Agricultural Production and Indian History

AGRICULTURAL PRODUCTION AND INDIAN HISTORY

Edited by
DAVID LUDDEN

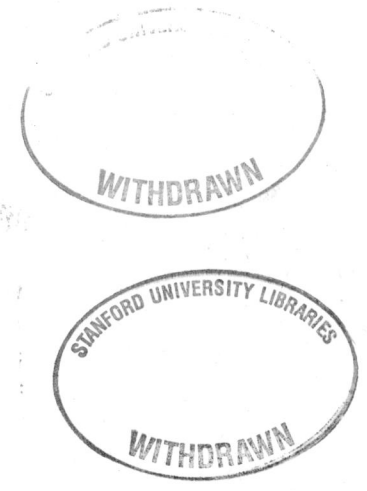

DELHI
OXFORD UNIVERSITY PRESS
OXFORD   NEW YORK
1994

Oxford University Press, Walton Street, Oxford OX2 6DP
Oxford New York
Athens Auckland Bangkok Bombay
Calcutta Cape Town Dar es Salaam Delhi
Florence Hong Kong Istanbul Karachi
Kuala Lumpur Madras Madrid Melbourne
Mexico City Nairobi Paris Singapore
Taipei Tokyo Toronto
and associates in
Berlin Ibadan

© Oxford University Press 1994

ISBN 0 19 563268 0

Typeset by Rastrixi, New Delhi 110070
Printed in India at Rekha Printers Pvt. Ltd., New Delhi 110020
and published by Neil O'Brien, Oxford University Press
YMCA Library Building, Jai Singh Road, New Delhi 110001

Contents

General Editors' Preface	vii
Acknowledgements	ix
Introduction: Agricultural Production and Indian History DAVID LUDDEN	1

DYNAMICS OF GROWTH

Dynamism and Enervation in North Indian Agriculture: The Historical Dimension ERIC STOKES	36
Property Structures, Demography and the Crisis of the Agrarian Economy of Colonial Bombay Presidency VASANT KAIWAR	53
Canal Irrigation and Agrarian Change: The Experience of the Ganges Canal Tract, Muzaffarnagar District (U.P.), 1840–1900 IAN STONE	114

PROCESSES OF COMMERCIALIZATION

Growth of Commercial Agriculture in Bengal—1859–1885 BINAY BHUSHAN CHAUDHURI	145

Expansion of Commodity Production and 182
Agrarian Market
A. SATYANARAYANA

Dimensions of Dependence 239
SHAHID AMIN

FORMATIONS OF SOCIAL POWER

A Typology of Agrarian Social Structure in 267
Early Twentieth-Century Bengal
SUGATA BOSE

Malguzar and Peasants: The Narmada Valley, 302
1860–1920
T.C.A. RAGHAVAN

Regional Dependence and Rural Development in 345
Central India: The Pivotal Role of Migrant Labour
CRISPIN N. BATES

Annotated Bibliography 369

General Editors' Preface

This series focuses on important themes in Indian history, on those which have long been the subject of interest and debate, or which have acquired importance more recently.

Each volume in the series consists of, first, a detailed Introduction; second, a careful choice of the essays and book-extracts vital to a proper understanding of the theme; and, finally, an Annotated Bibliography.

Using this consistent format, each volume seeks as a whole to critically assess the state of the art on its theme, chart the historiographical shifts that have occurred since the theme emerged, rethink old problems, open up questions which were considered closed, locate the theme within wider historiographical debates, and pose new issues of inquiry by which further work may be made possible.

This volume looks at agricultural production in colonial India. Nationalists inevitably linked agricultural stagnation and peasant impoverishment to the high levels of revenue demand under colonialism. To the imperial mind, convinced of the modernizing power of the state, the problem lay with the peasants: they had failed to respond to the transformative impulses of the market. Subsequent studies have gone beyond such naive assertions. The questions of peasant rationality and peasant calculations need further exploration, but price studies clearly indicate that peasants did respond positively to market stimulus. The link between revenue and production is important, but it cannot be seen in isolation from a variety of other determinations.

Recent studies have analysed how the specific nature of

commercialization redefines the agrarian landscape of a region, alters the temporal rhythms of production, re-allocates peasant labour. Agrarian historians now seek to understand how ecological contexts shape systems of production. They have identified different agrarian regimes in wet and dry zones, in uplands and lowlands. Another concern has been the link between power and production. Historians have shown how distinct systems of agrarian property define the contours of the regional production systems; how the complex relations between credit, market and production are mediated by power structures.

The essays in this volume reflect these complex concerns of agrarian historians, and reveal the variety of local worlds within which the everyday routine of production was carried on. The introduction analyses the process through which colonial knowledge about production was textualized, discusses the shifting focus within agrarian studies, and brings out the general implications of the specific essays on different regions.

Acknowledgements

The publisher is grateful to the following for permission to reproduce copyright material:

Cambridge University Press for:

1. Eric Stokes, 'Dynamism and Enervation in North Indian Agriculture', from *The Peasant and the Raj*.
2. Sugata Bose, 'A Typology of Agrarian Social Structure', from *Agrarian Bengal*.

Journal of Peasant Studies for:

V. Kaiwar, 'Property Structures, Demography and the Crisis of Agrarian Economy of Colonial Bombay Presidency'.

Sage Publications for:

1. T.C.A. Raghavan, 'Malguzar and Peasants', from *Studies in History*, 1, 2, n.s.
2. B.B. Chaudhuri, 'Growth of Commercial Agriculture in Bengal', *Indian Economic & Social History Review*, 7. 1 (1970).

The authors for:

A. Satyanarayana, 'Expansion of Commodity Production and Agrarian Markets'.

Crispin Bates, 'Regional Dependence and Rural Development', from *Modern Asian Studies* 19, 3, (1985).

Any omissions in regard to copyright, if notified, will be rectified by the publisher in subsequent printings of this book.

Introduction

Agricultural Production and Indian History

DAVID LUDDEN*

Texts to direct agricultural production appear in the far east and far west of Eurasia from the middle ages, but not in lands from Istanbul to Dacca.[1] Though many old texts from India contain knowledge about farms and farming,[2] they do not depict agri-

---

* I am grateful for detailed comments from Neeladri Bhattacharya.
[1] On long-term agrarian characteristics of the Middle East and South Asia, see my 'Agrarian Systems, Land, and Labor', in *The Political Economy of Ottoman, Safavid, and Mughal Empires*, edited by Tosun Aricanli, Ashraf Ghani and David Ludden (forthcoming).
[2] For source compendia, see M.S. Randhawa, *A History of Agriculture in India*, New Delhi, 1980, and Irfan Habib, *An Atlas of the Mughal Empire: Political and Economic Maps with Detailed Notes, Bibliography and Index*, Delhi, 1982, pp. 73–79. To sample the uses to which old sources have been put, see also A.M. Shastri, *India as Seen in the Brhatsamita of Varahamihara*, Delhi, 1969, 261–91; S. Singaravelu, *Social Life of the Tamils: Classical Period*, Kuala Lumpur, 1966; S.P. Raychaudhuri, 'Some Aspects of Agricultural practices in Ancient India', and A.K. Bhattacharya, 'A Glimpse into Soil Sciences and Agriculture in Ancient India', in *Proceedings of the Symposium on the History of Science in India (Calcutta, 1961)* National Institute of Sciences in India, 1961, pp. 107–117, 136–144; G.P. Majumdar, 'History of Botany and Allied Sciences (Agriculture, Medicine, Abori-Horticulture) in Ancient India (c. 2000B.C. to AD 100), in *Studies in the History of Sciences in India*, edited by Debi Prasad Chattopadhyaya, New Delhi, 1982, vol. 2, pp. 365–411; E.C.G. Rao, 'Some Aspects of Ground Water Exploration as Revealed in the Story of Dakargaladyaya of Brhatsamhita', in G. Kuppuram and K. Kumudamani, *History of Science and Technology in India (in 12 volumes)*, Delhi, 1990, vol. 12, pp. 131–54; David Ludden, 'Archaic Forms of Agricultural Knowledge in South India', in *Meanings and Purposes of Agriculture*, edited by Peter Robb and Utsa Patnaik (forthcoming); and Francis

culture as a distinctive subject of study. Irfan Habib indicates what this may signify when he suggests that the intelligentsia would engage agriculture as a textual subject in proportion to its power over production.[3] It seems that old texts on farm technology in China and Japan, and on management and tenancy in Europe,[4] signify powers over agricultural production that literati did not exercise in south and west Asia. In India, records indicate increasing farm surveillance by authors of texts from the seventeenth century, but colonialism created agriculture as a subject of study, an object of observation, analysis, and ma-

---

Zimmerman, *The Jungle and the Aroma of Meats: An Ecological Theme in Hindu Medicine*, Berkeley, 1987. Among old sources, the *Brhat Samita* and *Krshi Sukti* and others prescribe good farming practice; poetic genres, as in Tamil, depict detailed knowledge of agriculture; ayurvedic tracts create a unified field of medicine and ecology; epigraphy records irrigation building and agricultural expansion; while the *Ain-i-Akbari* and similar texts measure prosperity and describe rural conditions, as do travel accounts by Ibn Battuta and others. But none can be categorized as a study of agriculture, when compared to texts used for instance in Joseph Needham, *Science and Civilization in China*, vol. 6, *Biology and Biological Technology*, Part II, *Agriculture*, by Francesca Bray, Cambridge, 1984.

[3] An ingenious use of records to deduce power relations in part from what the records do not say is Dipesh Chakrabarty, 'Conditions for Knowledge of Working-Class Conditions: Employers, Government and the Jute Workers of Calcutta, 1890–1940', in *Subaltern Studies II*, edited by Ranajit Guha, Delhi, 1983, pp. 259–310. Irfan Habib suggests an explanation for the character of pre-modern records on Indian agriculture with this formulation: 'The ruling classes (whether the Mughal nobility or the hereditary superior rural classes of zamindars) were, in essence, rent-receivers, who shared out the bulk of the agricultural surplus, extracted in the form of land tax (usually realized in money), among themselves. They might be interested in extending cultivation by such means as laying out large irrigation works, or laying out orchards, for fruits for their table or sale, but this was perhaps the sum total of their direct interest in agriculture.' 'Technology and Barriers to Social Change in Mughal India', *Indian Historical Review*, 5, 1, 1978, p. 153.

[4] In addition to Needham, *Science and Civilization in China, vol. 6*, see Fukaya Katsumi, 'Peasant Culture and Consciousness in Tokugawa Japan: The Life and Consciousness of Small Peasant Families', *Peasant Studies*, 17, 1, 1989, pp. 1–22; Sugita, Kurima. 'Terrestial Deities and Celestial Bureaucrats: Transformation of the State and Local Communities in the Asiatic Mode of Production in Japan', in *The Study of the State*, edited by Henri Claessen and Peter Skolnik, New York, 1981, pp. 371–85; and Conrad Totman, *The Green Archepelago: Forestry in Preindustrial Japan*, Berkeley, 1989.

nipulation by the social classes whose written legacy dominates the historical record.

Over decades after 1780, two kinds of intellectual effort that had been critical earlier were combined by the East India Company with new discipline. As the Company surveyed resources to be taxed, it built on Mughal assessments and on eighteenth century regional state records of taxes, prices, crops, and rights, which enabled the Company to dig beneath zamindars and poligars for data on agriculture, which the Mughals never did.[5] At the same time, Company surveyors trekked forest and field to describe, map, and measure the land. Beginning with James Rennell, the Surveyor General organized geographical data, to inform Company commercial and military operations by calculating India's productive capacity. Colonial power thus began to constitute India as a domain of production, disciplining labour for surveillance of the countryside. It thus began to create the legacy of writing on agricultural production within which historians of Indian agriculture work today.

COLONIAL KNOWLEDGE

The method and theory of the state's appropriation of agrarian resources conditioned textual constructions of agriculture. Colonial knowledge arose from efforts to cut away layers of mediation between the Company and farmers and to command information for the increase of state power. Thomas Munro and William Jones worked respectively in rural south and in Calcutta courts to end the Company's dependence on natives for essential information. The results were the ryotwari system and Hindu law codes, rules and measures for state inquiry independent of people outside Company discipline.[6] Agriculture emerged in this process as an object of scrutiny guided by colonial rules and measures. Eighteenth century Indian states had begun to cut

[5] See Habib, *Atlas*, for details on Mughal statistics.
[6] David Ludden, 'Orientalist Empiricism: Transformations of Colonial Knowledge', in *Orientalism and the Post-Colonial Predicament*, edited by Carol A. Breckenridge and Peter van der Veer, Philadelphia, pp. 250–78.

away layers of mediation between rulers and producers, to appropriate agricultural wealth more directly, but the colonial state pushed very much harder in this direction.[7]

We inherit evidence on agriculture in proportion to its success. We have most data where the state penetrated localities most minutely. Our records indicate how the state was able and motivated to produce texts on agriculture. Thus we have more data for British India than for native states; for ryotwari than zamindari areas; for cash crops and plantations than for food crops and peasant farms; for areas of state irrigation work than for those where irrigation was a local affair; and more for times and places of turmoil than for everyday farming. Most data concern issues that were most important for the colonial state and colonial capital: such as revenue, property law, marketing, and credit. Relatively few data concern issues most critical for farmers, such as subsistence strategies and resource management. Data increase over time and change form as a function of state power. In early colonial years, data is spotty and localistic. Then the Company standardized and serialized data on revenue and cash crops. Crown rule produced comprehensive data in uniform, codified terms that render all localities components of the imperial system. By 1900, Indian agriculture was widely documented by the state, and analyzed in addition by scientific and political organizations, sponsored by the colonial elite and devoted to the progress of India's rural masses.[8]

Colonial rules and measures codified agriculture inside political economy. Farming was construed as an enterprise and agriculture as a subject defined by input-output accounting. In this framework, many attitudes appear in the records of the Raj, yet across the nineteenth century, the dominant attitude shifts toward a greater emphasis on the poverty of farmers, the backwardness of farm practices, more concern to increase output, and more interest in state investigations and interventions. Even

---

[7] See Burton Stein's introduction to *The Making of Agrarian Policies, 1770–1900*, in this series.

[8] See David Ludden, 'India's Development Regime', in *Colonialism and Culture*, edited by Nicholas B. Dirks, Ann Arbor, 1992, pp. 247–87.

in the early years of Company Raj, political economy impelled officers to consider the impact of revenue collection on farm investment and output, as well as on social stability. True, the Company did not take as direct an interest in Indian agriculture as did the British in Ceylon, where irrigation and plantations preoccupy early colonial records.[9] The politics of revenue overshadowed the economics of production in the Company's relation to agrarian India.[10] Even so, the Company took itself to be the lord of an estate worked by farmers whose taxes the state understood as rent. Its leaders adopted an attitude toward farms consistent with English landlordism. Securing revenue meant to gather production data. Export crops—especially indigo and cotton—invited special attention. Inquiries toward improving production moved alongside military campaigns and revenue settlements. Francis Buchanan, for instance, studied irrigation, animal breeds, seed varieties, and farming techniques, with an eye to improvement.[11] Munro conducted a survey of the Ceded Districts that revealed the extent to which even poor farmers relied on markets.[12] The Royal Botanical Garden is credited with introducing potato and strawberry cultivation in Bihar in 1815.[13]

---

[9] See 'Ceylon. Topographical and agricultural description of the Maritime Provinces and of the Tanks' [1807–1808], 'Commissioners of Eastern Enquiry. Ceylon' [1820–1829], 'Topographical and geographical description of Ceylon, presented by Sir Alexander Johnson, chief Justice of Ceylon', [1834] Colonial Office Records, London.

[10] See Burton Stein, *Thomas Munro: The Origins of the Colonial State and His Vision of Empire*, Delhi, 1989.

[11] See Francis (Hamilton) Buchanan, *A Journey from Madras Through the Countries of Mysore, Canara and Malabar*, Madras, 1870, and *Journey through the Northern Parts of Kanara*, with introduction by S. Silva, Karwar. Also Benjamin Heyne, 'Correspondence to Capt. Mackenzie, Supt. of the Mysore Survey', National Archives of India, Foreign Miscellaneous Series, No. 94; Benjamin Heyne, *Tracts, Historical and Statistical on India* . . . , London, 1814; 'Official Letters and Public Orders from Government and from other Public Authorities relating to the Surveys and more particularly to the Survey of Mysore', National Archives of India, Survey of India (Dehra Dun) Records. For an overview, of this survey literature, see R.H. Phillimore, *Historical Records of the Sruvey of India*, 4 volumes, Dehra Dun, 1945–1954.

[12] Stein, *Munro*, p. 82.

[13] Randhawa, *History of Agriculture*, III, p. 43. See Lucile Brockway, *Science*

Though early Company records concentrate on revenue, many contain data on production; and some include detailed enumerations and descriptions of irrigation works, trees, animals, carts, plows, and other assets.[14]

With William Bentinck in the 1830s, the more direct interest in agriculture that dominates the colonial record began to take the ideological form which became institutionalized after 1870.[15] 'We rode the country too hard', Bentinck said, 'through over-assessment and arbitrary methods of collection', thus 'the most lamentable poverty.' Indian peasants are 'the most obedient subjects in the world,' he continued, 'and they cannot be too much protected and encouraged.'[16] By 1840, political economy and moral discourse had converged on the idea that good policy stimulates investment and growth, while over-taxation depresses farm income and output. Good policy brings prosperity. In 1869, when Lord Mayo argued for setting up agriculture departments, he indicated the instrumental implications of this imperial paternalism with this pronouncement:

For generations to come the progress of India . . . must be directly dependent on her progress in agriculture . . . There is perhaps no country in the world in which the State has so immediate and direct an interest in such questions . . . Throughout the greater part of India, every measure for the improvement of the land enhances the value of the property of the State. The duties which in England are performed by a good landlord fall in India, in a great measure, upon the government. Speaking generally, the only Indian landlord who can command the requisite knowledge is the state.[17]

---

and Colonial Expansion: The Role of the British Royal Botanical Gardens, New York, 1979.

[14] David Ludden, 'Agrarian Commercialism in Eighteenth Century South India: Evidence from the 1823 Tirunelveli Census', in *Merchants, Markets, and the State in Early Modern India*, edited by Sanjay Subrahmanyam, Delhi, 1990, pp. 215–41.

[15] See Ludden, 'India's Development Regime.'

[16] John Rosselli, *Lord William Bentinck: The Making of the Liberal Imperialist, 1774–1839*, Berkeley, 1974, p. 238.

[17] Elizabeth Manak, 'Formulation of Agricultural Policy in Imperial India, 1872–1929: A Case of Madra Presidency', Phd.d. dissertation, University of Hawaii dissertation, 1979, p. 27. [Available at the National Library, Calcutta.]

Stimulated by riots and famines, which indicated deep distress in agrarian India, a flood of texts ensued on agricultural improvement. By the 1880s—when, for example, Madras Government printed an *Agricultural Class-book* and held one of its many agricultural exhibitions,[18]—the state no longer simply taught Europeans how to manage farms in India, it rather saw itself as the source of all scientific agricultural knowledge.[19] Scientific technologies and institutional means to improve agriculture expressed imperial power. Irrigation projects—above all, perhaps, the Punjab canal colonies—embodied the modern Raj. So did agricultural science. In 1908, the *Moral and Material Progress Report* declared that 'thirty years of experience and record gathering since the institution of the Departments of Agriculture lead to the recognition that Government's primary task was to apply European scientific methods to Indian agriculture.'[20] In 1929, the *Report of the Royal Commission on Agriculture* emerged as a textual monument to empire, to make agrarian Indian appear as a field for progress and as a production system under scientific management.

AGRARIAN STUDIES

Though counter-movements did punctuate thinking about Indian agriculture, so that we should not imagine a monolithic

---

[18] *Report on the Madras Agricultural Exhibition, 1883*, Madras Government, Madras, 1883. William R. Robertson, *An Agricultural Class-Book for the Use of Schools in South India*, Madras Government, 1880.

[19] See Ludden, 'India's Development Regime', and for examples of this literature: T. Greenaway, *Farming in India, Considered as a Pursuit for European Settlers of a Superior Class, with plans for the construction of dams, weirs, tanks, and sluices*, London, 1864; Frederick Pogson, *A Manual for Agriculture in India*, Calcutta, 1883; M.S. Randhawa, *History of Agriculture*, vol. 3; William R. Robertson, *Reports on the Agricultural Conditions, Capabilities, and Prospects of the Neilgherry and Coimbatore Districts*, Madras, 1881; Eugene C. Schrottky, *The Principles of Rational Agriculture Applied to India and its Staple Products*, Bombay, 1876; and Motilal Kashal Chand Shah, *Principles of Agriculture for India*, Ahmedabad, 1888.

[20] Manak, 'Formulation of Agricultural Policy', p. 27.

and homogenous colonial history of thought about agrarian issues; the official, elite origins of agricultural texts in the colonial period left a permanent imprint on agrarian studies, which arose initially from debates about policy. Farming attained modern intellectual significance from its appropriation by the state.

In subordination to empire, agriculture changed as a subject of study with the state's changing power over knowledge, and imperial politics always conditioned agricultural knowledge. When Buchanan, a botanist, toured Mysore, for example, he collected evidence to show how Tipu Sultan had impoverished the country; thus Buchanan helped to justify Company expansion.[21] Munro bested his competitors by using evidence submitted to Parliament to show how his ryotwari system served Company interests and suited India's agrarian tradition.[22] Company critics after 1830 demonstrated with revenue data the damage done to farms and treasury by high assessments.[23] Within decades, famines brought down upon the Crown the same critical scrutiny and condemnation that Buchanan turned on Tipu and its critics turned on the Company. As the first famine commission compiled evidence, nationalists used official data to show that British rule had drained away the wealth of India's farmers, reducing their ability to invest, even to survive.[24]

Economic nationalism used data on agrarian conditions against the Raj. Nationalists built empirical artifacts from official statistics and joined debates about policy, using measures of progress, as others had done before. Like their predecessors,

---

[21] Marika Vicziany, 'Imperialism, Botany, and Statistics in Early 19th Century India: The Surveys of Francis Buchanan (1762–1829).' *Modern Asian Studies*, 20, 4, 1986, pp. 625–660.

[22] See Stein, *Thomas Munro*, and Ludden, 'Orientalist Empiricism.' One outcome was to establish the lasting image of India as a land of peasant farmers—raiyats—who worked the land from time immemorial in self-sufficient village communities and who had suffered centuries of oppression by revenue contractors and tyrants, a theme elaborated by James Mill, Charles Metcalfe, and many others.

[23] See David Ludden, *Peasant History in South India*, Princeton, 1985 and Delhi, 1989, p. 116ff.

[24] Bipan Chandra, *The Rise and Growth of Economic Nationalism in India: Economic Policies of the Indian National Leadership, 1880–1905*, New Delhi, 1966.

they approached farming as a subject of policy debate. Dadabhai Naoroji, Romesh Chandra Dutt, and Mahadev Govind Ranade challenged the empire's pretensions and put its legitimacy permanently at stake in agrarian studies.[25] But their nationalism engaged agriculture only at points of policy debate, and one in particular, taxation. Before 1929, though opposition to taxes sustained rural Congress agitations,[26] party resolutions expressed an interest in agriculture that stopped at taxation.

Congress agreed with government that agricultural progress required programs such as catalogued by the Royal Commission.[27] Amid oppositions to empire, therefore, a unified paradigm for agrarian studies stretched across the political spectrum. Unity emerged from official data stressing commerce, growth, and state projects, and from ideas about development stressing rural poverty, encapsulated by a Congress declaration in 1947 that Indian poverty 'is essentially a rural problem.'[28]

---

[25] A debate very much like that between authors in Chandra, *Economic Nationalism* and contemporaries like S. Srinivasa Raghavaiyangar (*Memorandum of the Progress of Madras Presidency during the last forty years of British administration*, Madras, 1893, rpt New Delhi, 1988) continues today: see Irfan Habib, 'Studying a Colonial Economy—Without Perceiving Colonialism', *Modern Asian Studies*, 19, 3, 1985, pp. 355–82.

[26] David Hardiman, *Peasant Nationalists of Gujarat: Kheda District, 1917–1934*, Delhi, 1981.

[27] Congress resolved in 1901, for instance, 'that the Government should be pleased to bestow its first and undivided attention upon the department of agriculture and adopt all those measures for its improvement and development which have been made in America, Russia, Holland, Belgium, and several other countries so successfully in that direction.' A. Moin Zaidi, editor, *A Tryst With Destiny: A study of economic policy resolutions of the Indian National Congress passed during the last 100 years*, New Delhi, 1985, p. 34.

[28] 'Though povery is widespread in India, it is essentially a rural problem, caused chiefly by overpressure on land and lack of other wealth-producing occupations. India, under British rule, has been progressively ruralised, many of her avenues, of work and employment closed, a vast mass of the population thrown on the land, which has undergone continuous fragmentation, till a very large number of holdings have become uneconomic. It is essential, therefore, that the problem of the land should be dealt with in all its aspects. Agriculture has to be improved on scientific lines and industry has to be developed rapidly in its various forms . . . so as not only to produce wealth but also to absorb people from the land . . . Planning must lead to maximum

Unity also formed around a shared assumption that the state and science could end poverty. Nationalists challenged the empire to address poverty and argued that a state to represent India's interests would be necessary for progress. This left open space for collaboration in development work, even for those like Gilbert Slater, who in the twenties, as first Professor of Indian Economics in Madras, still saw hope for progress under the Raj.[29] Development work built links among foreign and Indian colleagues in academic fields—economics, biology, soil science, medicine, and engineering—with a shared scientific sensibility that lent further unity to agricultural texts that multiplied rapidly after 1900.

Universities and development agencies created professional bonds across political affiliations and intellectual space for inquiry at a broad intersection of imperial and national thinking on agriculture. Science dominates that intersection, perhaps even defines it, because the more a text claims scientific status, the more it appears to lie outside politics. In 1901, for example, Congress praised European scientific agriculture and urged government to adopt recommendations by John Voelkar, Consulting Chemist of the Royal Agricultural Society, who toured India in 1890. Congress repeated its praise of scientific agriculture often. Even Gandhian critics of western industrialism left its agricultural application unscathed.[30]

That science would solve India's agricultural problems remained a cornerstone of development ideology in India after 1947. Scientific institutions acquired a long history of invention and intervention in agricultural production. Research on crops of most concern to capitalists, like sugarcane, made India an international laboratory. In 1912, for instance, Madras govern-

---

employment, indeed to the employment of every able-bodied person.' Zaidi, *Tryst*, p. 72.

[29] Gilbert Slater, *Southern India: It's Political and Economic Problems*, Madras, 1936, pp. 24–9.

[30] For instance, J.C. Kumarappa, whose *An Economic Survey of Matar Taluka*, Ahmedabad, 1931, was followed by prodigious Gandhian labor and his *Appendix to the Report of the Congress Agrarian Reforms Committee*, New Delhi, 1945.

ment opened a Sugarcane Research Station at the Agricultural College in Coimbatore headed by Charles Barber, a South African botanist, who with his successors—T.S. Venkataraman and Nand Lal Dutt—bred eleven varieties. In aggregate, such efforts altered India's means of agricultural production with state-sponsored science and institutions like the Imperial Council of Agricultural Research—a kind of power that seemed politically unproblematic before the Green Revolution and whose history describes remarkable continuity across this century.[31] (See *infra*, pp. 233–5, 248–9, 251–3.)

By 1920, however, development work by social scientists from India and abroad imbued rural India with national significance deeper than is evident from Congress agitations and resolutions. National identity found roots in village India. That the village defined traditional India was accepted long before 1900. In Munro's minutes and Charles Metcalfe's famous submission to Parliament; in texts by Mill, Marx, Baden-Powell, Weber; and in work by orientalists, ethnologists, administrators, census takers, and others; an image of India's village tradition was established that served as a baseline for studies of agrarian change under the Raj.[32] For scholars of rural development, the village became a system of production whose fate was that of India itself. Radhakamal Mukerjee—economics professor, honorary organizer of cooperative societies in Murshidabad, and the most influential agrarian scholar of his time—collected village data over a period of years, and in 1916, published *The Foundations of Indian Economics*. Following its publication, the village became a standard object of economic analysis, under supervision by Gilbert Slater, Harold Mann, G.F. Keatinge, E.V. Lucas, and Mukerjee, a trend encouraged by the Royal Commission.[33] In the twenties, Gandhi's philosophy also influenced

---

[31] Randhawa, *History of Agriculture*, III, pp. 328–41. See Carl Esek Pray, *The economics of agricultural research in British Punjab*, Ph.d. dissertation, University of Pennsylvania, 1978.

[32] Ludden, 'Orientalist Empiricism.'

[33] See Radhakamal Mukerjee, *The Foundation of Indian Economics*, London, New York, Bombay, Calcutta, and Madras, 1916, p. xxiii. Ramkrishna Mukherjee, *Six Villages of Bengal*, p. x. Gilbert Slater, *The English Peasantry and the*

village studies, as acknowledged by Slater and stressed by younger economists who, like J.C. Kumarappa, often combined Gandhian idealism with economic empiricism.[34] The fate of village India, assaulted by western commercialism and by imperial power, became a centerpiece of texts on agrarian India.

Economic studies in the twenties portray village India as moving along a path charted by Mukerjee. Wracked by a colonial destruction of traditional harmony and self-sufficiency, village hopes for the future depended on enlightened reform. Mukerjee had a deeper and subtler vision of village India than Slater and most Englishmen, to be sure; and a stronger sense, which Gandhi would elaborate, that village tradition could guide India's national regeneration. Yet such ideas did not shock economists like Slater, who admired Gandhi and himself concluded a history of English enclosures with a plea to reinstate collective ownership.[35] The idea that self-sufficiency could be reconstituted in village India by reformers did not captivate many authors during the commercially expansive twenties—even those like Kumarappa who would later be its champions. The empirical evidence on agriculture that multiplied in the twenties—in statistical atlases, banking inquiries, village

---

Enclosure of Common Fields (London, 1907). Note the convergence of Slater's conclusion to *The English Peasantry* and Mukerjee's 'India's Message to the West' which is the epigram to his *Foundations*.

'Summing up, therefore, the economic results of a whole mass of little revolutions under examination, we find increased population, increased production of all sorts of commodities, increased national resources for purposes of taxation and foreign war. The moral effects we find to have been increasing misery and recklessness, showing itself in increased pauperism and drunkenness. An increase of the quantity of human life is attained at the expense of the degradation of its quality.' [Slater, *English Peasantry*, p. 266]

'India stands for living Humanity as against inert matter; for more equitable distribution of wealth; for less luxury and more brotherhood; for less industrial conflict an more co-operation; for wealth as a means as against wealth as an end; and for finding happiness not in resltess self-serving but in the consecration of life to the welfare of Society and Humanity.' [Mukerjee, *Foundations*, pp. 459-61, 465-7 and frontispiece].

[34] See note 30. Also S. Subbarama Aiyar, *Economic Life in a Malabar Village: A Study of Rural Economic Organization* (Bangalore, 1925).

[35] Slater, *English Peasantry*, pp. 265-6.

studies, and in work like that by Malcolm Darling, sponsored by provincial governments representing substantial agrarian interests—revealed problems in agriculture amenable to reforms like the establishment of cooperatives advocated by Mukerjee and tax reductions pressed by Congress.

AGRARIAN HISTORY

In the context of the twenties, William Moreland wrote the first monograph on Indian agrarian history, *The Agrarian System of Moslem India*. A chasm separated Indian history and agriculture at the time. *The Cambridge History of India* barely touched even agrarian policy. Historians in that day studied states in the progress of civilization, and agrarian history emerged as field in Europe only as the state's progressive role became contested. In India, nationalists began the contest, which confronted Moreland as he worked for the Agriculture Department in the United Provinces, and wrote manuals to educate bureaucrats and landlords. Moreland's political and intellectual environment must have conditioned his historical study. He wrote as a reforming Raj strove to improve agriculture and the Royal Commission gathered evidence; and during the rise of village studies, which explicitly for Mukerjee and Gandhi, and implicitly for others, like Slater, raised an image of peasants working in isolation from states before British rule. Ideas of village self-sufficiency disrupted by the Raj, and farmers subjected to crushing taxation by the British, permeated texts at the time of his writing. Moreland surely saw U.P. landlords as a Muslim legacy that obstructed reform; and his research may well have been influenced by ideas like those of Lord Mayo, the father of Indian agriculture departments, expressed in a quote above.

Moreland put agricultural development into a context that still frames agrarian history. Even this agricultural officer saw peasant farming as being outside history, including the physical features of farm life and intimate social relations of farm work. Institutional power relations that connect states and peasant villages constitute his agrarian system and determine its history.

His agrarian system consists of landed property rights and the state powers that sustain them. Though ideas animate agrarian systems, power maintains them, so that progress requires systemic change and the exertion of progressive state power. Moreland's concluding chapter argues that the 'idea of agricultural development . . . was already present in the fourteenth century . . . but the political and social environment was unusually unfavourable to its fruition.' From the Delhi Sultanate through the Mughal Empire, he writes, 'two figures stand out as normally masters of the peasants' fate . . . the [revenue] farmer and the assignee.' Moreland sees India's agrarian system over centuries before British rule as 'a barren struggle to divide, rather than . . . to increase, the annual produce of the country', a 'legacy of loss, which Moslem administrators left to their successors and which is still so far from final liquidation.'[36]

The energy driving Moreland's history clearly arose from his engagement with the practical conduct of agricultural development. In the English tradition, he conceives an agrarian system as composed of property rights, within which farmers work the land. Property rights determine how productive resources get used in development. In this light, peasants have histories only in their engagement with states. Those histories move forward to the extent that states facilitate agricultural growth. Behind Moreland's history lies a claim that the colonial state had stimulated progress, unlike its predecessors; so his history can be read as a defense of the Raj. But the implications of his argument do not stop there. Moreland argues against a historical sensibility that makes village India a world in itself, and he constructs a framework that could potentially embrace all the social relations of production. He argues that development requires systemic change and indicates that change wrought by British rule remained insufficient for progress.

Moreland's book appeared in 1929, as depression and socialism broke onto Indian politics. Amid contending political forces in that day, the book's concern for the conditions of

---

[36] W.H. Moreland, *The Agrarian System of Moslem India. An Historical Essay with Appendices* (Cambridge, 1929: rpt. Delhi, 1968), pp. 205–6.

agricultural progress and its focus on systems of power in development give it a critical valence that would typify many later agrarian histories. We should read Moreland in the tradition of James Mill and Lord Mayo, but also in that of Karl Marx and of Irfan Habib, who has expanded Moreland's conclusions mightily since 1963. We should read Moreland beside imperial apologetics, as farmers fought imperial taxation, but also beside this All-India Congress Committee resolution, composed as Jawaharlal Nehru became President:

> the great poverty and misery of the Indian People are due, not only to foreign exploitation in India but also to the economic structure of society, which the alien rulers support so that their exploitation may continue. In order therefore to remove this poverty and misery and to ameliorate the condition of the masses, it is essential to make revolutionary changes in the present economic and social structure of society and to remove the gross inequalities.[37]

This formulation sets the stage for later agrarian histories of British India, which would connect exploitation by foreigners and by Indians. In the discourse on inequality in India in the twenties, orientalist, legal, and ethnographic work focused on caste hierarchy, viewed purely as a product of Indian tradition. Muslim, non-Brahman, and low caste politicians focused on other inequalities that appeared to arise from Indian tradition. Economists considered inequalities that arose from unequal distribution of economic benefits for city and village, manufacture and agriculture, foreign and domestic sectors; and those stemming from usurious profits, high rents, and low wages. As the Great Depression struck, however, nationalism secured a grasp on connections between economic and social inequality, between British imperialism and oppression inside India. Worker and peasant movements forced 'gross inequalities' of wealth and power onto the political agenda and linked them to empire so as to make struggles within Indian society part of the struggle against empire.

As activists made rural inequality a political issue, scholars revealed dynamics of inequality inside village India. Agricul-

[37] Zaidi, *Tryst*, p. 54.

tural production thus became a theme in Indian history that entailed not only trends in output but also social inequality and state power. Moreland and Mukerjee made steps in this direction. And in their tradition, the authors who made agriculture a subject for historical study were politically engaged scholars working in social sciences of development. Ramkrishna Mukherjee studied Bengal villages in the early forties and, like Radhakamal, anticipated a surge in village studies—this time by anthropologists, who began in the forties to appropriate village India for their discipline, as economists had done in the twenties. Unlike his contemporaries,[38] Mukherjee would not concede agriculture to economics or rural social organization to anthropology. He crossed disciplinary boundaries and construed village society historically. With economist P.C. Mahalanobis and anthropologist K.P. Chattopadhyay, he gathered data to reveal class differentiation in rural society that he explained by transformations under British rule. At the same time, a rural sociologist, A.R. Desai, expanded studies of agrarian class formation, which Mukherjee pursued in a Bengal microcosm, to embrace Indian nationalism. They were soon joined by S.J. Patel, who explained the emergence of India's population of landless labourers by class differentiation under British rule.[39]

HISTORICAL AGRICULTURE

The legacy of the colonial era inflects agrarian history today. Critical evaluations of colonialism still drive research, as scholars engage issues of national significance. History's policy relevance today may seem remote, but state interest in agricul-

[38] His most important contemporary, M.N. Srinivas, studied a Mysore village in 1948 for an Oxford degree and led a generation that made the Indian village a field for anthropological study. See *India's Villages*, edited by M.N. Srinivas (London, 1955).

[39] S.J. Patel, *Agricultural Labourers in Modern India and Pakistan*, Bombay: 1952. A.R. Desai, *The Social Background of Indian Nationalism* (Bombay, 1947). Also Rajani Palme Dutt, *India To-day* (Calcutta, 1940; revised and enlarged, 1947).

ture became more direct than ever after 1947; and the need to analyze agricultural trends in light of colonial and national policies added new force to historical study. Similarly, political forces emanating from the countryside since 1947 have stimulated histories of agrarian power and inequality. Building on its base in social sciences of development, agrarian history still strives to integrate data and theories on political, economic, and social change; and having arisen with intellectual challenges from the Left in the 1930s, agrarian history remains a discourse in the legacy of Marx, amid Left challenges to India's development regime.

Yet methods of study have changed dramatically. One key trend has been toward empirical studies of agrarian regions. Research has followed lines suggested by Daniel Thorner, who argued for regional studies by citing 'the classic agrarian problem, namely, the interrelation of the institutional framework on the one hand, with the level of output and the distribution of the product on the other.'[40] By focusing on regions, scholars can analyze influences exerted by natural resources, economic geography, social organization, landholding patterns, power structures, labour and credit, technology, demography and other variables. Historical sources also encourage regional studies, because regional languages and institutions shape the nature and distribution of records. Such regional research in addition connects history and development studies to form a unified field of agrarian scholarship.[41]

With his 'classic agrarian problem' in view, Thorner defined thirty-nine agrarian regions in India.[42] By mapping dominant crops, geographers have since defined agricultural regions in more detail, and they show India divided east and west, along a line that swerves around the north-south line of the eightieth

[40] Daniel Thorner, 'Agrarian Regions', in *Rural Sociology*, edited by A.R. Desai (Bombay, 1959), pp. 152–60.
[41] See for example, *Agrarian Power and Agricultural Productivity in South Asia*, edited by Meghnad Desai, S.H. Rudolph, and Ashok Rudra (Berkeley and Delhi, 1984). Also Donald W. Attwood, *Raising Cane: The Political Economy of Sugar in Western India* (Boulder, 1992).
[42] Thorner, 'Agrarian Regions.'

degree east longitude. To the east (and on the western coast of the peninsula) the climate is more humid, and the dominant food grain is rice; in the drier west and interior, away from the coast, wheat and millets dominate food grains. This division—inscribed by rainfall and drainage—is significant in agrarian history for millennia and lays a template for agricultural regions today, which emerge from mapping dominant crop combinations and broadly fit the pattern described by Thorner.[43]

Following the legacy of colonial and national historiography, many historians of India focus on cultural and political regions, and agrarian history has likewise been written primarily with regional revenue and property systems in view.[44] But regions defined in addition by dominant crops, ecology, social relations and technologies have attracted more attention in the last two decades. Essays collected here come from a body of work that strives to integrate an increasingly complex set of regional elements. Such complex agricultural regions had long histories before British rule and attained specific colonial identities. Each had traditions of governance that inflected their integration into British India. Their specific histories represent forms of participation in empire, while their comparison describes all-India patterns and trends. Each region adds something to the agrarian history of India as a whole. What holds regions together are dynamics of integration and interaction; what holds their histories together is our effort to understand processes of change

[43] See Jasbir Singh, *An Agricultural Atlas of India: A Geographical Analysis* (Kurukshetra, 19740, pp. 301ff.

[44] See David Ludden, 'Productive Power in Agriculture: A Survey of Work on the Local History of British India', in Desai, et. al. *Agrarian Power and Agricultural Productivity*, pp. 51–99. It is noteworthy that agricultural regions have played a minor role in the work of historians of India, in general, who follow the tendancy to concentrate on political and cultural regions depicted in *Regions and regionalism in South Asian studies: an exploratory study (papers presented at a symposium held at Duke University, April 7–9, 1966)*, edited by Robert I, Crane, Durham, 1967. Compare the much greater role of agriculure in regions as defined by geographers: especially R.L. Singh, *India: A Regional Geography* (Varanasi, 1971; 1989], pp. 22ff and Figure 1.10. This difference is explained by geographers' greater concern for agricultural development in modern India.

in production relations and living conditions in rural India as a whole.

Reaching toward this end, scholars have slowly shifted their perspective on conditions in agrarian India at the time colonialism took hold. Their efforts have produced a second trend in scholarship, toward increasingly detailed appreciation of historical interactions among large-scale political and economic forces, on the one hand, and small-scale agrarian societies, on the other. Until the 1970s, in the tradition of Moreland and Mukerjee, rural India in pre-colonial times appeared to be a unitary social formation, wherein a subsistence-oriented peasantry inhabited self-sufficient villages, unchanged for centuries, until British rule inflicted dramatic shocks upon them. Today, agrarian regions appear as dynamic and differentiated for many centuries before colonialism. So the baseline for studies of change in colonial and post-colonial times cannot be established by assuming pre-colonial stability and uniformity.

Rather than resting simply on peasant household production, pre-colonial agriculture involved large, complex social formations that organized production and distinguished regions by modes of resource control and utilization. As irrigated agriculture in Asia differed in its social organization from dry farming in Europe,[45] so did India's irrigated, dry, shifting, and pastoral regimes differ from one another. In India's agricultural regions, these land-use regimes competed, expanded, and succeeded one another. Struggles to control land, labour, water, animals, and other resources enmeshed peasant families in social units—lineages, clans, tribes, villages, kingdoms, sects, religious networks, and states—which defined communities and social powers in production. The social organization of settled farming differed from that of forest and pastoral production; likewise, short- and long-fallow farming, irrigated and dry farming, commercial and subsistence farming differed from one another. Such differences defined spatially segregated production regimes and regions, which interacted politically and commer-

---

[45] Francesca Bray, *The Rice Economies: Technology and Development in Asian Societies* (London, 1986).

cially, as they competed for control of the landscape and succeeded one another over time, as more intense forms of land-use expanded or jungle invaded farmland and irrigation works declined. Such trends in production entailed changes in population density and composition, in urbanization and economic differentiation, in forms of land and labour control, and shifts in social relations. The broad trend toward more intense land-use progressed as settled farming succeeded shifting cultivation and pastoralism, as farmers invested labour on the land, fought to control investments and output, and conquered pastoral and jungle people. More intense land-use and shortening fallow complicated land rights; investments in ploughs, animals, and irrigation engaged households in hierarchies of rights and powers. In wet rice regimes, the most intense land and labour control entailed the most intricate, ritualized social and political hierarchies.

Agricultural expansion, competition, and succession occurred within political systems that organized power in production and also sustained kingdoms, empires, cities, and markets. Thus the history of agricultural regimes involved transformations in the institutional framework of farming as well as in the volume of output. Pre-colonial India no longer appears to be a land of self-reproducing villages, barely connected to commercial networks and connected to states only by the extraction of produce in taxes. Articulations among markets, states, and farms now seem complex, variable, and central in the historical dynamics of agricultural expansion. Urban-rural, inter-village, inter-local, and inter-regional integration and differentiation characterized India before the Delhi Sultanate, and expanded in subsequent centuries of migration, agricultural growth, technological diffusion, and political change.[46] In seventeenth century India, many regions, some bigger than European kingdoms, generated state revenue in complex networks of commercial exchange, vital for local reproduction and state revenue alike.[47] In many

[46] See a case study in Ludden, *Peasant History* and references in David Ludden, 'World Economy and Village India, 1750–1900', in *South Asia and World Capitalism*, edited by Sugata Bose (Delhi, 1990), pp. 159–77.

[47] See Ludden, 'Agrarian Commercialism', D.A. Washbrook, 'Progress and

regions, eighteenth century agriculture depended on urban demand.[48] In such regions—particularly in wet regions near the coast—European commercial centers laid the foundations of colonialism. Near the coast, agrarian regions became deeply involved in worldwide operations of merchant capital. Company profits depended on agrarian expansion that sustained overseas trade. Here, colonial power made its initial impact.

In contrast to the older view that Company Raj revolutionized agrarian production, by imposing steep revenue demands, alien forms of property, and commodity production, we now see that external forces emanating from capitalist empire entered agriculture differently and with differing results, depending on place and time. The Company had limited power to alter agrarian life. Colonial wars encouraged alliance-building to raise revenue and lower administrative expenses. Native state treaties, Permanent Settlement, and Ryotwari settlements punctuated an extended political process, in which the Company sought military stability, secure revenue, cheap administration, and allies; so that revenue settlements had initially modest, conservative effects on agrarian structures, in a context where resistance and revolt set terms of negotiation.[49]

Initially, Company power expressed itself primarily—and in sharp contrast to its own self-representation—by increasing the power of social groups on which it conferred ownership rights. In dry areas of extensive cultivation, this cut into long-fallow farming and pastoralism, enabling settled farmers to make new territorial claims, and enabling the best-endowed among them to grab open land and obtain labour to farm it. In more intensely farmed wet regions, landowners sought rental returns from property and thus more control over labour. Colonial governance, as it spread to the 1870s, exerted power on systems of

---

Problems: South Asian Economic and Social History, c. 1720–1860.' *Modern Asian Studies*, 22, 1, 1988, pp. 57–96.

[48] C.A. Bayly, *Rulers, Townsmen, and Bazaars: North Indian Society in the Age of British Expansion, 1770–1870*, Cambridge, 1983, and *Indian Society and the Making of the British Empire* (Cambridge, 1988).

[49] See Ludden 'World Economy and Village India.'

agricultural production in a direction that was neither alien nor pressed at revolutionary speed: toward more intense land and labour use and more appropriation by settled farmers of land used by pastoralists and shifting cultivators, thus increasing the labour available to landowners and for agricultural commodity production.

Company policy alone could not revolutionize production. Too many variables intervened. Before 1840, warfare and then declining prices weighed against vast intensification of land use and commercial production. Declining prices after 1820 reflect slow synchronization of distant economies, but regional diversity remained so pronounced that there was not so much one agrarian history in India as many regional histories with common themes and connections. By 1840, a convergence gained momentum that marks India's passage into the world of industrial capitalism.

As Karl Marx argued at the time, industrial capitalism, embodied in the railway and steamship, transformed rural India. Agricultural commodity production pushed ahead everywhere. Indian cotton rushed into world markets with the blockade of the American south, in 1860. Though ephemeral in the Deccan,[50] the Civil War cotton boom launched a seventy-year expansion of rural investment in India that brought new social power into agrarian production relations, based on commodity production. Government propelled expansion with its own investments in infrastructure, to intensify land and labour use. Transportation and transaction costs declined as the railway spread, Central India was opened up, and the imperial administration grew. After 1870, interior regions entered the world economy and India's national economy at the same time, as workers moved into labour markets in India and around the world.[51] Industrial empire brought all regions into a unified, national political economy. Yet Indian systems of agricultural

[50] Sumit Guha, 'Some Aspects of Agricultural Growth in Nineteenth Century India', *Studies in History*, 4, 1, 1982, pp. 57–86, and *The Agrarian Economy of the Bombay Deccan, 1818–1941* (Delhi, 1985).

[51] See references in Ludden, 'World Economy and Village India.'

production retained regional forms. Not only because specific crops thrived in specific regions, but because commodity production rested on systems of power with regional dynamics instituted by colonial governance. Colonialism first impressed modes of capitalist accumulation on wet regions near the coast. These areas, dominated by paddy cultivation and rice, had long and deep connections to overseas trade, ample water supplies, dense populations, intricate labour specialization, and cosmopolitan elite strata sustained by rent, commercial capital, and political status. The colonial state forged alliances with these elites classes in Zamindari and Ryotwari settlements. By contrast, in the relatively sparsely populated, dry, interior tracts, colonialism confronted extensive land-use regimes in which farming concentrated on crops that withstand drought, above all, millets and pulses. In these areas, production relations were organized socially in militarized political systems. Here the British found their 'martial races' and 'yeoman peasants', in the Deccan, Rajasthan, Central India, the Mughal heartland, and Punjab. Here colonial conquest and revenue settlement weighed in favor of settled farming and farmers, enhancing their powers to effect the expansion of cultivation; it weighed against shifting and long-fallow farming, and above all, against pastoralism. In the dry interior, as opposed to the wet coastal areas, colonialism and capitalist power strove to supplant militarized forms of territorial authority that articulated connections among diverse and conflicting extensive land-use regimes in pre-colonial times.

At the outset of colonialism, in densely populated wet regions near the coast—from eastern U.P. and Bihar, throughout Bengal, and along both coasts to Gujarat—a relatively large, stationary population of landless and near-landless labourers was controlled by its subordination to landed elites in intricate hierarchies of power. To these traditional modes of power colonial settlements added property law and debt coercion as means to exploit labour for the expansion of commodity production. By contrast, in the dry interior—from the tip of the peninsula to Rajasthan, Sind, Punjab, where, in 1800, extensive farming mingled with a vast pastoral economy—the labour force moved constantly over short and long distance in the everyday conduct of subsistence, to work land, trade, fight, tend animals, flee drought, seek water, open and defend territory. Here, labour was kept in control primarily by the kinship units that did farm work, by tribal and jati alliances that held territory, and by clan and caste alliances that built expansive royal domains, like the Maratha, Rajput, and Jat. Colonial land law reduced the rights of pastoralists and shifting cultivators dramatically, and opened new realms for commodity production by settled farmers, who colonized land previously outside their reach, using the labour of groups who had lost traditional entitlements to land. Here, the process of labour control—shorn of its military modality—became recast in terms of family, caste, village, market, and private property entitlements to land and labour.

Everywhere in agrarian India, agricultural commodity production entailed a more intensive utilization of land and labour. But in the densely settled, wet regions, this emerged more often within domains and forms of power that existed before colonialism and evolved to generate commodity crops. While in more sparsely settled, interior regions, more innovative social powers to produce agricultural commodities were created by the invention of landed property rights and their allocation to settled farmers, who emerged with a new set of powers for controlling land as well as landless and land-poor workers. This contrast explains why dry interior regions have appeared more innovative and progressive to observers since the colonial period. This appearance was furthered by irrigation works built under the imperial state, despite the distress these entailed. And broadly speaking, with notable exceptions—the Krishna-Godavari delta and Brahmaputra valley, where new irrigation created new paddy land, and dry peninsular tracts, where over-exploitation destroyed the land—a comparison of India's dry interior with its wet coastal regions (which had such long histories of growth before 1800) reveals proportionately greater increase in productivity in the dry interior after 1800, and more radical change in the social framework of agriculture.

In this Volume

Essays collected here represent varieties of research conducted within debates about systems of agricultural production in colonial India. The three sets of essays overlap thematically, as they explore growth, commercialization, and power. The first set considers farming as a social process inside regions where forms of investment determine patterns of growth. The second set concerns some key commodity crops and commodity production generally. The third suggests how production systems generate India's interactive agrarian regions.

In the first essay, Eric Stokes treats a contrast between eastern and western Uttar Pradesh that could be extended to include Punjab and Bihar. He argues for the importance of property forms in determining agricultural growth, and by doing so puts himself in opposition to scholars who highlight demography, ecology, and culture. He concludes that landlord property in the east, bolstered by caste hierarchy and rent-seeking behaviour by landed classes, 'proved atavistic devices' that 'survived because even a minute fractional share in the joint patrimony validated caste status.' By contrast, *bhaiachara* tenures, which arose in dry-farming conditions, were 'most readily adaptable to maximize production under Indian *petite culture*.' (p. 241)

Stokes does not seek to measure productivity trends so much as to highlight institutional factors. To account for institutional forms, he reverts to a Ricardian shorthand, by saying rents emerged with land scarcity to create landlordism in the east, but not in the open lands of the west. He does not do justice to literature on the subject. In fact, colonial discourse seethed with debate about the productivity implications of property forms, and the idea that productivity would benefit from peasant investment permeated policy thinking. Radhakamal Mukerjee took institutional factors very seriously, and in 1936, Congress declared 'poverty, and unemployment and indebtedness of the peasantry [to be] fundamentally due to antiquated and repressive land tenure and revenue systems', so the 'solution to this problem inevitably involves . . . emancipation of the peasants from feudal and semi-feudal levies.'[52]

Yet this is an excellent essay to begin this volume, because it forges a strong link between property institutions and the production process, a link with political overtones that must concern students of India's agrarian history. For despite Congress resolutions, zamindari abolition in the end involved a long struggle that was waged at the very time when 'institutional reform was assigned secondary importance in the program for increasing agricultural production' in India's first Five Year Plans.[53] India's official development regime, as embodied in the 'green revolution', has tended strongly to privilege the technical determinations against which Stokes pits his argument, and to put institutions and power to one side as being of secondary consideration for development. Among scholars, this division between proprietary power and productivity is enshrined in the tradition of considering property in the analytical realm of politics and culture, and relegating farm practice to economics and agronomy, where farmers appear as social isolates. In the same vein, it is important that this essay implicates the state in social formations that condition the conduct of agriculture. By reference to Edmund Leach's formula that kinship is merely another way of talking about property, Stokes enunciates a position that merits reiteration in all discussions about the role of caste in the Indian economy and about legislation that would alter the role of caste as a determinant of economic power. Finally, Stokes deploys a comparative method that distinguished his research and which stimulated advances in agrarian history that I have referred to above.[54]

Vasant Kaiwar argues in the same vein as Stokes, for the critical role of property forms, but against Stokes' conclusion that peasant property stimulates growth. Like Stokes, he argues against those explanations of agricultural growth that rest on

[52] Zaidi, *Tryst*, p. 28.
[53] Francine R. Frankel, *India's Political Economy, 1947–1977: The Gradual Revolution* (Princeton, 1978), p. 57.
[54] For example, D.A. Washbrook, 'Country Politics: Madras 1880–1930.' *Modern Asian Studies*, 7, 3, 1973, pp. 375–521.

variables isolated from property forms: capital, demography, and technology. His essay situates the demographic regime of western India, which produced a fragmentation of landholdings and spiral of indebtedness, inside the institutional force of the colonial regime. It construes 'social-property relations' as a rich analytical field in which to see peasant labour, inheritance, and subsistence as activities conditioned by their institutional environment. In contrast to Stokes and other authors in this volume, Kaiwar draws his comparative framework from the world of capitalist agriculture, in which he juxtaposes India's colonial peasantry to peasantries in Japan and Europe. This invokes a persistent problem for historians: to what extent and effect did India undergo a transition to the capitalist mode of production?[55] Kaiwar argues that, despite vast commodity production on peasant farms, peasant property combined with the exploitation and distortions of imperialism to prevent capital investments in farming that would have enabled agrarian India to reap the growth potential inherent in capitalist social-property relations.[56]

Kaiwar does not consider regional dynamics and interactions in western India, as do T.C.A. Raghavan and Crispin Bates in later essays, though Kaiwar does indicate that variations in the regional context of farm production in western India, induced by such factors as state investment in irrigation, did indeed differentiate production systems. Such factors do not attain significance in his method of comparison. His method has the additional disadvantage of posing a counterfactual implication that Indian farms would have worked like those in Europe and Japan had their institutional setting been the same. By con-

[55] D.A. Washbrook, 'South Asia, The World System, and World Capitalism', *The Journal of Asian Studies*, 49, 3, 1990, pp. 479–508, rpt in Bose, *South Asia and World Capitalism*. Also Ludden, 'World Capitalism and Village India.'

[56] This argument can be found in many forms: e.g. Irfan Habib, 'Potentialities of Capitalistic Development in the Economy of Mughal India, *Journal of Economic History*, 29, 1, 1969, pp. 32–78. E.L. Jones, L. *The European Miracle: Environments, economies, and geopolitics in the history of Europe and Asia* (Cambridge, 1981); S.S. Sivakumar, 'Transformation of the Agrarian Economy in Tondaimandalam, 1760–1900', *Social Scientist*, 6, 10, 1978, pp. 18–39.

trast, other essays in this volume take the institutional environment of British India as their arena for comparative analysis, to consider regional particulars, the impact of state intervention, and the influence exerted by world capitalism through the colonial state and commodity markets.

Ian Stone engages a debate about large colonial irrigation projects that resonates with controversies surrounding Narmada River schemes today. The debate has changed substantially since Stone wrote this essay, when the main issue was a critical evaluation of claims made by defenders of the Raj for the progressive impact of colonial projects. Even so, independent India remained institutionally committed to large scale irrigation works; and—as Kaiwar would agree—the real cost of capital inputs, subject to state policy intervention, is a critical factor in determining production conditions in agriculture. Many authors before and after Stone emphasize the benefits of irrigation as a stimulant to increasing productivity, as A. Satyanarayana does for Andhra Pradesh below. But long before debates arose around the Narmada schemes, Elizabeth Whitcombe described the ecological and economic problems produced by colonial irrigation in north India, and thereby questioned their net benefits.[57] Stone seeks to put such dislocations into the con-

---

[57] Ian Stone, *Canal Irrigation in British India: perspectives on technological change in a peasant society* (Cambridge, 1984). Elizabeth Whitcombe, *Agrarian Conditions in Northern India: the United Provinces Under British Rule, 1860–1900 (v. 1)* (Berkeley, 1972). Imran Ali, 'Malign Growth? Agricultural Colonization and the Roots of Backwardness in the Punjab', *Past and Present*, 114, 1987, pp. 110–132. Imran Ali, *The Punjab under Imperialism, 1885–1940* (Princeton, 1988). G.N. Rao, 'Canal irrigation and agrarian change in colonial Andhra: a study of Godavari district, c. 1850–1890', *The Indian Economic and Social History Review*, 25, 1, 1988, pp. 25–61. G.A.L. Satya Rani, 'Commercialization of Agriculture and It's Impact on the Socio-Political Awakening in the Godavari and Krishna Districts of Madras Presidency, AD 1858–1914', Ph.d. dissertation, Indian Institute of Technology, Madras, 1989. For current social science perspectives on irrigation benefits, see B.D. Dhawan, *Irrigation in India's Agricultural Development: Productivity, Stability, Equity* (Delhi, 1988). Dhanajaya Ramchandra Gadgil, *Economic effects of irrigation: report of a survey of the direct and indirect benefits of the Godavari and Pravara Canals*, Poona, 1948. Krishna Bharadwaj, *Irrigation in India: Alternative Perspectives*, Delhi, 1990. Niranjan Pant, editor, *Productivity and Equity in Irrigation Systems*, New Delhi, 1984.

text of agricultural change stimulated by the state provision of low-cost irrigation water, and concludes that 'the overall impression is of a substantial contribution to wealth and security.' (p. 108)

Stone employs a type of aggregate cost-benefit analysis typical in development project assessments. He does not confront the question as to whether differences between his conclusion and that of Whitcombe derive from ecological or other variables that distinguish his area of study from hers. He does not trouble himself with the simple yet challenging proposition that major irrigation projects have different consequences and can generate even opposite results in different physical, social, and political circumstances.

Stone's accounting of irrigation benefits rests on inference about individual investment decisions by farmers, in a context that allows the use of low-cost irrigation water to increase productivity by enabling the application of more labour to higher value crops, to 'set off a complex series of repercussions which left virtually none of the region's inhabitants untouched.' (p. 108) Farm responsiveness to opportunities provided by state investments to grow higher value market crops with low user-cost irrigation water thus explains the productive impact of irrigation projects, for Stone. This and the subsistence security on which it rests provided the rationale for state irrigation projects in the nineteenth century, as it does today.

Stone's discussion leads into debates that have arisen since 1947 about implications of state infrastructure investment for agricultural development. One debate has barely begun. It concerns the conditions under which large irrigation works raise living standards for a large enough proportion of the population in effected areas for long enough periods of time to qualify as investments in sustainable development.

Another debate, about peasant behaviour, in which Stone locates his argument, was essentially settled by Dharm Narain, in 1965, when he demolished the proposition that peasant

---

Daniel and Alice Thorner, 'The Weak and the Strong on the Sarda Canal', in *Land and Labour in India*, (Bombay: House, 1962), pp. 14–20.

farmers do not employ the market rationality on which state growth-oriented projects rely. Using price and crop statistics, he proved a broad farm responsiveness to agricultural commodity prices in modern India.[58] Subsequent work has shown that in India as a rule, even peasant farmers devoted to self-reproduction and security respond sensitively to market signals.[59] But this leaves many questions open, including those posed by Kaiwar as to where capital to invest in farms originates and how it is applied. How do subsistence and commodity production interact historically? What alterations in production systems attend increasing commodity output? When, where, how, and to what extent does agrarian India become dominated by the social and cultural logic of commodity and subsistence production, respectively? And above all, for histories written during India's green revolution, who benefits from agricultural commodity production and state investments toward its expansion? How do commodity production and the commercialization of agrarian relations effect social relations of production and power in rural India? Does commodity-led growth improve living conditions?

Such questions will not be answered adequately for years. Binay Bhushan Chaudhuri, the Dean of Indian agrarian history, has explored them in Bengal throughout his career.[60] His essay here follows a useful line of attack, considering general factors that encouraged commercial agriculture in Bengal in the first

[58] Dharm Narain, *The Impact of Price Movements on Areas under Selected Crops in India, 1900–1939* (Cambridge, 1965).

[59] See for instance, S. Prakash, 'Crop Choice and Peasant Rationality in British Gujarat, 1850–1937, *Studies in History*, 1, 2, 1985, pp. 201–20.

[60] Binay Bhushan Chaudhuri, *The Growth of Commercial Agriculture in Bengal*, Calcutta, 1964; 'Agricultural Production in Bengal, 1850–1900: Coexistence of Decline and Growth.' *Bengal Past and Present*, 88, July-December, 1969, pp. 152–206; 'Rural Credit Relations in Bengal', *Indian Economic and Social History Review* 21, 1, 1975, pp. 105–65; 'Agricultural Growth in Bengal and Bihar, 1770–1860: Growth of Cultivation since the Famine of 1770.' *Bengal Past and Present* 95, 1, January-June, 1976, pp. 290–340; 'The Land Market in Eastern India, 1793–1940, I: The Movement of Land Prices, and II: the Changing composition of Landed Society.' *Indian Economic and Social History Review*, 12, 1–2, 1976, pp. 1–42, 133–67; 'Movement of Rent in Eastern India, 1793–1930.' *Indian Historical Review* 3, 2, 1977, pp. 308–390.

decades of Crown Rule, and tracing the commercial career of particular crops. Four points merit emphasis. First, no clear line separates subsistence and commercial crops. For growers of rice in Bengal or wheat elsewhere in India, a bumper crop during a generous monsoon would spell hard times, as much as it did farmers in Europe and America by the 1880s. Second, market dynamics pertaining to specific crops differentiate conditions faced by producers. Specific marketing channels for export crops put Indian farmers in the grip of market cycles determined by the specific position of particular crops in world markets. In this light, we see that despite general commercial expansion, some commodity crops declined, like mulberry and indigo, as others expanded, like sugar and jute. Crops also concentrated in particular areas, and their methods of cultivation and finance could be significantly different, so that whereas switching from jute to rice was possible for Bengal peasants in the 1870s jute slump, it was not possible for peasants in Bihar to abandon sugar and indigo financed by landlords and mill owners, as we can see in the essay by Shahid Amin.[61] Finally, we see in Binay Chaudhuri's essay the beginnings of the trend that brought all agrarian life into the turbulent ups and downs of world commodity markets, a condition firmly established in Bengal and elsewhere by 1900.

A. Satyanarayana describes this trend in twentieth century Andhra. Regional specialization, which also emerged with the expansion of jute into eastern Bengal, became the hallmark of Andhra after irrigation arrived in the Krishna-Godavari delta; when irrigated cash crops—rice, sugarcane, tobacco, tumeric, and chillies—expanded along the coast; as dry cash crops—cotton and groundnut—spread through the interior, driving down the acreage under drought-resistant subsistence grains and expanding the domestic market for coastal rice. This trend continued, Satyanarayana argues, from a combination of internal and external demand that differentiated the regional impact of

[61] On indigo, see Colin M. Fisher, 'Planters and Peasants: The Ecological Context of Agrarian Unrest on the Indigo Plantations of North Bihar, 1820–1920.' In *The Imperial Impact: Studies in the Economic History of Africa and India*, edited by Clive Dewey and A.G. Hopkins (London, 1978), pp. 114–31.

the great depression by the more drastic decline suffered by industrial crops on world markets.[62] Though this essay does not explore the mechanisms of commodity production in Andhra villages, it describes as well as any the extent to which commercial agriculture reconfigured the agrarian landscape and produced specialized regions of cash cropping.

Shahid Amin engages a debate about what this meant for farmers. His essay should be read alongside others, some of which appear in other volumes of this series. Tom Kessinger, Bruce Robert, and Donald Attwood have argued that expanding cash crop markets in India enabled enterprising small-holding peasants to invest on production frontiers to increase their income and savings, stimulating further investment and increase in their family enterprises. Attwood in particular stresses the role of the irrigation frontier for sugar farmers whose profits and investments rested securely on low-priced water provided by the state.[63] Against these arguments, David Washbrook has posed evidence that in the dry interior of Andhra and Tamil Nadu, rich peasants with disproportionate power over land and capital captured the lion's share of the benefits from expanding commercial production, and used control over marketing to enhance their local powers.[64]

In literature on the landlord regime of eastern India we find

[62] On the impact of depression, see Christopher Baker, 'Debt and Depression in Madras, 1929–1936', in Dewey and Hopkins, *The Imperial Impact*, 1978, pp. 233–42, and *An Indian rural Economy, 1880–1955: The Tamilnad Countryside* (Oxford, 1984); Dietmar Rothermund, 'A Vulnerable Economy: India in the Great Depression, 1929–1939', in Bose, *South Asia And World Capitalism*, pp. 305.

[63] Donald W. Attwood, 'Peasants versus Capitalists in the Indian Sugar Industry: The Impact of the Irrigation Frontier.' *Journal of Asian Studies*, 45, 1, (November 1985): pp. 59–80; and *Raising Cane: The Political Economy of Sugar in Western India* (Boulder, 1992). Bruce Robert, 'Economic Change and Agrarian Organization in 'Dry' South India, 1890–1940: A Reinterpretation', *Modern Asian Studies*, 17, 1, 1983, pp. 59–78. Tom G. Kessinger, *Vilayatpur 1848–1968: Social and Economic Change in a North Indian Village* (Berkeley, 1973).

[64] D.A. Washbrook 'Economic Development and Social Stratification in Rural Madras: The "Dry Region", 1878–1929', in Dewey and Hopkins, *The Imperial Impact*, pp. 68–82, and 'Law, State and Agrarian Society in Colonial India', *Modern Asian Studies*, 15, 3, 1981, pp. 649–721.

the most eloquent accounts of domination by landed and financial elites during the expansion of agricultural commodity production. Shahid Amin rightly argues, in the introduction to the book from which this chapter comes, that most pertinent literature concentrates on crop finance and marketing, rather than production; and that historians have yet to pay sufficient attention to social dynamics of commodity crop production itself. Amin attempts to rectify this by showing how, in the production process, 'the dice [were] loaded in all possible ways against the smaller peasants and in favour of the capitalists and the dominant agrarian classes.'(p. 195)

Sugata Bose argues that the legal regime of zamindari property in eastern India overlaid a regional pattern of dominant agrarian classes that conditioned surplus appropriation and commercial production in Bengal as a whole. He begins the introduction to his book on agrarian Bengal, reprinted here, by arguing against 'the jotedar thesis' posed by Ratnalekha and Ratna Ray, who studied the power of dominant tenants who arose within the zamindari structure as a dominant class during the expansion of commercial farming. Bose analyzes the structure of social-property in Bengal to show three distinct regions: a northern frontier, where the jotedars thrived; a peasant smallholding region in the east, where jute expanded; and a peasant smallholding region filled with landless labourers in the old delta. In the remainder of his book, he shows how these three regions experienced the progressive penetration of commercial production and with it mechanisms of surplus appropriation through rent and credit demands, which 'loaded the dice' against peasants and landless labourers. With partition, the dramatic political consequences of Bengal's regional pattern became apparent.

Two essays on western and central India, by T.C.A. Raghavan and Crispen Bates, complete this volume, by opening up more questions for further study and engaging debates touched on in previous chapters. In the same way as Vasant Kaiwar argued against Eric Stokes' account of peasant productive dynamism, Raghavan argues that landlords in the Narmada Valley did not retard the expansion of commercial farming, but rather drove it forward as they concentrated its benefits to increase and fortify inequalities between malguzars and peasants. In the Narmada wheat boom, moreover, we see another example of a particular market nexus with very specific regional consequences for rural India.

Bates enters the debate to argue, in agreement with Stokes, that the Narmada wheat boom rested on a 'super-exploitation' of peasants that killed agricultural dynamism and made the Narmada wheat boom short-lived, compared with the cotton boom in Berar. More critically, he argues that ups and downs in the commercial production of Berar and the Narmada Valley determined the livelihoods of adivasi people from Chhattisgarh and other regions of Central India. Though it is clear that adivasi workers dislodged from traditional land-use rights played a major role in agricultural expansion in north, east, and central India in the nineteenth century,[65] there is little work on this important subject. Bates shows that colonial settlements created a property system that broke tribal entitlements to land and expelled adivasis to seek wages for much of the year. Thus, like the plantation production of tea and coffee, which rested entirely on the labour of migrants from agrarian tracts,[66] the wheat and cotton regimes of western India depended on labour from other regions, which specialized in labour exportation.

No debate engaged in this volume is closed. Many more, indeed, must be opened up in light of new appreciations for the complexity of agricultural development, which arise with each passing year. Like work compiled here, my research concentrates on small regions and on their interaction with forces emanating from great distances over long periods of time; and like authors in this volume, I search for lessons from agrarian history about the dynamics of modern transformation in rural India.[67]

[65] Dietmar Rothermund, 'A Survey of Rural Migration and Land Reclamation, 1885', *Journal of Peasant Studies*, 4, 3, 1977, pp. 230–42.

[66] J.D. Moore, 'Plantation Development and Labor Response in Nineteenth Century Mysore', University of Pennsylvania dissertation, 1983.

[67] See references above. Also 'Ecological Zones and the Cultural Economy of Irrigation in Southern Tamil Nadu', *South Asia*, I (NS), 1, 1978, pp. 1–13; 'Patronage and Irrigation in Tamil Nadu: A Long-Term View', *Indian Economic*

I hope this volume indicates that much more detailed work on agrarian localities needs to be done, if we are to learn those lessons properly. Such research will be sustained for many decades by the vast quantities of local data unused in archives throughout India. Yet scholars must also somehow fit the countless little bits of agrarian history's puzzle together. Our challenge is to create a form of agrarian knowledge that is true to the diverse and subtle realities of local farm life and which generates intellectual power sufficient to inform Indian historical writing as a whole.

---

and *Social History Review*, 16, 3, 1979, pp. 347–65; and 'The Terms of Ryotwari: Semantics and Disputes over Property Rights in Madras Presidency, 1800–1885', in *South Indian Studies: An Anthology of Critical Essays and Recent Scholarship*, edited by R.E. Frykenberg and Pauline Kolenda (Madras, 1985), pp. 151–170.

Chapter One

Dynamism and Enervation in North Indian Agriculture: The Historical Dimension*

ERIC STOKES

In the west, when they talk of a Purbi *(literally someone from the east, an inhabitant of the middle or lower Ganges)* they automatically add the adjective dhila *meaning rather unenterprising. One cannot but agree with the epithet. We are a long way from the robust northern castes*—Gilbert Etienne, Studies in Indian Agriculture: the Art of the Possible

One of the well-worn problems which have long engaged observers of the agrarian scene has been the uneven growth performance of Indian agriculture in different regions. In the north of the subcontinent there is the obvious contrast between the eastern and western portions of the Indo-Gangetic plain. While Bangladesh, Bengal, Bihar, and eastern U.P. have apparently remained sunk in stagnation and depression, western U.P., Haryana and the Punjab exhibit all the untidy signs of entrepreneurial activity and dynamic growth. Is not the expansion as straightforward as the phenomenon itself? The agriculturally secure regions were the first to enjoy prosperity and the first to

---

* Taken from Eric Stokes, *The Peasant & the Raj* (Cambridge, 1978).

fall victim to over-population, so that the centre of dynamic growth moved progressively away from the deltaic and lower riverine areas to the more thinly-held tracts of upper India, where the *Pax Britannica* and canal irrigation acted like a forced draught behind agricultural expansion. In this way, over the course of the nineteenth-century Lakshmi, the fickle goddess of fortune, betook herself with uneven tread westward from the lush verdure of Bengal until she has come to fix her temporary abode on the Punjab plain between Ludhiana and Lyallpur. The explanation has been applied over a narrower geographical span. Historians have become accustomed to tracing back the decisive east-west shift in economic power and activity in the U.P. region to the railway age of the 1860s and 1870s.[1] It was then that the thriving economy of the Benares region, founded on the export of cash crops like sugar, indigo, opium, and rice, and backed by an important handloom textile industry and a great entrepot trading centre at Mirzapur, began to lose out to the new centres of manufacture like Kanpur and to the wheat and sugar producing regions of the upper Ganges-Jumna Doab.

Overpopulation remained the favourite explanation of the contrast. In the wake of his labours on the U.P. Provincial Banking Enquiry Committee of 1928–30, the celebrated Professor Radhakamal Mukerjee commissioned a series of village and district studies by his M.A. pupils at Lucknow. In a foreword to one of these officially published monographs, Bholanath Misra's *Over-population in Jaunpur*,[2] Mukerjee employed the biological analogy of the 'fruit fly' effect to argue the recessive effects of overcrowding. From 1891 numbers in Jaunpur district had remained stagnant, the result of high disease mortality rates and emigration as population pressed up against the fearful natural limits Malthus had postulated. The pressure of numbers

---

[1] Cf. F.C.R. Robinson, *Separatism among Indian Muslims* (Cambridge, 1974) pp. 59ff.
[2] B. Misra, *Overpopulation in Jaunpur*, Dept. of Agriculture Bulletin, no. 59 (Allahabad, 1932). For a striking exposition of the difference between East and West U.P., see also *Report on the Present Economic Situation in the U.P.* (Govt. of U.P., Naini Tal, 1933); Hailey Collection, India Office Records, MSS. Eur. 1. 230/29C.

had lowered the size of the average cultivating holding to 3.5 acres and had resulted in a *petite culture* increasingly turned back towards subsistence rather than export cash crops. Sugar cultivation had declined progressively since 1841, and stood in 1929 at half the former level, an illustration, Mukerjee declared, of 'the agricultural adjustment of a district which has now more mouths to feed than the existing system and standard of cultivation can afford.'

Mukerjee contrasted Jaunpur with the most thriving district of the western U.P., Meerut. Here the average cultivating holding was more than twice as large (7.8 acres), a clear proof that the proportion of poor subsistence cultivators was much lower. Yet in other respects the contrast with Jaunpur was difficult to press. Compared with Jaunpur's population density of 797 per square mile, Meerut ran at the high figure of 702, and in the western portion the land carried as many as 1,000 per square mile, almost the equal of some of Jaunpur's most congested tahsils. Was it, then, the opening of the Ganges Canal in 1855 and the steady subsequent expansion of the irrigated area that had endowed Meerut with the dynamism of self-reinforcing growth equivalent in its effects to the presence of an expanding frontier into fertile virgin land? Yet, curiously, the double-cropped area in Meerut was no more than 22 per cent of the cultivation by 1908, less than Jaunpur had achieved (26.5 per cent) by its elaborate, though doubtless more labour-intensive system of well irrigation. The cropping pattern, of course, differed. One contrasting feature was the proportion of the cultivated area devoted to sugar by the turn of the century, Meerut's (11 per cent) being nearly double that of Jaunpur (6 per cent). At this time sugar yielded in Meerut the relatively good profit of Rs 15 per acre. By the 1940s both districts had each devoted a further 2 per cent of the cultivated area to sugar. The most striking difference, however, lay in the main crop. Meerut put 30 per cent of the cultivated area under wheat, yielding an estimated average of Rs 15 per acre, while Jaunpur had 26 per cent under rice, which yielded roughly only Rs 9 per acre in 1908. But one advantage of lowering the acreage under sugar was that it extended double-cropping; when rice or maize was substituted in the *kharif* the land could be used again for a *rabi* crop.³

What, then, was the secret of Meerut's far higher living standard, and, even more important, of the thrust and drive in its economic life that contrasted so glaringly with the listless stagnation of Jaunpur? Mukerjee had one principal explanation but half suggested another. The most obvious was the development of urban and industrial occupations in Meerut district which had not only boosted income but also relieved the dependence of the population on agriculture significantly. The more extensive urbanisation of Meerut was no recent feature. Edmund White, the census commissioner, had as long ago as 1882 drawn attention to the marked contrast between the eastern and western districts of the province. The most striking difference lay between Gorakhpur and Saharanpur, with the most agricultural and the least agricultural population, respectively. But Jaunpur and Meerut fell only shortly behind, Jaunpur having 75.77 per cent of its population classified as agricultural and Meerut only 52.2 per cent. The gap continued to widen after 1882, the 1901 census figures being 77.4 per cent and 49 per cent, respectively. By the time Mukerjee wrote in 1932, he reported that 76 per cent of the Jaunpur population were supported by agriculture, 10 per cent by industry, and 5 per cent by commerce, while in Meerut the figures were 42 per cent agriculture, 21 per cent industry, and 6 per cent commerce. The contrasting situation had not been created because of the population of the eastern districts multiplying more rapidly and more recently. So far as the imperfect census statistics indicate, the population of the province increased only slowly from 1853 to 1881. But the increase was far greater in the west than the east, 15.8 per cent in the Meerut division as against 7.2 per cent in the Allahabad and Benares divisions. 'The Meerut Division, as might have been expected', commented White, 'shows the greatest increase. In no part of the province is agriculture so flourishing, and the wealth of the urban population so marked.' Apart from the

³ *U.P. District Gazetteers*, IV, *Meerut* (Allahabad, 1904) pp. 37ff; xxviii, *Jaunpur* (Allahabad, 1908) pp. 31ff.

Benares district itself, the Benares division was purely rural. 'The agricultural classes must have long ago multiplied up to the limits permitted by the size of their holdings.'[4] Temporary emigration provided, of course, an important access to secondary occupations, in Jaunpur a net figure of some 80,000 emigrants being returned at the three censuses of 1901, 1911 and 1921.

Mukerjee believed that, because of fractionalisation and fragmentation of holdings, there was chronic under-employment and idleness in Jaunpur. 'Low agricultural income and agricultural idleness thus often go together and as a result either the non-cultivating money-lending classes or the landless labourers, or both, grow at the expense of small properties and tenants.' In terms of non-agricultural pursuits, the people of Jaunpur had shown far less initiative to strike out on new lines than their brethren in Meerut.[5] This was the nearest he got to half suggesting another explanation and to touching on a question others had treated much more boldly and confidently—the issue of caste.

The British administrative mind had always reverted in its consideration of the agrarian problem to the mental shorthand expressed in the notion of fixed ethnic types. Agricultural performance, it believed quite simply, could be directly predicated on the 'tribe' or caste of the agriculturalist. Such a belief constituted the staple argument of such a leading authority as Malcolm Darling in his *The Punjab Peasant in Prosperity and Debt* (1922) and the numerous subsequent books he published describing his peregrinations on horseback through the old Punjab province. The argument survived in Meerut down to the time that C. Cooke, in 1940, revised the settlement of the district in the expiring days of the Raj.[6] Since independence, mention of caste has, understandably, been more muted; and, for sound political reasons, the Indian Government struck out caste as a category from the decennial censuses. It became for a time as

---

[4] *Census Report N.W. Provs. and Oudh, 1881* (Allahabad, 1881), pp. 27, 31, Table XII.
[5] R. Mukerjee in Misra, *Overpopulation in Jaunpur*, p. 7.
[6] *Meerut Settlement Report 1940*, pp. 15ff.

impolite and impolitic to raise the matter as it once was to mention national work characteristics in the discussion of the stagnation of the British economy compared with the German or French. Economists dislike such irrational and unquantifiable factors: witness Morris D. Morris's attempt to banish caste from economic history in his well-known article in 1967.[7] Agronomists have also been taken up with the revolutionary potential of the new technology in agriculture, with pump sets, fertilisers, and high-yielding varieties, and believe that whatever problem remains may be resolved into one of class and not caste.

Yet, there have always been voices raising the old cry. Professor Etienne, in his book *Studies in Indian Agriculture: the Art of the Possible*,[8] has contrasted the superior energy and purposiveness of the agriculturalists in his sample village in Bulandshahr district with their dispirited counterparts in the Benares district. Like Darling and generations of European observers before him, he ascribed the dynamic informing the agriculture of the western region to the dominance of the Jats and their fighting traditions, so that even in agricultural economics it would seem that the biblical adage holds true: the kingdom of heaven suffereth violence until now and the violent take it by force. Etienne drew on a long tradition testifying to the superior capacity for toil in the west. In 1892 the settlement officer (J.O. Miller) of Muzaffarnagar, the district in the upper Doab immediately to the north of Meerut, commented on how the people 'shrank from a descent to the level of the Purbiyas or inhabitants of the most easterly districts.'[9] They accepted that their higher living standards depended on uninterrupted hard work. In this joy in labour the Jat took the palm, yoking his oxen to the plough before the coming of dawn and returning to the village long after the sun had set. William Crooke gave the Jat cultivator pride of place in the frontispiece photograph of his book, *The*

---

[7] M.D. Morris, 'Values as an Obstacle to Economic Growth in South Asia, *Jl. Econ. Hist.*, xxvii. 4 (1967) pp. 588ff.

[8] G. Etienne, *Studies in Indian Agriculture: the Art of the Possible* (London, 1968). First published in French, 1966.

[9] *Muzaffarnagar S.R. 1892*, p. 23.

North Western Provinces; 'in the whole of India', echoed Darling for the Punjab, 'there is no finer raw material than the Jat.'[10]

Yet, the very men who constructed their analysis out of such solid images proved strangely qualified in their detailed application. Even in the early days John Lawrence, Jat panegyrist though he was, found himself bound to conclude in 1838 that it was the nature of the land rather than innate caste characteristics that appeared to determine the character of agriculture and the mental attitudes of the people. The easily won crops of the sandy rain irrigated (barani) soils of the Bahora tahsil in the later Gurgaon district rendered even the Jat careless, while the stiffer loam of neighbouring Palwal not only brought out the Jat's sturdiest characteristics but also transformed the thriftless cattle-keeping Meo.[11] It was the same in the Doab districts. The change in human attitudes worked by the profitability of canal irrigation was noted by Thomason along the Eastern Jumna Canal well before the Mutiny. Most prominent were the Gujars, traditionally addicted to 'grazing their own and stealing their neighbours' cattle, and leading the idle life they love.' As R.W. Gillan, the settlement officer of Meerut, observed in 1901.[12] 'The Gujars are creatures of circumstance. Give them a canal and teach them the profits of agriculture, and they work their villages like Jats. Put them in a tract like the Loni Khadir and they pay their revenue by stealing cattle and committing burglaries in Delhi.' The same was true of higher castes like Rajputs, Tyagis, and Brahmins, who were important as owners and cultivators, the Jats farming, in fact, only 30 per cent of the Meerut district. When reporting on the revised settlement of the Jalalabad pargana of Meerut in 1868, W.A. Forbes remarked: 'Jats are always a busy pushing race, but Tagas and Rajpoots curiously enough seem always influenced by any example they may have beside

---

[10] W.C. Crooke, *The North Western Provinces* (London, 1857).

[11] *Reports of Revision of Settlements under Regulation IX of 1833 in the Delhie Territory* (Agra, 1846), pp. 25–6.

[12] *Meerut S.R. 1901*, p. 10. Also J. Thomason, Minute on E. Jumna Canal, 1 Mar. 1848, *Selections from the Records of Govt. N.W. Provinces, Mr Thomason's Despatches* (Calcutta, 1856), 1, p. 390.

them.'[13] Etienne noted how, in Bulandshahr district. Brahmins scorned no form of field work.[14] Caste distinctions in the west had always been relatively weak. In the 1870s Charles Elliott, the celebrated revenue authority, found that as far east as Farrukhabad caste had no influence over rent rates. Pradhan has noted how, in the 1950s, in the Jat villages of Muzaffarnagar the Jats engaged for sport in wrestling matches with the lowly sweeper.[15]

How far was this malleability of caste type and responsiveness to economic stimuli absent in the eastern district? Denzil Ibbetson, in his classical treatise on Punjab castes, argued that the people and culture of the Punjab and the contiguous western districts of the U.P. had never come under the influence of Brahmanical religion, and hence caste distinctions had never been ritualised among them, as in the eastern districts. Blunt, in the 1911 Census, used a related argument borrowed from Risley, that, as hypergamy was practised westwards, so caste distance increased eastwards. Did this mean that the higher castes' disdain for agriculture and hence their dependence on a socially inferior tenant and predial labour class, increased as one proceeded eastwards? Here, again fixed stereotypes collapse on closer inspection. C.E. Crawford, the settlement officer of Azamgarh in 1908, found that, while castes with a secondary occupation like Kayasths, and in some measure Brahmins, made indifferent cultivators, 'the Chhattri and the Bhuinhar are on the other hand both industrious and capable.' But to the north, in Gorakhpur district, Brahmins were so numerous that they had been driven much closer to the soil, owning a quarter of the district in 1891 and, together with Bhuinhars, cultivating some 18 per cent (almost exactly the same proportion as Brahmins and Tyagis in Meerut). In Azamgarh it was found that the caste prohibition against Brahmins handling the plough had been relaxed so far as to apply only to the actual operation of plough-

[13] *Meerut S.R. 1874*, p. 39.
[14] Etienne, *Studies in Indian Agriculture*, p. 80.
[15] C.A. Elliott in *Sels. Recs. Govt. N.W.P.*, 2nd ser., II, no. 4 (Allahabad, 1869). M.C. Pradhan, *The Political System of the Jats of Northern India* (Oxford, 1966), p. 50.

ing itself. Brahmins carried out every other agricultural task, even to yoking the oxen to the plough.[16] Plainly, to employ again Edmund Leach's well-known tag, the constraints of economics were prior to the constraints of morality and law.

If human nature proved adaptable to circumstance in this fashion was there perhaps some constraint it could not overcome? The most obvious was the pressure of population on the soil, causing excessive subdivision and fragmentation of holdings. Mukerjee, in 1931, pointed to the fact that the average cultivating holding in Jaunpur was half that in Meerut. The position in Gorakhpur looked even worse. In the Deoria subdivision (later district), where population clustered at 1,100 to the square mile, tenant holdings averaged apparently only 0.65 acres, according to the figures reproduced by another student of Mukerjee.[17] In practice, however, it was impossible to discover actual cultivating holdings out of the multitude of scattered plots into which they were fragmented, and extensive double-counting clearly occurred in census taking. Using plough areas or units, the settlement officer for eastern Gorakhpur in 1919 had a very different story to tell. Even in the overcrowded Haveli pargana of Deoria, he estimated that as much as 55 per cent of the cultivation was made up of holdings of 8–10 acres or more. In the very worst *tappas* of Deoria no more than 27 per cent of the cultivation was taken up by holdings of 4 acres and under, while the average for the subdivision came out at only 16 per cent.[18] This impression that the larger part of the cultivated area comprised holdings of an economic size received striking confirmation thirty years later in the All-India Rural Credit Survey carried out in 1951–2, on the eve of zamindari abolition. In the Survey's *District Monograph on Deoria*[19] a sample of eight villages showed that, although the average cultivating

---

[16] E.A. H. Blunt citing authority of H. H. Risley in *Census of India, 1911*, xv, U.P., pt I (Allahabad, 1912), p. 21, C.E. Crawford in *Azamgarh S.R. 1908*, p. 16. J.R. Reid, *Azamgarh S.R. 1881*, p. 86.

[17] J.K. Mathur, *The Pressure of Population: its Effects on Rural Economy in Gorakhpur District*, U.P. Agric. Bulletin, no. 50 (Allahabad, 1931), p. 9.

[18] *Gorakhpur (Tahsils Padrauna, Hata, and Deoria) S.R. 1919*, p 19.

[19] *All-India Rural Credit Survey: District Monograph on Deoria* (Bombay, 1958).

holding was 3 acres and 66.9 per cent of families held under this amount, nevertheless as much as 69.5 per cent of the holdings area consisted of holdings averaging 8.5 acres[20] and families cultivating more than 8.5 acres held over 40 per cent of the sown area. Of course, holdings in Meerut were unquestionably larger, but the disproportion with the eastern districts was nothing like so great as the misleading statistical averages used by Mukerjee and his pupils (in company with many eminent authorities) suggested. The closest comparable figures we have for Meerut show that in 1940 32 per cent of cultivators had possessed holdings of under 5 acres and occupied some 12 per cent of the cultivated area, while 36 per cent of cultivators held over 10 acres and occupied 60 per cent of the cultivated area.[21] The Jats were far from exempt from the problems of fractionalisation and fragmentation of holdings resulting from their inheritance customs; and even in the Jat homeland the Government of the Punjab took legislative powers in the 1920s and 1930s to effect consolidation of holdings.

Was there, then, some institutional constraint inherent in the system of tenure itself? Here one might venture an hypothesis. Parallel to the much stronger grasp of the higher castes over ownership and cultivating rights in the eastern districts was the contrast in tenures. As late as 1940 52 per cent of proprietary holdings in Meerut were classified as *bhaiachara*, while in the eastern districts joint zamindari and imperfect *pattidari* prevailed overwhelmingly. The vagaries of official classification and the distortions which sale and partition had inflicted on the fundamental tenurial forms made such comparisons valuable merely as broad indicators of the contrast in underlying structure. Even then it may be urged that the distinction among different forms of joint tenure is immaterial, merely signifying whether the land revenue was paid according to ancestral shares (joint zamindari and pattidari) or according to the amount of land held (bhaiachara). But this was not the really vital distinction. For the essence of true bhaiachara was the exclusion

[20] Ibid., pp. 15–16.
[21] Meerut S.R. 1949, p. 17.

of landlord forms within the village or neighbourhood community. Theoretically speaking, the tenure admits of no tributary tenant or quasi-tenant body, and proprietor and cultivator are identical.[22] Baden-Powell noted that bhaiachara peculiarly characterised Jat communities, and that except for the sense of supra-village 'tribal' union the tenure was indistinguishable from ryotwar. Now there seem good grounds for believing that ryotwar tenures throughout India were the tenurial form natural to regions of insecure agriculture where land was plentiful and hands few, just as landlord forms, whether zamindari, pattidari, or 'landed *mirasi*', were the products of regions of secure agriculture where population pressed on the land and generated a quasi-rental surplus.[23] Bhaiachara was the dominant tenure of Haryana, and it is reasonable to suppose that the Jat communities brought the tenure with them into the richer Ganges-Jumna Doab. When Sir Henry Elliot settled Meerut in the mid 1830s he found nothing in the nature of a tenantry existing in the western or Jat parganas, and hence found it impossible to construct standard rent rates on which to frame the assessment, and in the 1860s W.A. Forbes encountered the same difficulty.[24] Whiteway also found this condition persisting among the Mathura (Muttra) Jats at the end of the 1870s, at least in the insecure tracts west of the Jumna. In the Rohtak district (of Haryana) Martin Gubbins had reported in 1839: 'Rent rates do not exist in this

---

[22] In practice there was, of course, always a minority of temporary and hereditary 'non-proprietary' cultivators, who usually held land at the same rates as members of the proprietary body. In 1827 in Khanda, a village in the Kharkhaudah pargana of Rohtak district, G.R. Campbell classified the population as follows:

Proprietors	Hereditary cultivators and new settlers	Engaged in 'industry'	Trade	Labourers
1,273	204	74	71	6

I.O.R. Board's Collections. I,/4/1215, fo. 414.
[23] See above, pp. 7, 54, 88.
[24] *Meerut S.R. 1874* Resolution, Revenue Dept., 10 June 1880, p 8. Also H.M. Elliott in *Selections from the Reports of the Revenue Settlement of the N.W. Provinces under Regulation IX, 1833* (Benares, 1862–3), 1, pp. 179ff. Also *Meerut S.R. 1991*, pp. 15ff.

district—land being in greater abundance than hands to cultivate it—proprietors are generally glad to let their free lands to non-proprietors at the same rates of revenue which they pay themselves.' As late as 1910 two-thirds of the arable land in Rohtak was still being cultivated by proprietors and the remaining one-third was only notionally held by tenants. The settlement officer said that many of the latter 'were in the position of villagers who subsisted by taking in each other's washing. There is no real tenant class. Owners who exchange plots for temporary convenience in cultivation, and men who take a little free land from their fathers are all recorded as tenants.'[25] Sixty per cent of land ownership remained in Jat hands, a proportion that rises still higher if the non-Jat area of Jhajjar, added after the Mutiny, is excluded. In Haryana, the word for farmer and Jat was synonymous, suggesting that the economic basis for the social and ritual distinctions of caste did not exist within the cultivating communities. At the same time, agricultural conditions were such as to require and reward unremitting labour, neither breeding the fatalism of the semi-desert nor the careless indolence that Malcolm Darling believed had corrupted the Jats of the fertile Naraingarh tahsil of the Ambala district, where landlordism and Rajput life-styles had crept in.[26] In Rohtak, there was no weakening of the fibre. Here Darling thought human toil could be carried no further, even the Jat women rising before 5 in the morning, grinding the corn, performing all manner of field work, and not ceasing their labours until late at night.[27]

How did the bhaiachara form survive in the Doab? Sir Henry Elliot noted in 1836 how, in the Meerut district, Rajputs, Pathans and Sayyids, 'being too indolent and proud to cultivate much themselves, inclined to pattidari tenures.'[28] In other words, overlord castes, whenever opportunity offered, tended to retain a portion of the village lands in their own hands as their *sir* or home farm and to leave the remainder to be cultivated by others

[25] *Rohtak S.R. 1910*, p. 11.
[26] M. Darling, *Rusticus Loquitur* (Oxford, 1930), p. 68.
[27] Ibid., pp. 98, 103.
[28] H.M. Elliott, *Sels. Reps. Rev. Settl. N.W.P. Reg. IX, 1833*, II, p. 190.

in return for tributary payment or rent. Landlord forms were repressed among the Jat communities by the high pitch of the differential revenue assessment imposed upon them by the Begam Sumru and the successor British power down until the 1860s. When at length the demand was eased and the railway and canal irrigation gave promise of a new prosperity, the balance of advantage between rental income (or, more strictly, *malguzari* profits) and the profits of direct farming swung decisively in favour of the latter, assisted no doubt by the increasing rent restrictions of British tenancy legislation. Hence, there was little temptation to develop a rentier landlord class among the Jats.

In the eastern districts the regression of the proprietary castes back into cultivation probably began much earlier. Indeed, Buchanan Hamilton commented on the direct participation of Brahmin groups in agriculture at the beginning of the century.[29] The later decades of the nineteenth century witnessed a general quickening of the movement of proprietary castes to extend their cultivation by the enlargement of sir as well as by taking up land on tenancy.[30] At some points, overswollen Rajput proprietary communities reproduced the outward lineaments of the bhaiachara tenure, the most remarkable examples being the bighadam *mahals* of the Dobhi taluq in Jaunpur and of the neighbouring Deogaon and Belhabans parganas of Azamgarh, where the entire village lands were claimed as sir. Yet the slow, enforced descent from lordly status and the gradual shift from close joint landlord forms towards a type of ryotwari (bighadam) were hardly the conditions to breed dynamic entrepreneurship in agriculture. The old tenurial forms proved obdurate, not simply because of the struggle that had to be fought with lower-level cultivators to regain cultivating possession but because, for the higher castes, the old landlord tenures continued to possess considerable short-term social and economic

---

[29] Cited Montgomery Martin, *The History, Antiquities, Topography, and Statistics of Eastern India* (London, 1838), II, p. 452.

[30] *U.P. Board of Revenue Administrative Report 1903–4* (Allahabad, 1905), p. 18, para. 42. Also ibid., 1882–3 (Allahabad, 1884), Divisional Reports, p. 26, quoted above in Paper 9, p. 211.

advantages. Wherever possible Rajput, Brahmin and Bhuinhar remained a petty landlord class, driven close to the soil but still continuing to exploit it through lower caste sub-tenants and predial labourers.[31]

These tendencies were not absent from similar castes in the west. Not all Rajput groups in the upper Doab had improved themselves out of recognition. Mukerjee's M.A. students readily resorted to the old stereotypes of thakur ease and indolence when they came to examine Rajput villages like Khiwai in the Meerut district.[32] Nor, because of the almost equally intense pressure of population on the land, did the Jats wholly escape landlordism and preserve the pristine character of their bhaiachara tenures unimpaired. In the Doab, unlike much of Rohtak, the Jats had always made use of substantial Chamar predial labour and had never enjoyed the same almost exclusive tribal monopoly of population.[33] Moreover, there was not a straight choice between the roles of rentier and farmer. The fact that the largest average cultivating holding was 40 acres meant that the ownership of malguzari rights continued to offer itself as a means of obtaining additional income as well as of social and political consequence. (The estimated rental assets of the U.P. doubled between 1899 and 1929 while the revenue demand rose by only 12 per cent). The bhaiachara tenure, despite appearances, had never been an egalitarian structure and had always

---

[31] See above, p. 79.

[32] Report of Kunwar Bahadur, *U.P. Provincial Banking Committee Report 1929–30*, II (Allahabad, 1930), p. 228.

[33] Pargana caste statistics, given in *Meerut S.R. 1874*, Introdn. pp. 6–7, show that in the district as a whole Chamars (197,273) outnumbered Jats (145,524) in the total district population of 1,273,676, but in the western Jat parganas (Baraut, Chaprauli, Kotana, and Barnawa) Jats outnumbered Chamars more than two to one (56,152 to 23,408). In Rohtak at this time there was still no proper agricultural labour class and Chamar numbers remained proportionately small: Chamars in 1881 numbered 50,081 compared with Jats, 182,776 in a total district population of 533,609. *Punjab Gazetteers, Rohtak District 1883–4*, pp. 81, 125, 128 and Tables IX and XV. But in the rural population Jat preponderance was even more striking. In the Gohana pargana in 1839 the crude census showed Jats as forming as much as 80 per cent of the population. *Sels. Reps. Rev. Settl. Reg. IX, 1833, Delhie Territory*, no. 1 (Agra, 1846), p. 81.

spanned large discrepancies in the size of holdings. In formal terms, it could readily break up into a scatter of small zamindari properties. Evidently, in some villages Jats were demoted to become tenants of others, as in Edalpur in Mathura district.[34] Ownership was also intimately linked with the system of cash-crop production, marketing, and the provision of credit, in a skein so complex and variegated that commissions of inquiry never succeeded in disentangling it. How far this complex structure, along with greater urbanisation and a more substantial agriculture-based industry, supplied one of the keys to Meerut's superior economic drive must remain an important but unanswered question. One fact seems clear. In the eastern districts landlords and superior agriculturalists constituted the main source of agricultural credit, whereas in Meerut the professional moneylender and trader held a large share of the business. No doubt rural credit and usury remained an important buttress for landlord elements to shore up their position in the east.[35]

Ingrained attitudes associated with caste cannot be swept aside, but as a local kinship system caste was, in Leach's words, merely another way of talking about a system of property relations. What is suggested in this paper is that, while in the agrarian domain pressure of population and subdivision of holdings were tending to approximate conditions in the eastern and western U.P. more closely, the traditional tenurial expression of property rights was one of the factors helping to keep them on different paths. The owner-cultivator holding was the form natural to the Jats and bhaiachara the most readily adaptable tenure to maximise production under Indian *petite culture*.[36]

[34] *U.P. Pror. Banking Enqu. Rep.*, ii, Evidence, p. 250. For increase of rental assets, A.A. Waugh, *Rent and Revenue Problems* (30 Aug. 1934). Hailey Collection, I.O.R., Mss. Eur. E, 220/29C.

[35] Written evidence of Mohan Lal Sah, ibid., p. 51. Also pp. 153, 162, 175.

[36] The British classification of tenures was necessarily crude and inconsistent. Rarely did tenures conform to their 'ideal types' and the imperfect *pattidari* villages of the Punjab bore only the most superficial resemblance to those of the U.P. eastern districts just as did the western *bhaiachara* to the bighadam. The real test was the extent to which owner and cultivator diverged or coincided. Cf. the despairing cry over the confusion of terms by B.H. Baden-Powell, *The Indian Village Community* (London, 1896), pp. 353ff.

In the east the joint zamindari and imperfect pattidari tenures proved atavistic devices, landlord structures that had outlived their role but survived because even a minute fractional share in the joint patrimony validated caste status for other purposes.[37] Among the Dobhi *thakurs* the joint rather than the nuclear family remained overwhelmingly predominant. Among congested communities conditions inexorably drove the dominant castes into similar *de facto* owner-cultivator holdings (though ownership in this sense might mean the beneficial ownership equated with occupancy tenant right), but generally the slower institutional descent from landlord status failed to generate an answering entrepreneurial drive in the newer role of farmer.[38] In caricature it produced the squireen rather than the green revolutionary. S.P. Sharma has concluded,[39]

Some of the social differences between Rajput and Jat groups can still be found if we examine the Thakurs of Senapur [in the Dobhi taluq

---

Also H.M. Elliot, 31 Aug. 1836: 'The fact is, whatever definition may be given, it will not apply equally to every Zillah. Some little peculiarity will, perhaps, be found which would exclude each tenure from the limits which had been assigned to it; what one man includes under *zamindaree* another calls *putteedaree*, and one includes under *putteedaree*, what another calls *bhyachara*.' *Sels. Reps. Rev. Settl. N.W.P., Reg. IX, 1833*, 1, p. 186.

[37] S.P. Sharma, 'Marriage Family, and Kinship among the Jats and Thakurs of Northern India: some comparisons', *Contributions to Indian Sociology*, new ser., 7 (1973), pp. 81ff. Also see above, p. 42.

[38] The extent of owner-cultivation varied considerably in the eastern districts. In Jaunpur *sir* and *khudkasht viz*. proprietary cultivation, was estimated at 18.59 per cent in 1906 compared with 42.78 per cent in Meerut in 1901. Yet in Azamgarh, the district adjoining Jaunpur on the north, the area of sir and Khudkasht ran as high as 41 per cent in 1908, of which rather more than one-quarter was subject. *Azamgarh S.R. 1908*, pp. 14–15. The difference lay in the extent of cultivating rights relative to proprietary rights. In Meerut at the turn of the century Jats owned some 27 per cent of Meerut district and cultivated 30 per cent. There are no strictly comparable figures for Azamgarh, but in Belhabans pargana, for example, where the dense Bais Rajput communities owned 89 per cent of the land and sir amounted to 46.2 per cent 'the Rajputs hold less of the land than might be anticipated; for Ahirs hold nearly as large an area [as cultivators], and Brahmans, Lunias, and Chamars all cultivate a large acreage.' *District Gazetteers U.P.*, xxxiii, *Azamgarh* (Allahabad, 1911), p. 206.

[39] S.P. Sharma, 'Marriage, Family and Kinship', p. 83.

of Jaunpur] and the Jats of Basti [pseudonym of a village 15 miles south west of Delhi on the Gurgaon border] in contemporary context. Until the later part of the nineteenth century the Thakurs of Senapur, though they were the undisputed landlords of the village, did not personally cultivate the land because they thought of themselves as rulers rather than as workers of the land. All the agricultural land in their possession was worked by hired labour. It is only during the past two or three generations that the Thakur has been shamefacedly involved in agriculture. The Jats on the other hand have always been farmers, deeply rooted in the soil. In Basti Jat women and children work with their menfolk in the fields. According to a local saying: 'The Jat's baby has a plough handle for a plaything.'

Chapter Two

Property Structures, Demography and the Crisis of the Agrarian Economy of Colonial Bombay Presidency

VASANT KAIWAR[*]

Some of the most influential arguments about the failure of dynamic development in Indian agriculture during the colonial period have focused on the scarcity of capital, the absence of suitable technology, and demographic growth. This paper assesses the role of those limitations, but contends that they describe the impasse of colonial Indian agriculture rather than explain why it happened. It is necessary to explain why available capital was not used productively, why formal potentials for technical improvements were not introduced into production, and why demographic growth should not have stimulated innovations rather than cause technological stagnation. To do so requires bringing into central focus the social-property relations. The paper explores the manner in which peasant possession of land and other means of subsistence limited productive utilization of capital and technology, triggered a certain demographic regime and, in turn, disrupted further developmental possibilities. The paper is based on a detailed inquiry into the conditions prevailing in Bombay Presidency, an area roughly coinciding with today's Western Indian states of Maharashtra and Gujarat.

[*] Department of History, State University of New York at Albany, Albany, NY 12222. An earlier version of this paper was presented at the New York Asian Studies Conference at Cornell University, November 3–4, 1991. I wish to thank Sucheta Mazumdar for her time and insights. It is a pleasure to acknowledge the generosity of Robert Brenner, S. Ambirajan, Christopher

Introduction

In the Western states of Maharashtra and Gujarat capitalist development in agriculture has taken off in a significant way. Much of this development has occurred in the post-1947 period. During the late colonial period, however, this area went through a profound crisis manifested in falling yields, famines and epidemics. This paper explores the causes of that agrarian crisis.

By 1760, the most populous parts of this region had been occupied by the Marathas, with notable exceptions like Surat which was brought under East India Company control. Some expansion of tillage and irrigation had apparently been undertaken in Maharashtra under Maratha rule, but starting in the late eighteenth century a very disturbed period of political contention followed. Many areas in both Gujarat and Maharashtra were seriously devastated by wars, heavy taxation, famines and the ensuing depopulation. The northern Deccan district of Khandesh and parts of the Southern Maratha Country were most adversely affected. The heart of what later came to be called Bombay Presidency was created out of two major conquests by the East India Company—the districts of Gujarat in 1802-03, the Deccan and the Konkan in 1818 with the fall of the Maratha Poona Peshwa. The administrative unit thus created covered much of the western coast in a long narrow strip—nowhere was it more than 200 miles from the Arabian Sea [*Divekar*, 1982].

It is always tempting to consider the particular case study one is involved in as somewhat unique, and to overemphasise its diversity.[1] Bombay Presidency was not any more diverse than regions of similar size elsewhere in India. What makes this region interesting from the point of view of the long-term development of capitalism is the very high level of merchant capital in circulation long before the coming of the British. Of equal interest were tendencies for merchant capital in the earlier

---

Bayly and Peter Perdue in reading this paper critically. I am also indebted to Tom Brass, Alan Fogelquist, Richard Lachmann and Michael Soldatenko for their comments. The usual disclaimers apply.

[1] A good example of this is *Charlesworth* [1985: 10–16].

Maratha period to establish purely economic forms of control over the labour of peasant producers.[2] At the same time, peasant property forms came to entrench themselves.

The British colonial power accepted these long-term developments but proceeded to secure *individual* property rights in land even more firmly as a prelude to a system of direct revenue extraction from the peasantry—the ryotwari system of land settlement. Additionally, the state put in place an extensive railway system that brought into existence regional and national markets for bulk commodities like foodgrains and commercial crops. Finally, the colonial state introduced a variety of 'bourgeois' laws, e.g. laws of contract, and the Pax Britannica. It made market integration easier by developing a standard currency, weights and measures. All of these changes were fully expected to bring development to the agrarian system. Colonial officials and latter-day social scientists have been concerned to explain why that did not happen.[3] This essay will examine some of the explanations, note their shortcomings and offer an alternative.

ELEMENTS OF THE AGRARIAN CRISIS

Overall, agricultural growth in colonial India was quite unspectacular. Foodgrain production rose steadily from the mid-nineteenth century to about the World War I period, more or less keeping pace with growth of population. There appears to be a consensus that the turn of the century was a kind of watershed all over India. During the 1890s, small increases in foodgrain production were still being achieved, against a backdrop of unusually slow demographic growth [Mann, 1967: 322]. Once the period of famines and pestilence from about 1875 to 1918

---

[2] Some of this is captured in *Perlin* [1978].

[3] It is necessary to understand that we are dealing with a problem here that has exercised generations of practical administrators and scholars. My purpose is not to construct an arbitrary counterfactual question ('Why did development *not* take place?') but to show why certain expectations were not fulfilled, and how some of the explanations offered do not quite measure up against the historical evidence.

was past, population grew rapidly and food production could no longer keep pace. On the basis of production and area figures for crops given in the annual *Estimates of Area and Yield of Principal Crops in India*, economist George Blyn estimated that over the period 1891–1941 the yield per acre of foodgrain declined by 0.18 per cent per year, with the decline becoming particularly rapid (0.44 per cent per year) after 1921.[4] In Bombay, 'the disparity in trends (population and production) started about midway between 1911 and 1921', and by the 1920s per capita production was clearly deteriorating.[5] Even 'optimists' among economic historians—Charlesworth among them—acknowledge that the period after 1921 or so saw little per capita growth, as population growth grew rapidly and the depression hit the agrarian economy hard [*Catanach*, 1970: 181; *Charlesworth*, 1979; 1985]. Nationalist historians have tended to see the origins of the downturn occurring much earlier, in the last quarter of the nineteenth century [*Guha*, 1985a].

During the second half of the nineteenth century, there was a notable extension of cultivation in Bombay Presidency. This varied from region to region, depending on prior density of settlement and on the extent to which agriculture had been disrupted by wars and famines before the take over of Western India by the British.[6] There is no question that by the 1870s the supply of good quality arable land was near exhaustion through much of Western India. In some areas, hillsides had been denuded of shrub and forest cover. In others, marginal land previously left uncultivated was brought under cultivation.

There was severe damage to agriculture and in general to the ecology of the region from bringing into regular cultivation

[4] *George Blyn* [1966: 105, 171]; *S.C. Mishra* [1983: 173]; *W.J. MacPherson* [1972: 137]; see also *K.M. Mukerji* [1962]; *S. Sivasubromonian* [1960].

[5] *Blyn* [1966: 100]. Charlesworth dismisses Blyn's evidence suggesting among other things that even if the new land brought under the plough in the late nineteenth and early twentieth centuries was considerably less productive than the old land, 'it seems no difficult matter to have maintained or even increased per capita production over the early part of the twentieth century', [*Charlesworth*, 1985: 207]. I shall show that the qualitative evidence is overwhelmingly against such an assertion.

[6] For a detailed study of this, see *Kaiwar* [1989: 74–77].

poor-quality soils previously uncultivated. Damage could also result from common cultivation practices and from the overgrazing of inadequate and marginal pasturelands, a result of the growth of arable at the expense of pasture. Fully 38 per cent of the soil in the Deccan and 26 per cent of the soil in Karnataka was *highly* eroded by the end of the colonial period [*Basu and Puranik*, 1950: 482–84]. With the reduction of grazing to marginal land, roadside pastures and hillslopes with thin soil, cattle numbers and quality could not keep pace with the extension of cultivation. During the period, 1843–73, when agriculture was undergoing rapid expansion throughout the Deccan, cattle numbers actually declined by about 4.8 million [*Deccan Riots Commission, Report* (1875), henceforth *DRC*: 1,42]. Given the absence of famines in these decades, the reduction could only have come about as a result of the decline in the quantity and quality of pasture. When famine struck in the late 1870s, cattle mortality could be spectacular. In Shrigonda taluka (Ahmednagar district), between 31 July 1876 and 28 May 1877, the cattle population was reduced from 27,793 to 11,628 and in nearby Karjat from 21,586 to 5,584. Local shortages of pasture hurt all cultivators, rich and poor alike. For instance, in October 1876, with famine impending, 'well-to-do cultivators' in Sholapur sent their cattle to distant districts for grazing in charge of servants or family members. This desperate and costly measure did not succeed since it was reported that many of those people returned with only a small proportion of the cattle entrusted to them.[7] In the famines of 1896–1900, which affected all the districts of Bombay Presidency, the losses were staggering. Losses of plough cattle ranged from 25 per cent in Gujarat to 52 per cent in Karnataka (Southern Maratha country), with the Bombay Deccan losing 38 per cent. Gujarat lost 44 per cent of its milch cattle and 31 per cent of its young stock; Karnataka 41 per cent of its milch cattle and 7 per cent of its young stock; the Deccan 35 per cent of its milch cattle and 27 per cent of its young stock.[8]

This was accompanied by the disappearance of the nomadic

[7] *Bombay Revenue Department*, vol. 19 of 1877, Report by J. Davidson, cited Mishra [1982: 25].
[8] *Report on the Famine in Bombay Presidency 1899–1902*, [1902: 105].

cattle breeders, who in earlier times had been an important part of the rural economy of Western India. The availability of extensive free grazing in the forests and grasslands of the region had made it possible for these men to raise a strong breed of cattle, intended for sale to cultivators and kept separate from weak and sick animals.[9] As tillage expanded, these common resources were stripped away, fodder became scarce and expensive, and the nomadic cattle breeders were no longer able to continue their old profession. The remaining village commons became the indiscriminate breeding ground for a less sturdy breed of cattle, which moreover had to compete for fodder for at least six months of the year with 'aged and barren cows' and 'useless bullocks.'[10]

The stripping away of necessary forest cover undermined the ecological balance of rural Bombay Presidency. Colonial officials were well aware of the resulting problems. As one of them—A.G. Edie, Chief Conservator of Forests in the Bombay Presidency—noted in evidence to the Royal Commission on Agriculture, forests helped to retain moisture in the soil and regulate the flow of water in the streams. He observed that in forest-clad areas streams contained water all year-round, but in areas stripped bare the rivers were raging torrents in the rainy season and bone-dry the rest of the year. He also noted that land in the neighbourhood of barren hillsides was often flooded and in turn lost top-soil, a consequence in the first place of the deforestation of hillsides. On some hillsides, the top-soil had almost completely washed away as a result of both careless expansion of cultivation and overgrazing by animals.[11] In due course, this led to a lowering of the water-table and the consequent drying out of wells. In some parts of the Presidency, notably the Karnataka districts, tanks used in small-scale irriga-

---

[9] E.J. Bruen, written evidence to *Royal Commission on Agriculture, [RCA] Evidence Taken in the Bombay Presidency* (Bombay, 1927) [II:1: 399]; also *Department of Agriculture (DOA) Bulletin* [Bombay, 136 of 1927] [*Cattle Breeding in the Bombay Presidency* by E.J. Bruen].

[10] H.F. Knight, written evidence to *RCA* [II,1: 294].

[11] Evidence of A.G. Edie, Chief Conservator of Forests, Bombay Presidency to *RCA* [II,1: 146].

tion schemes silted up rapidly in the early decades of the twentieth century.

A very large proportion of the population remained dependent on agriculture throughout the colonial period. In 1928–29, in Olpad *taluka* in Surat district,[12] a fairly representative area of south Gujarat, 83.5 per cent of the work force was in agriculture, fishing or animal husbandry. Eighty two per cent of the total population was supported by these activities [Shukla, 1937: 69]. Notwithstanding this, food imports became a regular feature of Bombay Presidency's economic life. P.C. Patil, in oral evidence to the Royal Commission on Agriculture, pointed out that annual production per capita of cereals and pulses came to 713 lbs. (or about two lbs. per person per day). He worked out the per-capita requirement of cereals and pulses at 3.3 lbs. per day for a minimally nutritious diet, the chronic shortfall therefore being about 33 per cent. The calculation was made in 1913–14, a 'normal year.' In a year of serious drought, it would be even greater, probably approaching starvation levels unless large amounts of food were imported.[13] Though food was routinely imported into Bombay Presidency mainly from Sind, United Provinces or Madras [Watt, 1908: 1042], there is little evidence that the amounts were anywhere near adequate to meet the shortfalls indicated by Patil. In a 'normal' year in the early twentieth century, imports into Bombay Presidency (excluding Bombay city) varied from one-quarter million cwt. (1904–05) to about 2.5 million cwt. (1905–06).[14] In a famine year, this could double: in 1899–1900, four million cwt. of foodgrains were imported and in 1900–01, about 5 million cwt., but these amounts were really quite negligible in terms of meeting people's needs, especially in years when the crops almost totally failed over large areas of the Presidency. Gujarat—supposedly a pace-making region—was also importing food in the twentieth cen-

[12] *Taluka* refers to an administrative unit, usually a sub-division of a district.
[13] Patil, oral evidence, Q. 8017–23, to *RCA*, [II,1: 526]. Ironically, Mann visualised an increase in output of about 33 per cent from the adoption of improved methods of dry-farming. On mass starvation in Bombay Presidency, see *Klein* [1984].
[14] Twenty hundredweights (cwt.) make one ton.

tury though 55 per cent of its arable was under food crops [*Desai*, 1948: 58].

Despite the imports, the diet of the ordinary cultivator, and even urban workers, was woefully inadequate. Mukhtyar's research in Atgam in South Gujarat in 1930 indicated that ordinary cultivators suffered from chronic protein deficiency [*Mukhtyar*, 1930: 302–03]. Wages for agricultural labour between 1900 and 1920 did not keep pace with the rising cost of living (generally reflecting the rise in food prices). While the cost of living rose by 100 per cent or so, wages of agricultural labourers rose by 25–50 per cent [*Shirras*, 1924: 23–24]. Textile workers, generally transplanted from the hinterlands of Bombay and Ahmedabad, were not much better off. Expert evidence showed that the average weight of mill workers was around 99 lbs. and their consumption of cereals was lower than the amount recommended by the Bombay Jail Manual! To pay for this poor diet, mill workers spent nearly 60 per cent of their income on food.[15] Solvency data gathered in the Deccan villages of Jategaon Budruk and Pimpla Soudagar indicate that over time fewer cultivators were able to be solvent (i.e. meet their subsistence needs) entirely from cultivating their farms.[16]

I am proposing that this evidence taken cumulatively points to a decline in productiveness in the agricultural system. It also suggests that this decline affected ordinary cultivators in Bombay Presidency adversely. None of the elements of agricultural decline described above is decisive taken in isolation from the other elements. But together they make a clear case for a serious crisis of the system of economic reproduction of the peasantry.

This crisis created problems for the state whose income depended on revenue from the land. For instance, land revenue provided half the gross income of government for most of the

---

[15] This evidence was given to the *Royal Commission on Labour, 1931* [I,1: 317].

[16] See *Mann* [1917] and *Diskalkar* [1960, particularly the introduction to the volume by Mann] for evidence of agricultural decline in Pimpla Soudagar; *Mann and Kanitkar* [1921] on the original survey of Jategaon Budruk in 1917, and Mann's evidence to *RCA* [II,2: 16] where he provides information on the resurvey of the same village eight years later.

nineteenth century and as late as 1900 was the leading source of revenue contributing 40 per cent to the state's coffers [*Charlesworth*, 1982: 66]. However, this also limited the fiscal base of the state, especially as the agrarian crisis deepened in the twentieth century. Periodically, bad seasons and famines reduced the amount of land cultivated and hence the amount of revenue that could be collected from the peasantry [*McAlpin*, 1983: 124; *Charlesworth*, 1979: 120-1]. Arguably, the decline of the agricultural sector limited the capacity of the state to tax it for productive ends, i.e. the construction of necessary infrastructure and services needed for further development.[17]

It is a cliche of Indian historiography that industrial development did not draw population away from the countryside, thus relieving pressure on the land. It is a moot point as to whether this would, in itself, have solved the structural crisis of agriculture that I will discuss in this paper. Nonetheless, it is noteworthy that many people, not previously agriculturists, flooded into the agricultural sector during the colonial period.

Many of the erstwhile artisanal castes, e.g. weavers, potters and other artisans, had turned to agriculture as a primary source of living exerting greater pressure on the landed base. In Shukla's study of Olpad *taluka*, of the 842 persons whose names indicated they were of the weaver caste only 66 remained in their traditional profession, the rest having become agriculturists. Every other profession was more or less affected by industrial competition, if not as drastically as the weavers. Only carpenters, who made and serviced the wooden implements used by the cultivators throughout the colonial period, survived in any strength [*Shukla*, 1937: 59-60].

The pre-colonial economy that had supported some occupational diversity was being replaced by one much more purely agricultural in orientation. In the decade 1911-21, the population dependent on agriculture in all of India increased by 4,350,000

---

[17] It is hard to separate causes and effects here. Obviously, the economic performance was not unrelated to certain institutions and property structures put in place by the state or at least tolerated by it, but those in turn limited the capacity of the state to undertake an effective investment program.

and that dependent on industry declined by 2,150,000. This trend continued during the next decade, when population grew rapidly, thus greatly intensifying pressure on the land. In Bombay Presidency, the most industrialised area of British India, in the decade 1921–31, while population rose 15.1 per cent, that of Bombay City declined by 1.53 per cent [*Buchanan*, 1937: 130–31].

How did such a crisis come to pass? Scholars of Western India have, by and large, concentrated on one of the following: capital scarcity, that made it impossible to undertake new investments necessary to raise yields; the exhaustion of appropriate technology, that pushed available capital away from investment into speculative uses; and demographic growth that split farms into smaller units, encouraged the cultivation of more marginal land, discouraged specialisation and precipitated subsistence crises and economic decline. Are these arguments sufficient to *explain* the problems confronting the rural social-economy? Or is it necessary to focus on the social-property relations, the manner in which immediate producers established access to their means of subsistence, including land? Will an examination of the social-property relations yield a better, more inclusive, explanation of the *limits* to the productive uses of capital, and the adoption of available and suitable technology? And what is the relationship between access to property and the demographic regime in the rural social-economy of Western India? Finally, is it necessary to reorient inquiry towards the qualitative relationships between the major social agents—the peasantry, the capitalists and the state—in a specific historical context, and away from mere quantitative investigations of capital (in the sense of amounts of money), technology and demography (statistical studies of rates of population growth)?

CAPITAL CONSTRAINTS AND AGRICULTURAL GROWTH

Several Bombay officials thought the failure of agriculture to develop was due to the lack of capital. Harold Mann, Director of Agriculture of Bombay Presidency, giving evidence to the

Royal Commission on Agriculture in 1927 argued that the 'credit' which had been available to the *ryot* (peasant) in the past was for practical purposes exhausted. Contemporary Indian economists, like P.C. Patil, concurred with this reading citing, among other things, the problems of finding capital to set up co-operatives. Patil's empirical studies also led him to conclude that most agricultural production was for subsistence, and not for profit. Presumably the surpluses needed for investment and growth were not being generated within the agrarian sector.[18] However, the capital scarcity argument is problematic.

There is ample evidence that the spread of agriculture in the second half of the nineteenth century on to previously uncultivated land was largely financed by 'credit', i.e. money capital advanced as loans. In 1855, in Kolhapur district, the total amount of money capital available in the form of credit was estimated to be around Rs 3.1 million. In 1875, in a comparable area, Satara district, the total amount of money-capital of this sort was around Rs 15.3 million.[19] Throughout the Deccan, many of the moneylenders—nearly 40 per cent—had established themselves after 1850 when the expansion was in full swing. Investigative reports by members of the colonial bureaucracy found close links between the expansion of agriculture and the growth in the volume of money capital.[20] As the state's revenue demands on the peasantry fell after 1850, the volume of money-capital advanced by private interests increased. George Wingate, an official of the Revenue Department, reporting on a group of villages in southern Bombay Presidency in 1875 noted that total interest paid by villagers (Rs 10,932) exceeded the state's revenue demand (Rs 8,381). Loans taken by a group of

---

[18] For the capital constraint argument, see Mann's oral evidence Q. 3500–02, response to a question from Hubert Calvert, *RCA*, [II:1: 75]; also see Patil's oral evidence, Q. 7958–63 to *RCA* [II,1: 523].

[19] *Bombay Government Records*, [No. 8: 275]; *Bombay Revenue Department*, volume 8, 1875, Report of A. Wingate.

[20] The connections emerge not least of all in the reports submitted to the *Report of the Deccan Riots Commission*, henceforth *DRC*. Between one-third and two-thirds of the ryots were recorded as being involved in debt-relations [*Banaji*, 1978: 371].

251 peasants in Poona district went up from Rs 28,775 in 1860 to Rs 128,696 in 1875. It is reasonably clear that the agricultural expansion of the second half of the nineteenth century was financed by merchant-moneylending capitalists, and showed up symptomatically as the increased indebtedness of the immediate producer [DRC II, 1875: 2,39].

By the late nineteenth century, the supply of cultivable land was more or less exhausted. Occasionally, however, due to famine or epidemic, large amounts of land did become available. Such land was rapidly settled by peasant cultivators, with loans from merchant-capitalists. In the Gujarat districts of Kaira and Panch Mahals, the famine of 1899–1900 took a heavy toll on the population.[21] The former lost close to 30 per cent of its population in those two years. On the Mehlol estate in the Panch Mahals, where the government gave out large farms at low rents, tenants came from as far away as Charotar and Kanam in Kaira district. Some of them had access to large amounts of money-capital, as evidenced by the case of the *ryot* who invested Rs 5,000 on irrigation. Others had access to large amounts of interest free loans under the Agricultural Loans Act of 1884. In Mundgod Peth, of Kanara district in the Southern part of the Presidency, large tracts of rice land which lay uncultivated in 1923–24 following the influenza epidemic of 1918, were brought under cultivation by merchant capitalists who undertook the considerable expense of bringing the land under cultivation, hiring tenants, and building houses for them, with government assisting in the process by giving out the land on 'easy terms.'[22]

Although the merchant capitalists were seemingly willing to open up land for cultivation and give loans to facilitate this process, evidence given to the Committee set up to investigate the Deccan Riots in 1875 commented that capitalists were averse to a deeper involvement in agriculture: 'Profits do not return to it [the land] to cultivate and to manure.' The report said that these capitalists were unwilling to invest in long-term fixed

[21] R.M. Maxwell, Collector of Kaira, written evidence to RCA [II,1: 341].
[22] [*Bates*, 1981: 804]; R.M. Maxwell, Collector of Kaira, written evidence to RCA [II,1: 339].

capital.[23] *Prima facie* evidence suggests that capital was available for the extension of cultivation, though the situation was more problematical when it came to long-term fixed capital investments. The following sections will explore facets of this problem.

Sugarcane Production on the Canal Zones

The opening of major canal systems in the Deccan—the Nira Canal in the late nineteenth century, the Godavari and Pravara canals in the twentieth century—made new lands available for cane cultivation. It was not the high cost of sugarcane cultivation that limited it to a mere 60,000 acres in the Presidency, but the requirement for reliable, year-round irrigation. Even along the canals, water was not evenly available to all cultivators and thus only the best located cultivators could make the switch to sugarcane cultivation. At the best of times, profits of Rs 100–200 could be made on an investment of Rs 600 per acre [*Attwood*, 1984: 24–25; *Charlesworth*, 1979: 127; *DOA* No. 60: 33]. The outlay, considerable by the standards of Western Indian agriculture, included the whole range of production activities: land improvement, the purchase of good quality bullocks, heavier —often metal—implements, fertiliser (both organic and inorganic), seed-cane, not to mention 'recurrent' costs, e.g. purchase of irrigation water and labour power. Initially, capital in the form of advances was supplied by local moneylenders often in return for rights to market the crop. The rate of interest was high, around 20 per cent.[24]

---

[23] Report by the Collector of Nasik, in *DRC* [Appendix A:235–36]. These capitalists, however, remained active in the land market, buying and selling, and they also financed the rural elites *Shukla* [1937]; *Cheesman* [1982].

[24] For costs of production on the various canal systems, see *DOA Bulletin* No. 147 of 1928 (*The Organisation and Cost of Gul [Crude Sugar] Making in the Deccan Sugarcane Tracts* by P.C. Patil); on furnaces and power crushers, see *DOA Bulletin* No. 144 of 1928 (*Furnaces for Making Gul or Crude Sugar in the Bombay Presidency* by P.C. Patil); *DOA Bulletin* No. 139 of 1927 (*Sugarcane Mills and Small Power Crushers in Bombay Presidency* by P.C. Patil); also *Gadgil* [1948: 116–25]. On marketing methods, see [*Catanach*, 1970: 88]. Interest rates are also dealt with in Catanach, and Mann [1918: 121–6].

The profitability of sugarcane cultivation soon attracted urban moneylenders from Poona city. In 1891, the latter lent as much as Rs 25,000 a year to the *ryots* in one village without 'bond or security' because they grew sugarcane on canal land. The urban entrants pushed aside the local moneylenders by providing loans at 12 per cent.[25] The Bombay Provincial Cooperative Bank also began lending money to sugarcane cultivators. Not only did the entry of these urban sources ease any possible remaining constraints on capital availability, but it seems that in places credit became 'too facile.'[26] Waterlogging and salinity reduced the output of crops, and from the mid-1920s prices took a downward curve making it difficult to collect interest on loans. Outstanding loans mounted steadily. In the Nira Canal area, for instance, the increased from Rs 1.5 million in 1924 to Rs 3 million in 1926, before declining to Rs 1.8 million in 1930 [*ITB* I, 1932: 448].

Agents controlling money-capital responded to this crisis by withdrawing capital from the canal zones. Between 1919 and 1927, the area financed by the Bombay Provincial Cooperative Bank had increased from 1,200 acres to 3,000 acres, at which point nearly a third of the area under sugarcane was financed by it (and its local agencies). Thereafter, as prices declined and serious ecological damage from waterlogging set in, fresh loans were no longer available and the Bank's agents concentrated on recovering old loans. The area financed by the Bank and its agencies declined sharply in the next three years to 50, 100 and 400 acres respectively. Urban merchant-moneylenders cut their finances nearly in half giving no more than Rs 200–300 per acre and requiring that cultivators find their own start-up funds.[27]

---

[25] Report of the Commission Appointed to Inquire into the Working of the *Deccan Agriculturalists' Relief Act*, 1891-2 (Bombay, 1892), (App. I: 4–5); see also *Attwood* [1984, 1985]; *Indian Tariff Board, Evidence on the Indian Sugar Industry*, [*ITB* volume 1: 68–76]; *Guha* [1985a: 128–9].

[26] G.F.S. Collins, Registrar Co-operative Societies, Bombay Presidency, written evidence to *RCA* [II,1: 194].

[27] [*ITB* I, 1932: 447–48]; also oral evidence of Prahladji Govardhandas Q. 2695–6 in *BPBEC*. vol. II.

Cotton Production in Khandesh and Gujarat

Rich peasant households in Bombay Presidency had always had access to significant amounts of money. In the late nineteenth century, such households in Nasik and Khandesh districts had loans of around Rs 2,000. But with the expansion of cotton cultivation in the late nineteenth and twentieth centuries, large peasant households were able to get significantly larger amounts of money. For example, among the cotton-growers of the village of Umra in Olpad *taluka*, a Brahmin family and a Kanbi family had Rs 12,000 and Rs 10,000 on loan respectively. Ten Parsi families in the village of Karanj had borrowed Rs 21,000; 15 Rajput families in Mahmadpore were indebted to the tune of Rs 20,000. In the cotton-growing belt of the eastern villages of the same *taluka* the average debt per family in the late 1920s was about Rs 1,000, nearly three times that of people living in the less fertile coastal areas. Not only did cotton growers have access to large amounts of money, they also paid relatively low rates of interest, in some cases lower than those in the canal zones, between six to nine per cent. By contrast, the greater part of the loans in the poorer coastal tracts paid a much higher rate, 12 to 18 per cent [*BPBEC*, II; Shukla, 1937: 226–27]. In Khandesh, cotton cultivators borrowed liberally. A survey carried out in the 1920s found that the interest on over 67 per cent of the loans in Khandesh was below 12 per cent.[28] Much of the borrowing came either at the commencement of the agricultural season, ostensibly for the purchase of seed, cattle, and so on, or at harvest time to hire labor and carts to carry the produce to market.[29] In one sample, 54 per cent of the cotton growers in Khandesh reported that they relied on moneylenders for the elements of production, though as we shall see borrowed money was not by and large used to improve production.[30]

[28] *Indian Central Cotton Committee, General Report on Eight Enquiries into the Finance and Marketing of Cultivators' Cotton, 1925–28* [1929], Table IX.

[29] W.J. Jenkins, officiating secretary of the Indian Central Cotton Committee, written evidence to *RCA* [II,1: 448].

[30] This should not however, be taken to imply that the large cotton growers were subordinated to moneylenders in the sense in which *Banaji* [1978] describes the formal subsumption of labour to capital. This much should be

The volume of debt generally reflected relative conditions of economic vitality and distress, higher in the more vibrant areas like Khandesh, and lower in the economically depressed areas, like the east Deccan. An official compilation, based on various surveys in the 1920s,[31] showed the highest debt per family in Khandesh (Rs 685), followed by South Gujarat (Rs 551). The lowest figure was for the impoverished coastal tract of the Konkan (Rs 135), and next up from that was the so-called famine tract (non-cotton growing) of the eastern Deccan (Rs 245). In general, money-capital—in the form of loans and credit—was readily available in areas tightly integrated into market circuits. There was seemingly no technical limit to its availability. Where necessary, development could be funded by urban capital, the dichotomy of city and country being practically non-existent.

It is therefore erroneous to argue, as Mann and Patil did, that capital 'scarcity' hindered the progress of Bombay agriculture.[32] Rather the problem seems to have been the way capital was used. Much of the expansion of agriculture in the second half of the nineteenth century was on to marginal agricultural land. Further, as population grew within a system of partible inheritance, landholdings got smaller, and the installed capital base, e.g. wells, drainage channels, and so on, suffered a decline. Important investments in land improvement to counter long-term declines in soil fertility were neglected. All these factors merely aggravated the insecurity of small-property agriculture, leading to declining productivity, lowered profits, and either

---

clear from the very low, below average, rates of interest they paid, the easy access to large amounts of credit, and their complete independence when it came to marketing their crop. [*Indian Central Cotton Committee, General Report*: 15]; also *Dantwala* [1937: 112–13]. They also used these funds to reinforce their dominance of village society.

[31] The materials used included village studies, an intensive survey of five villages, reports and statements based on house to house surveys in 53 villages by persons especially entrusted to do so by the BPBEC, reports submitted by officers of the Revenue and Cooperative Departments [*BPBEC* I: 40.]

[32] Mann and Patil were not the only ones who believed this to be true. In bureaucratic circles in India to this day, capital scarcity arguments are almost a cliche. The crisis and its dimensions are fully spelt out in *Kaiwar* [1989]. See in particular Chapters two, three and four.

unproductive uses or withdrawal of capital from agriculture. Wealthier peasant households tended to take large loans from urban merchants at low interest rates, and to lend the money to poorer peasants at higher rates of interest, rather than undertake the more arduous processes of land consolidation and improvement. There is also evidence that some peasant households withdrew capital from agriculture and invested in trade or urban ventures, especially from the 1930s on.

Unproductive Uses of Money

For most peasants, loans were part and parcel of a subsistence strategy.[33] The Banking Enquiry Committee in Bombay Presidency discovered that the bulk of the loans taken by the cultivators were for 'unproductive purposes' [BPBEC: 43–52; Darling, 1947: 19]. A 1930 survey of 2 villages in Broach district found that much of the money owed by producers to *sowkars* (moneylenders) was either to repay old debts or on account of 'distress borrowing'[34]—46.2 per cent in the village of Khanpur and 31.2

---

[33] At the outset, I will emphasise that this kind of subsistence orientation is not to be confused with moral economy. Moral economists specify several mechanisms that would allow us to place a rural society under that rubric. Broadly speaking, they are: (i) village-level mechanisms to redistribute resources and/or land from rich to poor with the *explicit* intention of reducing disparities of wealth: (ii) obligations placed on the rich by the poor to make their wealth, or at least part of it, available to the entire village during times of crises; (iii) village-level arrangements for collective security, welfare of the old, the sick, the poor, and other handicapped persons; (iv) mechanisms to ensure labour supply to projects that enhance village level economic and social well-being. Above all, it presupposes a rural community not (or not yet) subject to capital's alienating tendencies. None of this applies to rural Bombay Presidency. Ironically, a subsistence orientation among family farmers can easily come to dominate in the absence of functioning mechanisms to insure against risks, etc. The absence of such mechanisms may make a subsistence strategy the only *prudent* course to follow. There is nothing to halt the free fall of a cultivator into ruin as a result of a speculative venture. On the moral versus political economy approaches, see *Scott* [1976] and *Popkin* [1979].

[34] There had been a partial failure of crops in Broach in 1928–9, but nonetheless the figures for unproductive borrowing discovered by the BPBEC

per cent in the village of Sajod. A further 29.1 per cent in Khanpur and 45.3 per cent in Sajod went into the construction of houses, marriages and other ceremonials. Only a small amount was spent on items immediately connected with the production process: 3.5 per cent in Khanpur and 2.1 per cent in Sajod on land purchases; 21.2 per cent in Khanpur and 21.4 per cent in Sajod on 'agricultural expenses.'[35] A survey of Matar *taluka* (in the Kaira [Kheda] district of Gujarat) produced much the same picture, with productive expenses well below those which might be broadly called social or subsistence borrowings [*Kumarappa*: 1944: 112; *Desai*, 1948: 202]. In Olpad *taluka*, heavily involved in cotton production for the world market, expenses associated with items related to the production process were likewise very small. Livestock (4.38 per cent), land purchase (7.18 per cent), implements (0.13 per cent), land improvements (0.26 per cent) occupied a minuscule per cent of total debt [*Shukla*, 1937: 231–37].

The Technical Impasse to Growth

Some Indian historians and economists have argued that 'traditional' agriculture in Bombay Presidency, and in the rest of India, had simply exhausted the productive potentials of actually available technology [*Mishra*, 1981; *Guha*, 1985a]. The technical-impasse (or techno-determinist) argument states that if investments of the productivity-enhancing kind were not made, it was simply because there were none left: 'Besides irrigation, and a few improved implements and seeds, [few] opportunities for investment [existed], causing a production threshold to be reached, once these investments had been made' [*Mishra* 1981: 259, 267]. With the exception of cotton and sugarcane, improved seeds, we are told, had ambiguous effects on total yields and

---

gives us an understanding of deeper structural problems. Remember that Broach district was situated in one of the most fertile areas of Gujarat.

[35] [*BPBEC* I: 49]. The purchase of land does not really qualify as productive, because some of it may have been speculative, fueled by population growth, and possible also by exhaustion of land parcels previously cultivated.

economic utility. For example, certain types of *jowar* and *bajra* seeds—the staple food of the people of the Bombay Presidency—raised yields but were unsuitable for fodder. Land and water management techniques such as irrigation works, embankments, terracing and so on either did not repay investments or were at best defensive measures, delaying further soil erosion. There were also real limits to sinking wells in the Deccan [*Guha*, 1985a: 116–17]. And since feasible techniques were limited to 'labour and locally-produced materials', agriculture did not exercise any multiplier effects on industry. In other words, the 'techno-economic background' determined the absolute level of feasible investments [*Guha*, 1985a: 195,196].

These arguments are far from convincing. It should have been in the interest of peasants to raise income from farming by maximising yields, at the expense of additional (if practically costless, because otherwise unemployed) family labour. P.C. Patil's study of several Deccan villages in the mid-1920s showed that yields varied widely in the same village (in the same season) sometimes being of the order of 10:1 in the case of foodcrops [*DOA* 149: 10,13]. It might be argued that yield variations were the result of differences in soil quality and rainfall. But, Patil noted that neighbouring plots of similar soil quality owned by different persons produced very different amounts of grain, one plot sometimes yielding more than twice as much as a neighbouring plot. Output on irrigated plots varied wildly. In the study cited above, the lowest yield of an irrigated field of *jowar* was 160 lbs. per acre and the highest yield was 1,080 lbs. per acre! Higher yields tended to be the result of careful tillage employing the ordinary implements found in the village, and a systematic utilisation of crop rotations, manures, and in some cases careful crop-watching [*DOA* 149: 1518]. The crucial elements seem to have been timing, care, and a systematic approach to the work of farming.

The average yield of *bajra* in the surveyed villages was 540 lbs. per acre, an extremely low figure considering the population density of the Deccan [*DOA* 149: 14]. In general, output tended to cluster around the minimum rather than the maximum productivity levels. At the very least, this suggests that in a reorganised (and rationalised) property system, there could have been a large increase in productivity as a result of a careful and systematic application of labour and existing resources, i.e. without any new technology in the conventional sense of the term.[36] One can argue, in a preliminary fashion, that if the peasants of Bombay Presidency did not employ labour and existing resources *systematically* to raise yields to the maximum, it must have been due to practical constraints other than either technical limits or shortages of investable funds. One possible explanation for the widely varying yields on plots owned by the same cultivator (a phenomenon noted by Patil and others) is that it was difficult, with widely scattered plots in a monsoon-dominated agricultural regime, to devote the same amount of time to careful preliminary tillage and crop-watching on the entire holding. These were necessary, along with weeding and other related activities, to get the maximum output from a plot of land. Cultivators may have tried to get a decent yield from one or two plots of land and let the others lie in a state of semi-fallow, getting what they could by the most cursory tillage, planting, weeding, and so on. Needless to say, this was a wasteful practice at a time when population was pressing on existing landed resources.

The fact is that in Bombay agriculture available technological resources were not deployed to raise yields. In Khandesh, remnants of the *phad* system, a complex village-level irrigation system using the run-off from monsoonal streams, indicated that there were potentials for improved agriculture. But this required village-level cooperation and communal organisation, which effectively overcame the limits of small-property agriculture.[37] But in the colonial period, as a result of far-reaching legal and economic changes, the social mechanisms for maintaining irrigation works and organising labor were lost. In consequence, officials noticed that large numbers of old weirs across rivers and *nullahs* (water channels) had been abandoned. Areas that previously grew irrigated crops now grew dry crops. At the

[36] See, for example, the argument in Robert Brenner [1986: 47, n. 13].
[37] Jenkins written evidence to *RCA* [II,1: 449].

same time, a large number of wells had gone out of use,[38] partly due to deforestation and the consequent lowering of the watertable. But in other cases, wells were not maintained as a result of their having been divided up between two or more heirs to a piece of property.[39] As noted, the extension of arable at the expense of pasture, led to a decline in the numbers and quality of work cattle, and in many places to serious soil erosion. Finally, rich manure which could have been used for production was often wasted or exported as fuel to the city. All of this conveys the general impression of a decline in the installed productive base of agriculture, rather than a maximal use of available technical means for improving agriculture.[40] In spite of this evidence, the economic historian Sumit Guha argues that the methods widely in use were 'adapted to local ecology and local needs' [*Guha*, 1985a: 116–17]. On the contrary, those methods led directly to the decline of the installed technical base, and to serious ecological damage that impaired soil fertility through erosion.[41] As to supplying 'local needs', the massive and growing rural poverty revealed by the 'solvency data' that Guha himself cites argue otherwise.

Another argument of the techno-determinists is that in the absence of viable new technology, specifically labour-saving machinery, small holdings could be expected to overwhelm larger ones. Thus, when a holding grew beyond a certain size, it would be handicapped by the need to employ increasing amounts of (expensive) wage-labour. Techno-determinists maintain that smaller holdings were superior in one sense—they could avoid hiring labour by superexploiting family labour. Under the prevailing technical conditions, earnings from cultivation were sometimes lower than those from wage labour,

[38] [Ibid.: 450]; oral evidence, Q. 7766–7772 [*RCA*, II,1: 500].

[39] Mann, in particular, points to these symptoms of decay in the agricultural system of Bombay.

[40] This much is implied in the techno-determinist argument. The argument would be otherwise meaningless.

[41] Guha himself points out that yields could have declined in the second half of the nineteenth century [Guha's review of M.B. McAlpin, *Subject to Famine*, in *Journal of Asian Studies*, 47.1 [1987: 237].

thus giving a section of wage labourers a 'larger surplus income' than big landowners! Given their 'desire to hold land', wage labourers would bid up prices until large landowners sold off portions of their land [Guha, 1985a: 236–37]. The small farm becomes, in a sense, the appearance form of the underdevelopment process whose roots lie in the absence of appropriate new technology.

Actually, if the first of the above conditions for the break-up of large holdings is to be taken literally, i.e. if we were to place a strict interpretation on it, then we would expect the average farming unit that emerged to reflect to some degree the optimal exploitability of the existing technology. Kingsley Davis, a demographer, has demonstrated convincingly that at the existing level of technology, the average cultivator (with a pair of bullocks) could cultivate a unit about 1.5 times the size of the average holding in 1941. This would imply that existing farming units were using actually available technology sub-optimally. As we shall see below in the section on *Fragmentation and Subdivision*, the tendency was for ever smaller farms to overwhelm larger farms reflecting the increasing pressure of population on the land,[42] rather than any particular optimal adjustment to technology. The problem with the techno-determinist position is that it represents the small farm as a simple functional adjustment to existing technology, which was further linked to the 'expensiveness' of wage labor. But it cannot really tell us precisely *why* the tendency should have been for uneconomic units to emerge—units that could neither supply a proper subsistence basis for the cultivator and his family, nor use the existing technology properly.[43] Moreover, it cannot tell us why produc-

---

[42] In India, between 1911 and 1931, the number of males in the agricultural and pastoral sectors increased from 71,834,000 to 79,874,000 or approximately at the same rate as the population. The number of acres per person diminished from 2.23 in 1891–92 to 1.90 in 1939–40. Though Gujarat experienced relatively rapid urban it still saw an absolute increase in rural population, from 2,140,306 in 1911 to 2,851,517 in 1941 [Choksey, 1968: 47]; see also *Census of India*, 1911 [volume VII] and 1941 [volume III].

[43] If technological decline and every smaller farms coincided, then it might be argued that technology still determined farm size. However, common sense and actual experience suggest that this explanation is absurd. A threat

tive units should have been so hopelessly fragmented. An overwhelming percentage of farms in Bombay Presidency consisted of more than one land parcel. Typically, even sub-economic farm units tended to consist of several fragments scattered around a village.

The contention that large farms were overwhelmed because they possessed no *competitive* advantages *vis-a-vis* smaller farms assumes, implicitly and illicitly, that large peasants were operating their farms as profit-making enterprises. But the historical evidence suggests that peasants, by and large, did not operate their holdings as capitalist undertakings, even in the intensely irrigated areas.[44] Production geared to achieving solvency (i.e. making ends meet) versus production geared towards maximising profits signal two entirely different modes of social reproduction.[45] In Bombay Presidency, subjective evidence and hard data indicate that it was the former motivation that drove most

---

to the integrity of a farm via subdivision or other mechanisms might dictate that a prudent cultivator would forego costly investments. Particularly, a cultivator might decide to forego those improvements to the farm that would only pay for themselves over a long period of time. This might more plausibly explain why in the long run there was no incentive to maintain the existing technical base.

[44] See chapter four of *Kaiwar* [1989], where the relative absence of specialisation on the irrigated farms on the Godavari and Pravara canals is discussed. Specialisation, and the resulting increases in productiveness are, as Brenner [1986] rightly points out, a necessary manifestation of the development of capitalism in agriculture, as it is in industry. In its absence, particularly in a situation where producers could potentially specialise as on the perennially irrigated lands, the strong presumption must be that they were not capitalist producers, though they often used marginally better technology.

[45] This might appear to be somewhat contradictory, because earlier I noted that peasants switched some of their resources into trade, moneylending and speculation, where they would be forced to compete with other agents controlling capital. But, in fact, the contradiction is easily resolved. *Agricultural production* was not fully disciplined to the needs of the commodity economy, but the money the rich peasants held could become profit-seeking capital *once it exited the realm of agricultural production*. Landholding gave, and still gives, peasants non-market access to their means of subsistence, provides inheritance to male heirs, support in old age, prestige in the village necessary to maintain a social niche, and so on.

peasants, large and small.[46] In these circumstances, it is hard to see—given the higher (value) output and earnings per acre of large farms and the further association between higher (value) output and solvency (but not necessarily profits)[47]—why large farms should be overwhelmed by smaller ones, unless there were social pressures operating to dissolve large units. Partible inheritance, powerfully reinforced by the state, would be an example of such a pressure operating on all landowners, large and small, which they would be powerless to resist.

Demographically Based Explanations for Economic Growth and Stagnation

The case for demographic growth as the autonomous cause of economic decline in the long run has tended to be more persuasive than either the capital constraint or techno-determinist arguments. Population in Bombay Presidency grew from 14 million in 1872 to around 23 million in 1941. The period from

---

[46] See for example, *ICAR* data *Cost of Production of Crops* volume II: 106ff; *DOA Bulletin* No. 149, 1927; see also [Gadgil and Gadgil, 1904]; [Gadgil, 1948: 26–7; 48–51; 64]. All these studies show that farms across a whole spectrum of size had similar accounting losses.

[47] Village studies by *Mann* [1917]; *Mann and Kanitkar* [1921]; *Mukhtyar* [1930], especially the latter's study of a south Gujarat village, found an explicit connection between higher output per acre and overall higher levels of income. Mukhtyar also found that the output per acre and value of output on larger farms were greater than on smaller farms. This might contradict some later [post-1947] arguments by Indian economists. But we must distinguish between 'large' farms that are simply the aggregate of many small strips versus those that are composed of larger than average fields. In the former case, there will be no economies of scale. The need to tend to many strips would also lower the labour intensity, and thus output, on any particular strip. Where the strips of fields were larger and more consolidated, as in Atgam in 1930, there were obviously economies of scale that benefited larger farms. The tendency in colonial Bombay was towards greater fragmentation over time—thus progressively annihilating such economies of scale. The observations of Indian economists from the mid-to-late 1960s onwards probably record this phenomenon, but draw some quite ahistorical conclusions from it.

about 1818 when the British established their hegemony in Western India to 1872 is now believed by historians to have been a period of fairly rapid growth, anywhere from 1.2 to 1.5 per cent per annum.[48] The population of Bombay Presidency might therefore have doubled during the years from 1818 to 1941.

The following is a summary of arguments linking demographic growth with the stagnation, if not decline, of Indian agriculture:

(i) Population growth in colonial India generated a great demand for land by people with no alternative source of employment, driving up the price of land so that farming became a 'deficit undertaking.' So much of the farmer's 'capital resources' was absorbed by rent (or the high price of land) that there was little left for efficient production [*Davis*, 1951: 218; *Ghosh*, 1946: 47 respectively].

(ii) As long as land was available to be brought under the plough, neither 'indebtedness' nor 'tenurial arrangements' constituted on obstacle to investment. The transformation of the prosperous investing peasant into a rack-renting landlord was quite simply a function of the exhaustion of the land-frontier [*Guha*, 1985a: 199–200].

(iii) Those who possessed land (and capital) found the income differential between organising cultivation and leasing land minimal. An increase in the number of tenants encouraged landlords to lease land in small parcels, and thus militated against the exploitation of economies of scale.[49]

(iv) Even without the rent-nexus, the declining per-capita surplus accruing to peasants from self-cultivation of smaller parcels of land resulted in fewer savings, and thus a reduction in 'the capital applied to land.' That, in turn, put a stop to the process of growth [*Mody*, 1982: 238].

If population growth encouraged disinvestment and therefore economic decline, it also served relatedly to 'disarticulate'

---

[48] See the chapter on 'Population' in *Guha* [1985a]; see also *Choksey*, [1968]
[49] [*RCA* II,1: 529], Written evidence of B.S. Patel, Professor of Agricultural College, Poona.

the market in agricultural commodities, causing a contraction of the commodity economy itself. The trade in foodgrains declined in the long run. This argument is put forward by economist Ashoka Mody [1982: 239–43], who proves the argument by comparing the amount of foodgrains traded for comparable harvest years in two periods: the first, pre-1920 and the second, post-1933. The second period figures give the effects on trade of the post-1920 rise in population, and would therefore illustrate the adverse consequences of demographic growth on the market economy. Mody's figures show a significant drop in the volume of foodgrain traded in the latter period, anywhere from 4 to 54 million hundredweights for roughly comparable harvest years. This contraction of foodgrain trade took place against the backdrop of a continually expanding transportation system. In 1891, there was one mile of railway per 90 square miles; in 1940, one mile of railway per 25 square miles [*O'Malley*, 1941: 238–39; *Davis*, 1951: 39]. Obviously, transportation was not sufficient to ensure the continued development of the commodity economy. Rather, along with the contraction of foodgrain trade, there was an increased amplitude of price swings, in line with local harvest conditions.[50]

A manifestation of this contraction of the market economy was the tendency of peasants to store the surplus harvests of good years to be used during bad seasons [*BPBEC* II: 295]. It was also manifested ironically in the tendency of farmers in the most privileged tracts—the perennially irrigated zones in the Ahmednagar district, for example—in the late 1930s to grow their own food [*Gadgil*, 1948: 140–41]. This was a rational defensive mechanism but it represented a deduction from the overall productive potential, *via* specialisation, of the economy.

There was also a change at the all-India level in cropping

---

[50] [*Mody*, 1982: 265]. The crucial barrier to the development of the commodity economy was, of course, the incapacity of the agrarian economy to improve its productiveness. On this question, see *Smith* [1959: 157–58] who argues correctly that improvements in commercial organisation and transportation, while important to the development of a market were not decisive in assuring a surplus in Japan. It rested upon 'steady improvements in farming methods.'

PROPERTY STRUCTURES AND DEMOGRAPHY / 79

patterns. In Greater Bengal and the Central Provinces, there was a decline in land growing non-food crops, and only a marginal increase in the United Provinces and in Punjab. Only Bombay and Madras Presidencies experienced anything like a notable increase in areas devoted to non-foodgrains, and here too the evidence is ambiguous. Mody states that in Bombay the prospect of importing food actually encouraged peasants to switch from *bajra*—a coarse millet—to the production of groundnut for export, a case of specialisation for the world market [*Mody*, 1982: 255–57]. But, in fact, over the period from 1903 to 1940, area sown to foodgrains as a percentage of the total cultivated area was reduced by a mere five per cent, from 75.8 per cent to 70.6 per cent. Cotton, the most valuable export crop, also experienced some decline during this period [*Mishra*, 1981: 159–60]. Thus, both production and circulation appear to have been adversely affected by the growth in population.

These arguments, however, appear to be somewhat one-sided. There are numerous studies which attempt to demonstrate exactly the opposite relationship between population and economic development as the ones argued above. Boserup, for example, claims that population growth has, on occasion, been a stimulus to technological development and organisational innovations.[51] Clifford Geertz [1968: 69–82] has argued, in the case of Java, that a rapidly growing population was countered by an ever more productive agriculture based on intricate co-operation and innovative organisation. In the long-term, such innovations did not lead to capitalist growth but over two decades or more in the nineteenth century when the Javanese population grew by over two per cent per annum, there was no evidence of a decline in the productiveness of agriculture, as was certainly the case in twentieth century Bombay.

At a more general level, the futility of according to the demographic phenomenon such an overwhelming determining power should be obvious. If population growth *per se* had the

---

[51] *Boserup* [1970]. The argument is modified in her later work but the gist of it remains that population is not automatically a depressant to technological innovation [*Boserup*, 1981: 141ff].

effects attributed to it, then a transition towards capitalist development should have been impossible just about anywhere. After all, the rate of growth of population in the Indian subcontinent in the nineteenth century was slower than in Europe or North America and just about the same as that of Japan [*Visaria and Visaria*, 1982: 522]. There were districts in Bombay Presidency, e.g. Khandesh in the north Deccan, with abundant fertile soil and a relatively low population level through much of the nineteenth century. One can see no *a priori* reason why development did not occur there, and exercise upward multiplier effects on the economies of surrounding districts. More particularly, one might ask why such a vast majority of the additions to the population base in Bombay should have become landholders or tenants.[52]

THE SOCIAL BASIS OF ECONOMIC DEVELOPMENT OR GETTING AWAY FROM DEMOGRAPHIC AND TECHNICAL EXPLANATIONS

The arguments that I have outlined above pose the problem of economic stagnation/decline rather than *explain* the causes of the economic impasse of the rural economy of Western India. To do so requires a re-orientation of the inquiry towards the social basis of economic development or underdevelopment, as the case may be. In the first place, we need to look at the socio-cultural practices underlying the demographic regime of Western India, and from there proceed to examine the property relations, whereby the rural producers established rights to the possession of land, and their means of subsistence and production.

In some ways, the arguments that tend to see India underdevelopment as the outcome of exogenous movements of population and technology are not different from 'neo-Smithian' tendencies in European historiography. In this latter mode of

[52] Omvedt notes that this is by no means a universal outcome. In countries like England, and Japan, people were systematically excluded from land by primogeniture. Complex rules of entail prevented estates from being split up, and so on [Omvedt, 1980].

argumentation, certain external stimuli, e.g. the rise of population or the appearance of new productive forces, act as both the necessary and sufficient conditions 'to detonate a process of growth' [Brenner, 1986: 35–36]. The Indian story, at least as told by Mody, Guha, Mishra and others, was one in which population growth went beyond being a stimulus to becoming a drag on development, and in which new (appropriate) productive forces were not available to induce growth.

The theoretical underpinnings of the techno-determinist and demographic-determinist complex of arguments are as follows: the characteristic forms within which production was organised (e.g. the peasant farm) are purely techno-economic relations,[53] functional expressions of a certain level of technology. The autonomous development of technology would impel peasant households to adopt it if it increased their profits, thus changing the relations of production and effecting a transition from a pre-capitalist to a capitalist organisation of production. This type of explanation reduces property relations to a technical organisation of production, taking for granted what needs to be demonstrated, namely that peasant farms were in fact capitalist. In Bombay Presidency, at least, the evidence suggests quite the reverse: that agricultural producers were, by and large, not capitalists, that their strategy of reproduction, including those with abundant land and other resources, was really to ensure the solvency of the family. This might mean running the farm, year in and year out, as deficit undertakings, seen from a capitalist perspective. Clearly, direct producers in agriculture in Bombay Presidency felt no urgency to make profits, to adopt the best available techniques, or to maximise through innovation the price/cost ratio. They did not seemingly face the ultimate sanction of being put out of business or going bankrupt as a result of being unsuccessful on the market. In a situation where land provided the only secure economic asset, peasant families would naturally be most reluctant to part with it or would seek

---

[53] Guha uses this phrase, but the reductionist view of socio-economic reproduction as a purely/largely technical activity is common to Mishra and Mody as well.

to get it back if they lost it by one means or another and were prepared to pay a premium price to do so. Officials correctly noted this phenomenon but misattributed it to some irrational 'desire for possession' or 'vain pride in ancestral possession.'[54]

The rest of this paper will examine the economic dynamic of agriculture in colonial Bombay Presidency, relating it to the property system and the ensuing subdivision and fragmentation of peasant farms. I have argued against the notion that these latter phenomena can be explained merely by the limitations of pre-modern technology. Instead, I propose to link them to demographic growth within a system of partible inheritance. Demographic growth, in turn, has to be related to the lowered barriers to entry to the means of subsistence, given full and formal expression by the *ryotwari* system of Western India, and the ensuing intimate link between production and biological reproduction that developed in the family economy of rural Bombay Presidency. Population growth was, therefore, not a natural happening, a movement external to the social-property relations of Bombay agriculture.

TOWARDS AN EXPLANATION OF THE ECONOMIC DYNAMIC OF BOMBAY AGRICULTURE

The Social Underpinnings of Population Growth in Bombay Presidency

Underpinning, or at least making possible, the rapid population growth observed in colonial India was the high fertility of the rural population. Census reports show that it remained fairly constant around 33–39 per thousand in the twentieth century. But high as these figures are, they appear to be something of an underestimate. According to demographers, the birth rates in colonial India were closer to 45–50 per thousand. The revised estimates for Bombay Presidency indicate a birthrate of 59.9,

---

[54] The former quote is from H.H. Mann [*RCA*, II,1, Q. 2941]; the latter from written evidence given by G.K. Chitale to *BPBEC* [II: 397].

48.1, 55.4, and 49.3 per thousand respectively for the decades ending 1901, 1911, 1912 and 1931 [Davis, 1951: 68–69; Visaria and Visaria, 1982: 508–09]. Though the death rate began a secular, decline in the 1920s, birth rates continued at the old level. Up to 1920, the death rate was quite frequently higher than the birth rate, but thereafter the birth rate was consistently higher than the death rate. The latter fell from 30.2 per thousand (five year average for the year ending 1915) to about 22.3 (five year average for the year ending 1945). When this happened the situation was ripe for a major squeeze on the agricultural sector [Davis, 1951: 33,36–37].

Village and *taluka* studies confirm these general observations made at the all-India and regional levels. In Atgam, surveyed by Mukhtyar in 1926–27, the birth-rate was 40 per thousand and the death rate 36 per thousand [Mukhtyar, 1930: 241]. In the immediate aftermath of famines and epidemics, when out-migration by males in search of work and mortality among women of child-bearing years could be expected to exceed normal levels, the total number of births nevertheless remained high. In Atgam, for example, during the disastrous decade 1900–1910, the total number of births was around 500, compared to the 'normal' rate of 700 per decade [Mukhtyar, 1930: 50–51]. In Olpad *taluka*, five year averages for births in the years 1891–1900 (36.1 per thousand) and 1901–1910 (34.3 per thousand) were only marginally lower than birth rates in other more normal decades: 1911–20 (42.4 per thousand) and 1921–30 (40.3 per thousand) [Shukla, 1937: 49–50]. In the Deccan, in the famine years 1900 and 1901, the birth rates per 1000 married women, ages 15–39, were 193 and 173 respectively, but in 1902 it was up to 236, about the average for normal years [Guha, 1985a: 178]. These particular observations tie into the larger picture of high fertility. Apparently, fertility levels were not adjusted to disastrous circumstances, i.e. there seems to have been little restriction on marriage and biological reproduction during and immediately following a severe contraction of the rural economy.

Functionalist arguments put forward by demographers link high birth rates in India to high death rates, the former seen as a necessary insurance to maintain family size at a level suitable for production given massive infant and child mortality [*Blake and Davis*, 1956: 211–235; *Visaria and Visaria*, 1982: 482]. But these arguments miss the point. In the twentieth century, the real threat was not labour scarcity but an increasing population more or less overwhelming a stagnant productive base, consequent high levels of underemployment, and sub-optimal use of existing technology. The actual fertility may have been below the biological maximum, mainly as a result of practices such as the ban on widow remarriage among the upper castes and secondary sterility resulting from early marriage and child-bearing among a malnourished population, but it was nonetheless very high. The 1931 Census states that in the Bombay Deccan 4.04 children were born per family. The corresponding figure in Gujarat was 4.06 [*Visaria and Visaria*, 1982: 32; *Census of India, 1931*, IX: 32; *Sovani*, 1942: 69–78]. Given the tendency of the Census data to understate fertility the figures may have been much higher. The historical evidence suggests that the Indian peasantry, even when faced with being overwhelmed by poverty, was unable, or quite unwilling, to make adjustments in their patterns of biological reproduction. In fact, all strata of the peasantry acted in a manner that suggested a systemic pressure to marry and start up families as soon as possible. Caste and class-based studies show very little differences in the reproductive rates among married women [*Davis*, 1951: 74–75].

We can place these fertility figures in perspective by comparing them with another peasant society, that of Tokugawa Japan. Here in the eighteenth century, birth rates were in the mid to low twenties per thousand; in the first half of the nineteenth century, they declined to the high teens and low twenties. Death rates were correspondingly low and quite close to the birth rates so that between 1800 and 1850, the Japanese population grew by only one million [*Hanley and Yamamura*, 1977: 211]. The number of children born per family averaged between two and three, or about half the level of Bombay Presidency, in the late nineteenth and early twentieth century. In a famine year, the mortality rate was only 26 per thousand [*Hanley and Yamamura*, 1977: 302–03], or fully one-third below that of Bombay Presidency in a 'normal' year in the late nineteenth century!

In Bombay Presidency, the high fertility rate was a direct consequence of the institution of universal marriage. The Census data show that in the Deccan, among women aged 15 to 39, 84.8 per cent were married, 12.4 per cent widowed, and only 2.8 per cent remained single. In the age group 20–29, according to the Census of 1921, unmarried women were a mere 2.7 per cent, indicating that not only was marriage universal, but also that it occurred early [*Guha*, 1985a: 180]. Child-marriage was common, particularly among the upper castes, making it possible for reproduction to begin early, thereby ensuring the longest possible span of reproduction [*Mukhtyar*, 1930: 68–69].

As a matter of fact, while the Indian pattern was not untypical of Asian peasant societies—'almost universal marriage and at a very young age' [*Wrigley*, 1969b: 90]—it was by no means the case in Europe or Japan.[55] In Western Europe, both late marriage and enforced spinsterhood were common. In Languedoc, for instance, the age at first marriage in the eighteenth century was 25 years for women, and 29 years for men [*Le Roy Ladurie*, 1974: 302]. In Northern Europe, a high proportion of women, 40–60 per cent, never married [*Hanley and Yamamura*, 1977: 315]. In Tokugawa Japan, the first marriage was widely postponed beyond the age at which reproduction could feasibly begin, and at times of *worsening* economic conditions, the number of marriages occurring in a village *dropped* and along with that the number of children born [*Hanley and Yamamura*, 1977: 227, 248–49]. Unlike India, where women often bore children until menopause, the average age for last birth in Japan was 32 to 33 [*Hanley and Yamamura*, 1977: 216]. In a similar vein, marriage was far from being universal in Tokugawa Japan. In a

---

[55] One wonders if Wrigley's observation about an Asian pattern would hold true for pre-colonial India. My work on pre-colonial Western India suggests that marriage was by no means universal in rural society, and that there were elaborate rituals involving married couples designed to limit fertility. D.R. Gadgil, in his introduction to *Sovani* [1942] makes the following tantalising observation: 'In India, as a whole, the process of social disintegration has largely upset the older checks [to fertility] during the last sixty years while nowhere have newer forms such as contraception made any significant advances.'

system governed by primogeniture, the percentage of the population consisting of wives of persons other than those in the main line of descent was very low—zero to four per cent—and generally the higher number coincided with years of economic opportunity. To limit population further, villagers in Tokugawa Japan practiced both abortion and sex-sélective infanticide, so that in many places the male-female ratio was as high as 1.24. In general, it varied between 1.05 and 1.21, compared to Bombay Presidency where it was between 1.04 and 1.06 [*Hanley and Yamamura*, 1977: 238–39,252]. In the 'frontier' district of Khandesh, which was settled by migration as well as population growth among the local peasantry, the sex ratio (male-female) varied from 1.02 to 1.04 [*McAlpin*, 1983: 235–36], suggesting that entire peasant families migrated and took possession of vacant land.

In general, there is no convincing evidence that the peasantry in Western India, or in the rest of India, made any systematic effort to curb household formation or family size, though famines and shortages—the latter evidenced by the growing insolvency of the immediate producers—made it clear to them that there was a serious crisis in the old patterns of reproduction. On the other hand, the peasantry of Tokugawa Japan, or for that matter Northern Europe, were clearly able to regulate their population in line with available resources and job opportunities in the agrarian sector. What accounts for this difference? Obviously, functionalist arguments do not work. It is, therefore, crucial to consider the relationship between access to property in the means of subsistence (most notably land in the agricultural sector) and the peasantry's strategies of demographic reproduction.

The Relationship between Access to the Means of Subsistence and Demographic Growth

The key to the relatively low fertility in Tokugawa Japan was the clear effort to limit the number of households in each village, and the number of persons in each household, carried out in part by the villagers themselves and in part by the state. The number of childbearing couples in each household was limited to one or none and the total number of households rose only slowly over time. The custom restricting marriage to the main line of descent was connected to primogeniture, whereby only one son could count on inheriting from his father. The others had to leave the village to seek work, or remain at home 'unmarried and subservient to a brother' [Hanley and Yamamura, 1977: 250–51].

The effort to limit family size was facilitated by the institution of adoption. Hanley and Yamamura cite this as 'one of the major reasons why the pre-modern Japanese were able to limit family-size in a society in which the continuation of the family was of the utmost importance, both economically and socially.' In one case, in Fujito, of the 105 families for which details are available, 56 (53 per cent) adopted males to carry on the family line. Hanley and Yamamura explicitly contrast this practice with the prejudice against adoption in India. In Japan, 'any male in the family, adopted or not, is eligible to become head, to carry on the family name and business' [Hanley and Yamamura, 1977: 232]. The village was able, therefore, to regulate the rate of demographic growth by delaying or preventing marriage, by selective infanticide and adoption, thus making population responsive to economic opportunity. When employment opportunities increased, whether in agriculture or outside, so did the number of marriages, as in the early nineteenth century. When job opportunities decreased, marriages followed suit [Hanley and Yamamura, 1977: 248–50].

In parts of Western and Northern Europe, rather like in Japan, the most widely employed check to fertility was delayed marriage, a degree of enforced spinsterhood, and rigid controls on unmarried parenthood (in Languedoc, only two per cent of children were born out of wedlock) [Le Roy Ladurie, 1974; Seccombe, 1983: 39]. These controls were especially enforced when the land situation was tight, as in the eighteenth century. But this was possible because a de-facto system of primogeniture existed. Seccombe found that though the laws of inheritance varied from place to place, in practice 'peasant heads and seigneurs both overwhelmingly agreed that it was desirable, above all, to keep the main land parcel—the main means of production—intact, avoiding subdivision where possible' [*Seccombe*, 1983: 39]. Patriarchal controls over household formation were facilitated by the co-residence of several generations in the peasant household [*Medick*, 1976: 303–04]. In some cases the village council regulated the lives of young people without access to land through the institution of domestic service, 'continuously redistributing young adults *en masse* to the more productive holdings of the local area' [*Seccombe*, 1983: 39].

The opening up of the demographic regime in Europe with the break-down of the patriarchal peasant household, and the proliferation of small/marginal peasants, structurally dependent on wage-labour (both in agriculture and industry), requires similarly to be situated in a particular socio-economic reality. Once access to a sufficient land-parcel to meet the needs of simple reproduction was no longer the key to family formation, proletarianised peasants in Europe could marry early, indeed were forced to marry early to assemble an industrial work-team. Patriarchal restrictions on family formation were no longer effective. 'Beggars-marriages'—between partners who had no considerable inheritance or dowry—became more common [*Medick*, 1976: 303–04]. The material base—the exploitation of family labour—also encouraged high fertility. Even under worsening economic conditions, a 'retreat' into a 'more restrictive, traditional marriage pattern' was an impossibility for these marginal peasants/proto-industrial workers. This was because the adult proto-industrial worker was unable to reproduce himself individually, *especially* under worsening economic conditions, when he depended upon the labour of the entire family. As a result, 'a fluctuating process of economic expansion went along with a relatively constant process of demographic expansion' [*Medick*, 1976: 305]. This is borne out by the fact that it was among these marginal peasants/proto-industrial workers that the European population rose most dramatically in the eighteenth and nineteenth centuries [*Seccombe*, 1983: 34–35].

To sum up, then, limiting access to the means of subsistence (through primogeniture or the impartible land base) went hand in hand with restrictive marriage patterns in pre-industrial

Japan, Northern and Western Europe. This made for a low rate of population growth, in line with economic opportunities. The continuation of the family line and the provision of labour in the agrarian economy were taken care of by the institution of adoption (Japan) and by a fairly sophisticated system of circulating young adults through the productive households of a local area (Europe). These practices mitigated the pressure on each household to ensure its own labour supply. The break up of the system, by whatever means, also gave rise to a different kind of demographic regime.[56] In this new demographic regime, the establishment of a family was *made possible* by independent access to the means of subsistence, i.e. where the means of subsistence were no longer controlled by the patriarch. On the other hand, the establishment of a family was *made necessary* by the breakdown of social arrangements for the supply of labour to the household production unit. Production and reproduction became intimately linked [*Levine*, 1983: 9–34; *Levine*, 1987: 94–159].

I am trying to suggest that population growth, and its negative impact on the resource base, should not be seen as a natural happening. There is a tendency in the Indian historiography to attribute to population growth a kind of autonomous movement and to assume that it had a necessary causal link to agricultural stagnation/decline. But population growth, or more particularly the demographic regime, is intimately connected with the way access to property in the means of subsistence and the supply of labour-power to the producing unit is socially organised. Economic growth, stagnation, or decline resulting from such population growth ought to be connected to larger structures of production mediated through the household economy. Comparative studies of different peasant societies are necessary to demystify demographically-centered explanations for economic performance.

On the face of it, the pattern of reproduction of the Western Indian peasantry corresponded to the proto-industrial pattern.

[56] It is beyond the scope of this paper to investigate this topic. I am merely trying to locate the 'nature' of the household and the small-property agrarian economy in Bombay Presidency in a comparative historical perspective.

I will argue that this was so because there was no barrier to entry to a land-parcel and because there were no social mechanisms for supplying labour-power to the peasant household.

Both the Hindu (and Muslim) laws of inheritance, and the actual practice of the peasantry in Western India, gave all adult males a right to a land parcel. All the evidence is that individuals did claim a share, however small, in the family inheritance—land plus the instruments of production. No doubt, much squabbling, litigation, and loss of productive time followed from this arrangement.[57] This kind of individual access to landed property (and more generally the means of subsistence) was made possible by the removal of barriers to entry formerly posed by the joint family. As late as 1861, observers remarked that however divisible in theory the possessions of a family might have been, they were, in practice, rarely distributed among the individual members of the family. Twenty years later, the joint family was already breaking down. Partitions of the family estate were becoming more common, and families lasting more than two generations were very rare. By the 1930s, it was very common for a family to be dissolved on the death of a father [*O'Malley*, 1941: 340–41; *Davis*, 1951: 173] or when the sons reached maturity. This kind of change, no doubt, was the outcome of the land settlement introduced by the British wherein property had to be registered in the name of the individual head of household for the purposes of land revenue payments, and so on. There is also some evidence of a breakdown of old-regime mechanisms of organising a labour-supply to the farmers which tided over periodic shortfalls of labour. In many villages, vestiges of this kind of 'co-operative' labour, which in Gujarat was known as *sondhal*,[58] still survived, but it must have been far more prevalent.

---

[57] Regarding the loss of productive time, Patil cites an instance, which must not have been untypical, of a large peasant who had to cut back on production, because there was a family dispute, and litigation, surrounding a division of the land [*DOA Bulletin* No. 149]. The sons, when they came of age, could demand their share of the land though the father was still alive.

[58] *Shukla*, [1937: 129]. Ravinder Kumar also points out that the breakdown of labour organisation made it difficult to maintain any technical appurtenan-

A further factor to consider is the process that the Indian economist, Amiya Kumar Bagchi has called 'deindustrialisation' [*Bagchi, 1976*]. Without commenting on the appropriateness of this term, it is relevant to note the impact this had on the agrarian sector. In the general economic context of a decline of traditional crafts due to competitive pressures from the industrialised West, those who in the past followed other professions tended to move into agriculture, where lowered barriers to entry made it possible for them to gain land. In one *taluka* in South Gujarat, 82 per cent of those who had taken up occupations other than the traditional ones had become agriculturists. Examples abound from *taluka* and village studies of artisans giving up their erstwhile occupations to go into (a marginal sort of) agriculture. The *ghanchi* (oilman) who found his castor oil replaced by kerosene; the *kumbhar* (potter) who found his earthenwares replaced by metal utensils; the *dheds* (weavers) who lost their profession due to imported cloth, often found themselves among the ranks of the sub-economic landholders [*Shukla, 1937: 160–61; Choksey, 1968: 60; Mukhtyar, 1930: 118*].

These two conditions—free access to the means of subsistence and the break-up of the social organisation of labour supplies to the individual producing unit—constitute the material pre-conditions in which a particularly dynamic kind of demographic regime could be expected to come into place. The break-up of the joint household made it *possible* for individuals to start a family as soon as they could legally come into the inheritance of a piece of land; the breakdown of any kind of social organisation of labour—coerced or otherwise—made it *necessary* for the farmer to assemble a 'work-team' out of the family unit itself, i.e. to exploit for most purposes the labour-power of women and children, depending on an uncertain labour market for the shortest possible time. Naturally, marriage and biological reproduction were integral to this kind of family economy, and ensured a high birth rate even, and especially, during crises of the rural economy. By causing a multiplication of small, uneconomic units of production, such a system of property rela-

---

ces requiring the labour of many persons [*Kumar, 1968: 258–59*; see also *Ludden, 1984: 76*].

tions could, and I am arguing *did*, precipitate a crisis of the agrarian economy, which on a superficial reading appears as a crisis caused by a population/resource scissors.

The relationship between having children and the need to assemble a work-team for the farm is quite clearly brought out in a number of inquiries. Harold Mann, referring to the problems of educating the children of the peasantry, said that when a boy, or girl, reached the age of eight, he/she was taken away from school to herd sheep or cattle.[59] Quite consistent with the labour-requirement argument, large farms and large families coincided. Large families, especially those with many sons, tended to cause the break-up of the farm into many fragments, and households with small children tended to have great difficulties making ends meet [*Gadgil*, 1948: 67; *Mann*, 1917: 143].

It is only in this property-production matrix that the correlation between population growth and the large observed increase in small holdings makes sense.[60] Again, it is only in this general situation that the historian Charlesworth's observation about quantitative growth in production being spent mainly on feeding more people is appropriate [*Charlesworth*, 1985: 207]. Otherwise, remarks of the kind we have criticised whether based on impressionistic accounts of peasant subjectivity, or technically and demographically determined limits to growth are meaningless.

Subdivision and Fragmentation

The Royal Commission on Agriculture in India defined subdivision as the process by which 'the land of a common ancestor was distributed among his successors ... usually in accordance with the laws of inheritance, but sometimes effected by voluntary transfers among the living by sale, gift or otherwise.' Fragmentation, while related to subdivision, refers to a different process, one in which 'land held by an individual comes to be

[59] H.H. Mann to *RCA* [II,1: 74].
[60] H.F. Knight, written evidence to *RCA* [II,1: 292]; *Mishra* [1982: 32]; *Guha*, [1985: 236–37].

scattered throughout a village in small plots or fields and does not constitute a contiguous block' [cited *Shukla*, 1937: 82].

The subdivision of holdings during the colonial period is illustrated by the example of Pimpla Soudagar. In 1817–18, when the village came under British control, the total number of landholders was 42, the average size of the holding being about 17.5 acres. For a while, subdivision proceeded slowly: there were 52 landholders in 1829–30, 54 in 1840–41. Between 1840 and 1915 subdivision accelerated, the number of landholders rising to 156, the average size of the holding being correspondingly reduced to about five acres [*Mann*, 1917: 43–46]. This trend continued after 1915: in 1947–48, when the village was resurveyed, the number of landholders had gone up to 176, with a further decline in the size of holdings. There was a vast increase in the number of sub-economic holdings between 1771 and 1947–48. In 1771, there was only one holding in the 5–10 acre range, while there were 34 such holdings in 1914–15, and 33 in 1947–48. Below the five acre level, there were none in 1771 or 1817, but 93 in 1914–15 and 113 in 1947–48. Though this statistical increase may have been a function of better, or different, recording methods, I think it also illustrates the dual impact of population growth, and reduced barriers to entry into the ranks of landholders. Similar effects were noticed in Jategaon Budruk. Here, at the below-20 acre level, there were only three holdings in 1790, eight in 1817, but 113 in 1917 [*Fukazawa*, 1982: 201].

In Atgam, records for the early colonial period are unfortunately not available but there was, nonetheless, a great increase in the total number of holdings from 219 in 1900–01 (the year after the famine), average size 14 acres, to 431 in 1926–27, average size 7 acres [*Mukhtyar*, 1930: 110–12]. Not all of this increase was due to population growth. Part of it was the result of the break-up of large holdings due to the splitting up of families, mortgages falling in, and/or new entrants of artisanal and urban background becoming landowners by purchasing portions of larger holdings.[61]

---

[61] Guha mentions this phenomenon for the Deccan. In Gujarat, according to Bates, the great famine of 1899–1900 had a dislocating effect on older

All of these tendencies are illustrated by Mukhtyar's study of the Kikla Panchia family holdings. In the 1870s it was still a joint family of fifty with 'extensive landholdings.' By 1900, the original family had split into five branches, which owned 247 acres in 80 plots. Mukhtyar surveyed the village in 1926–27 and found that the original 247 acres were held by 53 separate owners, a dramatic increase since 1900. Interestingly, only fourteen were from the original stock, the remaining 39 being 'outsiders.' Of the 39 outsiders, 37 acquired the land as repayment for debt, one held a plot as a mortgage, and the other received his plot as a 'gift.' The fourteen families from the Kikla Panchia stock owned no more than 104 acres, down from 247 acres in 1900, and those acres were now fragmented into 258 plots, again a vast increase since 1900 [*Mukhtyar*, 1930: 116]. The 'gift' plot suggests that land was often used in distinctly 'pre-capitalist' ways. When the members of the Kikla Panchia family sat down to divide the land around 1900 or so, they did so in a way as to secure to each 'holdings of equal fertility and assessment.' At the end of the process, there was one plot left, and not knowing how to dispose of it, it was given away as a gift to a Dhed, 'who had served the family as a village policeman' [*Mukhtyar*, 1930: 117].

While the process of *morcellement* went on at the bottom end of the landholding scale, there was no corresponding accumulation at the top. In Gujarat, in 1916–17, 89 per cent of landholders had land parcels below 15 acres, and these holdings accounted for 51 per cent of cultivated acreage. In 1942–43, 90 per cent of landholders had less than 15 acres, but the area held in these subeconomic holdings had increased to 56 per cent of the total. There was simultaneously a decline in economic holdings, i.e. those above 15 acres [*Desai*, 1948: 107–09; *Fukazawa*, 1982: 201]. In East Khandesh, the percentage of cultivated land held in units up to 15 acres was 24.8 1916–17, 28.9 in 1921–22 and 49.1 in 1947–48. Holdings of 100 acres and above declined from 19.1 per cent of cultivated land in 1916–17 to 16.6 per cent in 1921–22

---

landlord families, and allowed smaller landholders, persons with money, an opportunity to purchase portions of their estates [*Bates*, 1981: 804ff].

and 6.5 per cent in 1947–48. Similar trends operated in every district of the Deccan [*Guha*, 1985b: 236]. Small holdings gained at the expense of large holdings, with concomitant effects on the potentialities of exploiting actually usable economies of scale.

Small cultivators tried very hard to piece together strips of land which would yield them an economic holding but their efforts were futile. In Atgam, the average owned holding was 7.2 acres, while the average cultivated holding was 8.3 acres, still well below the 15–20 acres suggested as a minimum subsistence holding. In Olpad *taluka*, the results were broadly similar, average owned acreage being 7.7 acres, and average cultivated acreage being 11.6 acres [*Desai*, 1948: 109–10]. The manner in which these somewhat larger holdings were put together actually shows how the problem of uneconomic holdings was compounded rather than solved by the renting of land by peasants. Often, the most marginal small-holders, unable to cultivate their land, leased it to other, less worse-off peasants. Such land was often of poor quality [*Mann and Kanitkar*, 1921: 45; *Mishra*, 1981]. Cultivators were sometimes able to get land on lease from larger peasants, or moneylenders. Nonetheless, the problem of creating a reasonably solid, i.e. contiguous, block of land remained. This brings us to the pervasive problem of fragmentation of land.

Taken in conjunction with what I have already said about the subdivision of properties, the evidence presented below points, in my opinion, to one of the *key causes* of the tendency for agricultural productiveness to stagnate during the phase of expansion, and thereafter to decline in the face of a surging population.

In colonial Bombay, holdings were not only subdivided, but they were also intensely fragmented. A survey of Olpad *taluka* in the mid-1930s revealed what was a fairly typical state of affairs. On average, holdings contained anywhere from 4 to 6 fragments, with larger holdings being made up of a large number of fragments [*Shukla*, 1937: 100]. George Keatinge, while Commissioner of Agriculture in Bombay Presidency, recorded a case in Kanara district in which 52 acres, sufficient for '3–4 rich holdings, or 6–8 fair ones', was in fact divided into 50 holdings in 139 fragments. To make matters worse, the strips were cultivated by all the holders on a rotating basis.[62] Land with any kind of commercial potentiality became intensely fragmented—more so than the usual levels. Thus, in Ratnagiri district, on land growing coconuts individual fragments could be as small as 1/160 of an acre (about 30 square yards). Each tree could be an individual holding, or for that matter be divided among several holders [RCA, General Report: 134]. In some cases, fragmentation could go so far that it rendered some of the fragments uncultivable. This sometimes happened on very valuable, i.e. irrigated, land [Keatinge, 1921: 210–11]. On the old canals in Poona district, fragmentation was intense, and in many cases field shapes and sizes made *effective* irrigation impossible.[63] Some of the fields were ribbon-shaped, others triangular or trapezoid; in some cases, fields were mere strips 20 feet wide, but running the length of an entire field, which made thorough tillage all but impossible. Fragments belonging to an individual might be scattered in every direction. When land became fragmented, so did the ownership of appurtenances on it, e.g. wells. If, fortuitously, the owners could co-operate to maintain the well, it might still be used. More often, it appears, none of the co-sharers wanted to incur the expenses involved causing the well to be abandoned.[64]

In areas of intense fragmentation, large holdings held no particular advantage of scale. They were, generally, split up into many fragments, and the individual fragments tended to be of approximately the same size as the fragments on small holdings.[65] Moreover, village maps showed many examples of cul-

[62] *Keatinge* [1921: 195]. George Keatinge was the Director of Agriculture in Bombay Presidency before Harold Mann, and made careful observations of Bombay agriculture during his tenure. It is not known how widespread this practice was but it seems to have been a particular problem in the southern part of the presidency.

[63] Inglis, oral evidence Q. 5291 to *RCA* [II,1: 241].

[64] *RCA* [II,2: 162]; *Keatinge* [1921: 221]; *Mann* [1917: 31].

[65] This might account for Mann's contention—based on his observations in Jategaon Budruk, a Deccan village—that large holdings had no advantage over small. That is not universally true. Relatively unfragmented large holdings, of the kind found in parts of Gujarat or Khandesh, were actually much

tivators with far-flung plots, sometimes as far as 3/4 mile from each other. Even where the distance separating individual plots was somewhat less, the absence of proper paths of communication meant that a cultivator would have to cross other people's fields, or waste a considerable amount of time skirting innumerable fragments. If a cultivator needed more land, it could only be acquired by renting, or purchasing, from other holders or moneylenders, who acquired small pieces of cultivators' holdings as mortgages fell in. This accounted for the fact that cultivation, often involving a certain amount of renting of land, was more fragmented than ownership [Mann, 1917: 51–53; Keatinge, 1921: 202–03]. Perhaps not surprisingly, the interventions of moneylending capitalists in the rural economy increased the fragmentation of land. Typically, their holdings tended to consist of a larger number of fragments than holdings of corresponding size owned by agriculturists. Peasants, confronting a subsistence crisis, often either mortgaged or sold parts of their holdings. Such land tended to 'circulate' through the hands of moneylenders in strips, rather than stay in their possession and become consolidated over time.[66]

Orthodox thinking blamed subdivision and fragmentation on the operation of Hindu (and Muslim) laws of inheritance whereby land was equally divided among claimants, each further splitting every field so as to have 'every superior and inferior patch among them', rather than the 'whole of various sections of it.'[67] However, it was not the abstract operation of the law, but the particular historical circumstances in which the potentialities of the law were exploited that created the problems associated with fragmentation and subdivision. During the pre-colonial period, and for a generation or so after the transition

---

more productive than smaller holdings. But where fragmentation had gone very far, as they did in older settled districts, then obviously Mann's observation holds some force. As noted, fragmentation increased over time obliterating economies of scale, quite independently of subdivision. On fragmentation and plot size, see Keatinge [1912: 40]; Charlesworth [1985: 236].

[66] Keatinge [1921: 206]. This last contention is borne out by various studies of landowning by sowkars: DARA Papers, [volume 2: 79,207]; Catanach [1970: 184–86].

[67] Quotes from Charlesworth [1985: 232]; Mann [1917: 47] respectively.

to colonial rule in Bombay, the joint-family system regulated the division of income from land. Essentially, the income from land was divided among the family members according to their contribution to cultivation and size of the family. But the evidence available suggests that all males cultivated the ancestral holding jointly.[68] With the disintegration of the joint family, the land came to be equally divided into individual shares, the sharers 'preferring the tangible benefits of equal shares in all the fields with varying quality to the invisible economic advantages of a compact holding.'[69]

Once the system of partible inheritance was set in place—and reinforced by the state's direct collection of land revenue on individual holdings—the logic of subdivision and fragmentation could work against the accumulation of land, or at any rate the continuation, much less formation, of compact farms. To the extent that large farms could sustain large families, the end result would be to subdivide larger farms, making scattering rather than accumulation the normal dynamic within this kind of rural economy.

In the general economic setting of Bombay Presidency, with frequent drought and famine, landless labourers were often the worst hit as employment opportunities literally dried up. Practical experience therefore taught peasants to try and lay claim to any land they were entitled to. This response, rational in the short-term, simply made land a much sought after item. To some historians, this might appear as peasant 'tenacity', land hunger or the 'preference' of the peasantry for cultivation [*Shukla*, 1937: 87; *Guha*, 1985a: 196]. It might therefore appear that the subjectivity of the peasantry contributed to underdevelopment, but in fact this would be an exceptionally facile interpretation. The

---

[68] The *narwa* system in the Patidar villages of Gujarat gave shares in the income from land to various families, who had them cultivated with diverse forms of tenant and bonded labour. This system was broken up in the second half of the nineteenth century by the property settlements in Gujarat, and slowly undermined the greater Patidars, who were thereby subjected to the same pressures to subdivide and fragment the land, as the kunbis of Maharashtra [*Hardiman*, 1981: 37–40; *Bates*, 1981: 777].

[69] *Desai* [1948: 111–12]; see also *Davis* [1951: 173].

subjectivity of the peasantry was formed within certain sociopolitical conditions of opportunity (access to the means of subsistence) and constraint (lack of good jobs outside agriculture, shortages and famines). To argue otherwise is seriously to invert the causal connection between the peasantry and underdevelopment.

Fragmentation and Agricultural Productiveness

One of the great problems associated with fragmentation and the open-field system was the inefficiency of cultivation and the use of labour-time. Add to this the fact that the village was often some distance from the fields, and one gets an idea of factors rendering difficult, if not impossible, the optimal use of working time. Some of these are listed below:

(i) Fragmentation prevented a farmer (even if he had wanted to) from living on his holding; this meant that he was only able to visit his various fragments to perform labour needed for the current crop. As a respondent to the Royal Commission on Agriculture noted, the farmer and his family could not afford to make permanent improvements to their fields. There were so many small fragments that the cost of doing so would be prohibitive and there was the ever-present possibility of further splitting of the fields.[70]

(ii) A considerable amount of land was wasted in borders, boundary marks, and so on. Small fragments left after fields were divided up often went out of cultivation, or were reduced to producing coarse grass [*Keatinge*, 1921: 71; *Pardi Taluka Economic Inquiry*: 16; *Desai*, 1948: 112].

(iii) Crop watching, to prevent damage from animals and birds, was tricky and often unprofitable due to the scattered nature of the fields. In some cases, crop watching was so expensive that it reduced the economic viability of otherwise productive farms.[71]

[70] G.H. Desai written evidence to *RCA* [II,2: 162–63]; *Keatinge* [1921: 71].
[71] G.H. Desai, to *RCA* [II,2: 162–63]; *DOA Bulletin* [No. 149: 15ff].

(iv) Boundary disputes, and litigation within families and between neighbours, were endemic and cut into productive time.[72] In some cases, it seems disputes were resolved by the various claimants establishing temporary rights to pieces of the property. In this revolving property rights system, there was no incentive to improve the land.

(v) Fragmentation prevented orderly organisation of 'labour and capital', thereby frequently preventing second crops from being grown. It also discouraged, in practice, innovative crop patterns. Open fields over which animals were allowed to stray imposed upon the cultivators a uniform pattern of cropping. Innovative rotations and specialisation would have constituted, in themselves, powerful productivity-enhancing features in a pre-modern economy. These were ruled out by the system of letting cattle loose on the village lands after the harvest was in.[73] Fodder crops, which could have provided nutritious feed for cattle, as well as replenishing the soil, could not be systematically grown in this system. Rather, cultivators had to grow crops which simultaneously served as both food and fodder, at the risk of reducing the yield of the food-crop.[74]

(vi) One indirect consequence of fragmentation was that any crop which required 'high culture'—i.e. careful cultivation—could not be produced. Mann felt that fragmented holdings were responsible for the decline of wheat cultivation. Good quality, heavy ploughs, proper cattle and careful irrigation, necessary for the high standard of cultivation required for wheat, were simply beyond the cultivator who had to tend to several plots.[75]

---

[72] *Keatinge* [1921: 71]; *DOA Bulletin* [No. 149: 15].

[73] *Desai* [1948: 112]; see also *Calvert* [1936: 376] on the negative effects of this kind of open-field, strip farming.

[74] [*RCA General Report*: 135]. Much of what came to be classified as fodder crops by the Bombay Administration was really grass—much of it coarse and not very nutritious.

[75] In the Haveli taluka, Poona district, for example, wheat declined from 14,467 acres in 1885–86 to 7,550 acres in 1914–15 [*Mann*, 1917: 96–97]. This decline coincided with a boom in the world market demand for wheat. The reasons for that should by now be quite clear.

(vii) I have elsewhere shown the poor condition into which the existing technical base had fallen in late colonial times [*Kaiwar*, 1989: 63–189]. There is no question that the reluctance to sink wells, or to maintain existing ones, had a great deal to do with fragmented holdings.[76] The tools used by cultivators were suited to 'small and scattered holdings.' Iron ploughs, in an improved dry-farming system, could yield good results, but they were too heavy to carry from field to field, and too burdensome for the poor quality draught cattle, which came to dominate Bombay agriculture, especially after the famines of the late nineteenth century.

(viii) Up to a point, and apart from other considerations, fragmented holdings served to discourage agents controlling capital from direct intervention in production. Any attempt to create a large, consolidated farm would require the cooperation of several landowners, and anyone of them could spoil the continuity of the land surface. Unless, conditions were exceptional, capitalists would naturally avoid direct involvement [*Mann*, 1917: 53–54]. But this would not prevent them from buying and selling small fragments setting in motion an inflationary price spiral in a crowded agrarian economy. This diverted income from 'profits of enterprise' (or potentials thereof) into rents and interest.[77]

(ix) The state provided 'external economies' by constructing large-scale irrigation works in certain districts, but fragmented fields made it impossible to use the resources optimally. Fields were divided into shapes and sizes that made proper irrigation impossible. Since canal irrigation was an expensive proposition in the Deccan, and water had to be stored and transported from giant reservoirs

[76] This was not only true of Bombay where the cost of sinking a well was high, but also in the Punjab, where it was not [*RCA, General Report*: 343–44; Darling, 1947: 128]. This would suggest that the prime consideration against sinking a well was not cost *per se*.

[77] Calvert [1936: 222] argued that wherever fragmentation of land had gone far, such tendencies existed. He cited the case of Belgium.

high up in the Sahyadris, the Irrigation Department tried to minimise losses due to evaporation by only keeping open sections of each canal for short durations (generally ten days for each section). As each section was opened, there was a mad scramble on the part of a multitude of small producers. Water was 'lavishly used and freely wasted to the detriment of the current crop and often to the permanent damage to the soil' [*Keatinge*, 1921: 82]. High-yields were realised in sugarcane cultivation but the result was rapid water-logging and salinity.

This formidable list of problems in maintaining the productiveness of agriculture as fragmentation (and subdivision) advanced with growing population suggests that the non-availability of new (appropriate) technology *per se* was a secondary detail. The key to the problem of agriculture in Bombay Presidency lies in understanding the relationship between the property system (ownership and inheritance), production and demographic reproduction. These effectively imposed tremendous pressures on individual peasants to take whatever land and other instruments of production they were entitled to on reaching maturity. This unleashed a process of subdivision and fragmentation of agricultural land. In the absence of alternative, non-agricultural avenues of secure employment for most, this process was not checked by the out-migration of a sufficient number of people from the rural sector to ease pressure on the land. The demographic regime of an individually owning and cultivating peasantry further aggravated the pressure on the land. This vicious cycle, set in motion in the mid-nineteenth century, shows no signs of abating in the Western Indian countryside. The only exceptions are where peasant cooperative societies in the post-1947 period have been able to stem the process of subdivision and fragmentation, or where the state has stepped in and imposed consolidation in return for productive inputs.

Conclusion

The evidence from Bombay Presidency suggests that the growth that occurred in the second half of the nineteenth century took the form of a multiplication of units of production with a built-in tendency towards a stagnant or declining productiveness in the agricultural sector. Absence of primogeniture and early if not universal marriage were already part of the rural social culture before the arrival of the British. The full potential of these practices were held in check by customs that slowed or prevented subdivision and fragmentation of land and controlled demographic reproduction. There were also significant barriers to landownership for a whole range of rural castes before the arrival of the British.

Once British rule was established in Western India, not only were these checks on property subdivision removed but the barriers to landownership also went by the wayside. Full peasant property rights with freedom to inherit and subdivide land were entrenched and a certain demographic dynamic followed. I have described the dynamic and its associated economic consequences. Since colonial bureaucrats confidently expected their *ryotwari* property settlement to generate sustained development, it follows that the disastrous results of their policy came as an unintended consequence. Debates within bureaucratic ranks on the advisability of introducing primogeniture and giving capitalists full access to land show that officials realised the need to restructure property relations in the countryside. They also realised that to do so would effectively proletarianise masses of property-owning peasants and probably initiate a full scale class struggle against the British Raj. By and large, the state did little in the way of constructive intervention to change the property system.

As time went by and population pressures on the land intensified, many peasants became structurally dependent on wage labour to make ends meet. But many if not most of them also owned small plots of land on which they and their families raised all or some of their food needs, reducing the pressure to become fully proletarianised, thus slowing the development of a labour market. During the colonial period agriculture remained shielded from the competitive pressures of the marketplace. Symptomatic of this was the coexistence in Bombay Presidency of a wide range of techniques, output, productivity of land and labour over a long period within the same ecological zones. There was also a tendency over the long-term for existing —and therefore, presumably, viable—technology to fall into disuse. There appears to have been no systemic pressure generally to adopt the best methods possible, or for that matter the best methods available.

As noted, producers apparently aimed to reduce dependence on the market by growing as many of their subsistence requirements as possible. They also made sure that they did not become overly dependent on the market for the means of reconstituting the production process. In intensely irrigated areas, where specialisation was eminently feasible, immediate producers grew their own food, and raised their own draught cattle, thus constricting the scope of specialisation in neighbouring areas, which might have led to development. Economist Dharm Narain's study of the impact of price on economic behaviour showed that the 'Indian farmer devotes a significant proportion of his land to raising his own requirements; and in his decisions in regard to the uses to which he puts it he is insensitive to price' [Narain, 1965: 4]. Only the 'surplus' acres were sown with a view to selling the produce on the market. Many of the 'factor inputs' were not purchased on the market. Where possible, the implements of production were appropriated through non-waged relations involving fixed payments —hardly a modern concept of wage labour [Gadgil, 1948: 52–53].

In this kind of agrarian system population growth could have a 'disarticulating' effect on the market for foodcrops, causing it to contract. If fully capitalist social-property relations— minimally expressed as the *compulsion* to commodify all production—had been established in colonial Bombay Presidency, population growth could not have had a 'disarticulating' effect on the internal market. The rising demand for food should have caused an increase in price which producers would have interpreted as a sign to produce more food for the market, at higher output per acre. Analysts like Guha and Mody attribute overwhelming determining power to population growth, without attempting to explain why it occurs, and its specific effects with reference to particular agrarian socio-economic systems. Historically, economic development has accompanied population growth in some areas while in others it has stopped or reversed growth.

It confuses the issue to argue that 'if agriculture is not to be a roadblock, then other arenas of the economy must develop to absorb capital and labour' [*Charlesworth*, 1985: 298]. Arghiri Emmanuel rightly shows that there is an 'advanced' agriculture and a 'backward' agriculture; that it is a myth to see agriculture as generally backward, and industry as such as advanced. Advanced agriculture supplies abundant food, raw materials and labour to industry, while absorbing, indeed being the biggest market for, both consumer and producer goods [*Emmanuel*, 1974]. The momentum for economic transformation can be, and has historically been, generated from within the agrarian sector —the roadblock cannot be removed by the exogenous development of industry, and the import of industrial technology into agriculture. The dissolution of the barriers to modern economic growth in agriculture requires a transformation of the social-property relations within which the reproduction of agriculture takes place. During the colonial period the state's policy was to support peasants' rights to land making it difficult for agencies controlling capital to expropriate them fully. Equally important, the state failed to develop workable policies to enforce a system of primogeniture. On the other hand, peasants themselves, except in an odd case here and there,[78] did not or could not exercise the necessary discipline to prevent the progressively debilitating subdivision and fragmentation of land. The results were predictably disastrous.

---

[78] The Charotar Patidars are a case of controlled fertility and access to land. See, *Clark* [1983].

REFERENCES

Attwood, D.W., 1984, 'Capital and the Transformation of Agrarian Class Systems: Sugar Production in India', in M. Desai, S.H. Rudolph and A. Rudra, (eds.), pp. 20–50.

—— 1985, 'Peasants versus Capitalists in the Indian Sugar Industry: The Impact of the Irrigation Frontier', *Journal of Asian Studies*, vol. 45, No. 1, pp. 59–80.

Bagchi, A.K., 1976, 'Deindustrialization in India in the Nineteenth Century: Some Theoretical Implications', *Journal of Development Studies*, vol. XII, No. 2, pp. 135–64.

Banaji, J., 1975, 'India and the Colonial Mode of Production', *Economic and Political Weekly*, vol. 10, No. 50, pp. 1887–92.

—— 1976, 'Chayanov, Kautsky and Lenin: Considerations Towards a Synthesis', *Economic and Political Weekly*, vol. 11, No. 40, pp. 1594–1607.

—— 1978, 'Capitalist Domination and the Small Peasantry in the Deccan Districts', in *Studies in the Development of Capitalism*, Lahore: Vanguard Publishers, pp. 351—428.

Basu, J.K. and N.B. Puranik, 1950, 'A Pattern of Land Use Planning of the Dry Areas of the Bombay State', *Proceedings of the National Institute of Science in India*, vol. 16, no. 6.

Bates, C.N., 1981, 'The Nature of Social Change in Rural Gujarat: The Kheda district, 1818–1918', *Modern Asian Studies*, vol. 15, No. 4, pp. 771–821.

Baviskar, B.S., 1980, *The Politics of Development. Sugar Co-operatives in Rural Maharashtra*, New Delhi: Oxford University Press.

Blyn, G., 1966, *Agricultural Trends in India 1891–1947: Output, Availability and Productivity*, Philadelphia: University of Pennsylvania Press.

Bois, G., 1978, 'Against the Neo-Malthusian Orthodoxy', *Past and Present*, No. 80, pp. 55–69.

Department of Agriculture, Bombay, Bulletin No. 60 of 1914, *Sugarcane: Its Cultivation and Gul Manufacture* by J.B. Knight, Bombay.

—— No. 136 of 1927, *Cattle Breeding in the Bombay Presidency, Principles and Progress* by E.J. Bruen, Bombay.

—— No. 139 of 1927, *Sugarcane Mills and Small Power Crushers in the Bombay Presidency* by P.C. Patil, Bombay.

—— No. 144 of 1927, *Furnaces for Making Gul or Crude Sugar in the Bombay Presidency* by P.C. Patil, Bombay.

—— No. 147 to 1927, *The Organization and Cost of Gul (Crude Sugar) Making in the Deccan Sugarcane Tracts* by P.C. Patil, Bombay.

—— No. 149 of 1927, *Studies in the Cost of Production of Crops in the Deccan, No. 1—Crops in the neighbourhood of Poona* by P.C. Patil in collaboration with T.G. Shirname and T.B. Pawar, Bombay.

Report of the Bombay Provincial Banking Enquiry Committee 1929–1930, Bombay: Government Central Press.

Boserup, E., 1965, *The Conditions of Agricultural Growth: The Economics of Agrarian Change under Population Pressure*, London: Allen and Unwin.

—— 1981, *Population and Technological Change, A Study of Long-Term Trends*, Chicago: University of Chicago Press.

Brenner, R., 1982, 'The Agrarian Roots of European Capitalism', *Past and Present*, No. 97, pp. 16–113.

—— 1986, 'The Social Basis of Economic Development', in J. Roemer, John (ed.), pp. 23–53.

Buchanan, D.H., 1934, *The Development of Capitalistic Enterprise in India*, New York: The MacMillan Company.

Calvert, H., 1936, *The Wealth and Welfare of Punjab*, Lahore: Civil and Military Gazette Press.

Cassels, W.R., 1862, *Cotton: An Account of Its Culture in the Bombay Presidency*, Bombay: Bombay Education Society's Press.

Catanach, I.J., 1970, *Rural Credit in Western India 1875–1930*, Berkeley, CA: University of California Press.

Census of India, 1881, Operations and Results in the Presidency of Bombay, Bombay.

—— *1891: Bombay Presidency, vol. VII*, Bombay.

—— *1911: Bombay Presidency, vol. VII*, Bombay.

—— *1931: Bombay Presidency, vol. IX*, Bombay.

—— *1941: Bombay Presidency, vol. III*, Bombay.

Charlesworth, N., 1972, 'The Myth of the Deccan Riots of 1875', *Modern Asian Studies*, vol. 6, No. 4, pp. 401–21.

—— 1979, 'Trends in the Agricultural Performance of an Indian Province: the Bombay Presidency, 1900–1920', in K.N. Chaudhuri and Clive Dewey (eds.), pp. 113–40.

—— 1982, *British Rule and the Indian Economy, 1800–1914*, London: MacMillan Press.

—— 1985, *Peasants and Imperial Rule: Agriculture and Agrarian Society in the Bombay Presidency, 1850–1935*, Cambridge: Cambridge University Press.

Chaudhuri, K.N. and Clive Dewey (eds.), 1979, *Economy and Society, Essays in Indian Economic and Social History*, New Delhi: Oxford University Press.

Chayanov, A.V., 1966 (transl.), 'The Peasant Farm Organization', in D. Thorner, B. Kerblay and R.E.F. Smith (eds), pp. 29–270.

Cheesman, D., 1982, 'The Omnipresent Bania: Rural Moneylenders in 19th Century Sind', *Modern Asian Studies*, vol. 16, No. 3, pp. 445–42.

Chevalier, J., 1983, 'There is Nothing Simple about Simple Commodity Production, *Journal of Peasant Studies*, vol. 10, No. 4, pp. 153–186.

Choksey, R.D., 1968, *Economic Life in the Bombay Gujarat, 1800–1939*, Bombay: Asia Publishing House.

Clark, A., 1983, 'Limitations of Female Life Chances in Rural Central Gujarat', *Indian Economic and Social History Review*, vol. 20, no. 1, pp. 1–26.

Dantwala, M.L., 1937, *Marketing of Raw Cotton in Indian*, Bombay: Longmans, Green and Co.

—— 1948, *Hundred Years of Indian Cotton*, Bombay: Orient Longmans.

Darling, M., 1977, *The Punjab Peasantry in Prosperity and Debt*, Delhi: Manohar.

Davis, K., 1951, *The Population of India and Pakistan*, Princeton, NJ: Princeton University Press.

—— 1955, 'Institutional Factors Favouring High Fertility in Underdeveloped Areas', *Eugenics Quarterly*, vol. 2, No. 1, pp. 33–39.

Davis, K., and J. Blake, 1956, 'Social Structures and Fertility: An Analytic Framework', *Economic Development and Cultural Change*, vol. 4, Vol. 3, pp. 211–35.

Desai, M.B., 1948, *The Rural Economy of Gujarat*, Bombay: Oxford University Press.

Desai, M., S.H. Rudolph and A. Rudra, 1984, *Agrarian Power and Agricultural Productivity in India*, Berkeley, CA: University of California Press.

Dewey, C., and A.G. Hopkins (eds.), 1978, *The Imperial Impact: Studies in the Economic History of India and Africa*, London: University of London, The Athlone Press.

Diskalkar, P.D., 1960, *Resurvey of a Deccan Village: Pimple Saudagar*, Bombay: The Indian Society of Agricultural Economics.

Divekar, V.D., 1982, 'Western India', in D. Kumar and M. Desai (eds), pp. 332–51.

Drake, M. (ed.), 1969, *Population in Industrialization*, London: Methuen.

Dubois, Abbe J.A., 1924, *Hindu Manners, Customs, and Ceremonies*, (Translated and edited by Henry K. Beauchamp, 3rd Edn.), Oxford: Clarendon Press.

Emmanuel, A., 1974, 'Myths of Development versus Myths of Underdevelopment, *New Left Review*, No. 85, pp. 61–82.

Findlay Shirras, G., 1924, *Report on an Enquiry into Agricultural Wages in the Bombay Presidency*, Bombay: Government Central Press.

Fukazawa, H., 1982, 'Western India', in D. Kumar and M. Desai (eds), pp. 177–206.

Gadgil, D.R., 1948, *Economic Effects of Irrigation. Report of a Survey of the Direct and Indirect Benefits of the Godavari and Pravara Canals*, Poona: Gokhale Institute of Politics and Economy, Publication No. 17.

Gadgil, D.R., and Gadgil, V.R., 1940, *A Survey of Farm Business in Wai Taluka*, Poona: Gokhale Institute of Politics and Economics, Publication No. 7.

Bombay Presidency, *Gazetteers of The Bombay Presidency*, 1879, Ahmedabad, vol. IV, Bombay.

——— 1877, *Surat and Broach*, vol. IX, Bombay.

——— 1880, *Khandesh*, vol. XII, Bombay.

——— 1884, *Ahmednagar*, vol. XVII, Bombay.

——— 1885, *Poona—3 parts*, vol. XVIII, Bombay.

——— 1885, *Satara*, vol. XIX, Bombay.

——— 1884, *Sholapur*, vol. XX, Bombay.

——— 1884, *Belgaum*, vol. XXI, Bombay.

——— 1884, *Dharwar*, vol. XXII, Bombay.

Geertz, C., 1968, *Agricultural Involution: The Process of Ecological Change in Indonesia*, Berkeley and Los Angeles, CA: University of California Press.

Ghosh, D., 1946, *Pressure of Population and Economic Efficiency in India*, New Delhi: Indian Council of World Affairs.

Guha, S., 1985a, *The Agrarian Economy of the Bombay Deccan, 1818–1941*, Delhi: Oxford University Press.

—— 1985b, 'Some Aspects of the Rural Economy in the Deccan', in K.N. Raj *et al.* (eds.), pp. 210–46.

Hanley, S., and K. Yamamura, 1977, *Economic and Demographic Change in Preindustrial Japan 1600–1868*, Princeton, NJ: Princeton University Press.

Hardiman, D., 1981, *Peasant Nationalists of Gujarat, Kheda 1917–34*, Delhi: Oxford University Press.

Indian Central Cotton Committee, 1929, *General Report on Eight Enquiries into the Finance and Marketing of Cultivators' Cotton, 1925–28*, Bombay.

Indian Tariff Board, 1932, *Evidence on the Sugar Industry*, 2 vols., Calcutta.

Imperial Council of Agricultural Research, 1938, 1940. *Report of the Cost of Production in the Principal Cotton and Sugarcane Tracts in India*, Calcutta.

Kaiwar, V., 1989, *Social-Property Relations and the Economic Dynamic: The Case of Peasant Agriculture in Western India, ca. mid-Nineteenth to mid-Twentieth Century*, (Ph.D. dissertation, UCLA).

Keatinge, G., 1912, *Rural Economy in the Bombay Deccan*, London: Longmans, Green and Co.

—— 1921, *Agricultural Progress in Western India*, London: Longmans, Green and Co.

Klein, I., 1984, 'When the Rains Failed: Famine, Relief and Mortality in British India', *Indian Economic Society History Review*, vol. 21, No. 2, pp. 185–214.

Kumar, D. and M. Desai (eds.), 1982, *The Cambridge Economic History of India, vol. 2: c. 1757-c. 1970*, Cambridge: Cambridge University Press.

Kumar, R., 1968, *Western India in the Nineteenth Century*, London: Routledge and Kegan Paul.

Kumarappa, J.C., 1944, 'Handicrafts and Cottage Industry', *Annals of the American Academy of Political and Social Sciences*, vol. 233, pp. 106–13.

Le Roy Ladurie, E., 1974, *The Peasants of Languedoc*. (Translated and with an Introduction by John Day). Urbana, IL: University of Illinois Press.

Levine, D., 1987, *Reproducing Families: The Political Economy of English Population History*, Cambridge: Cambridge University Press.

—— 1983, 'Proto-Industrialization and Demographic Upheaval', in L.P. Moch and G.D. Stark, (eds.), pp. 9–34.

Ludden, D., 1984, 'Productive Power in Agriculture: A Survey of Work on the Local History of British India', in M. Desai, S.H. Rudolph and A. Rudra, (eds.), pp. 51–99.

McAlpin, M.B., 1983, *Subject to Famine*, Princeton, NJ: Princeton University Press.

MacPherson, W.J., 1972, 'Economic Development in India under the British Crown, 1858–1947', in A.J. Youngson (ed.), pp. 140–72.

Mann, H.H., 1917, *Land and Labour in a Deccan Village (Study number I)*, London and Bombay: Oxford University Press.

—— 1918, 'Economic Conditions in Some Deccan Canal Areas', Bombay Co-operative Quarterly, vol. 2, No. 4,

Mann, H.H., 1967, *The Social Framework of Agriculture: India, the Middle East and England*, Bombay: Vora and Co.

Mann, H.H., and Kanitkar, N.V., 1921. *Land and Labour in a Deccan Village (no. 2)*, London and Bombay: Oxford University Press.

Medick, H., 1976, 'The Proto-industrial Family Economy: The Structural Function of Household and Family During the Transition from Peasant Society to Industrial Capitalism', *Social History*, vol. 1, No. 3, pp. 291–315.

Meillasoux, C., 1983, 'The Economic Bases of Demographic Reproduction: From the Domestic Mode of Production to Wage-Earning', *Journal of Peasant Studies*, vol. II, No. 1, PP. 50–61.

Mishra, S.C., 1981, Patterns of Long-Run Agrarian Change in Bombay and Punjab 1881–1972, (Ph.D. Thesis, University of Cambridge).

—— 1982, 'Commercialization, Peasant Differentiation and Merchant Capital in Late Nineteenth Century Bombay and Punjab', *Journal of Peasant Studies*, vol. 10, No. 1, pp. 3–51.

—— 1983, 'On the Reliability of Pre-Independence Agricultural Statistics in Bombay and Punjab', *Indian Economic and Social History Review*, vol. 20, No. 2, pp. 171–90.

Moch, L.P. and G.D. Stark, (eds.), 1983, *Essays on the Family and Histor-*

—— 1983, 'On the Reliability of Pre-Independence Agricultural Statistics in Bombay and Punjab', *Indian Economic and Social History Review*, vol. 20, No. 2, pp. 171–90.

Moch, L.P. and G.D. Stark, (eds.), 1983, *Essays on the Family and Historical Change*, College Station, TX: University of Texas at Arlington Press.

Mody, A., 1982, 'Population Growth and Commercialisation of Agriculture: India, 1890–1940', *Indian Economic and Social History Review*, vol. 19, Nos. 3–4, pp. 237–66.

Mukerji, K.M., 1962, *Levels of Economic Activity and Public Expenditures in India, a Historical and Quantitative Study*, Bombay: Asia Publishing House.

Mukhtyar, G.C., 1930, *Life and Labour in a South Gujarat Village*, London: Longmans, Green and Co.

Narain, D., 1965, *Impact of Price Movements on Areas Under Selected Crops in India, 1900–1939*, Cambridge: Cambridge University Press.

O'Malley, L.S.S. (ed.), 1941, *Modern India and the West*, London: Oxford University Press.

—— 1941, 'Mechanism and Transport', in L.S.S. O'Malley (ed.).

Omvedt, G., 1980, 'Migration in Colonial India: The Articulation of Feudalism and Capitalism by the Colonial State', *Journal of Peasant Studies*, vol. 7, No. 2, pp. 185–212.

Perlin, F., 1978, 'Of White Whale and Countrymen in the Eighteenth Century Maratha Deccan: Extended Class Relations, Rights, and the Problem of Rural Autonomy Under the Old Regime', *Journal of Peasant Studies*, vol. 5, No. 2, pp. 172–237.

Popkin, S., 1979, *The Rational Peasant*, Berkeley, CA: University of California Press.

Raj. K.N., *et al* (eds.), 1985, *Essays in the Commercialization of Agriculture*, New Delhi: University of Oxford Press.

Rao, V.K.R.V., *et al.* (eds.), 1960, *Papers on National Income and Allied Topics, vol. 1*, Bombay: Asia Publishing House.

Report on the Census of the Bombay Presidency taken on the 21st February 1872, 1875, Bombay.

Report of the Deccan Riots Commission, 1875, Bombay.

Report of the Indian Famine Commission, 1880, London.

Report of the Commission Appointed to Enquire into the Working of the Deccan Agriculturists' Relief Act 1891–2, 1892, Calcutta.

Roemer, J. (ed.), 1986, *Analytical Marxism*, Cambridge: Cambridge University Press.

Royal Commission on Labour in India (Whitley Commission), Report and Evidence, 1931, Calcutta.

Scott, J.C., 1976, *The Moral Economy of the Peasant*, New Haven, CT: Yale University Press.

Seccombe, W., 1983, 'Marxism and Demography', *New Left Review*, No. 137, pp. 22–47.

Selections From The Records of the Government of Bombay (New Series)—No. 157, Deccan Agriculturists Relief Act Papers and Proceedings, 6th April 1877 to 24th March 1880, 1881, Bombay.

Shukla, J.B., 1937, *Life and Labour in a Gujarat Taluka*, Bombay: Longmans, Green and Co.

Sivasubromonian, S., 1957, 'Estimates of Gross Values of Output for Undivided India—1900–1 to 1916–7, in V.K.R.V. Rao, *et al.* (eds.), pp. 231–43.

Smith, T.C., 1959, *The Agrarian Origins of Modern Japan*, Stanford: Stanford University Press.

Sovani, N.V., 1942, *The Population Problem in India; A Regional Approach*, Poona: Gokhale Institute of Politics and Economics, Publication No. 8.

Talati, R.P., 1941, 'Damaged Lands in the Deccan and their Classification, *Indian Journal of Agricultural Science*, vol. II, No. 6, pp. 951–77.

Thorner, D., B. Kerblay and R.E.F. Smith, 1966, *A.V. Chayanov on the Theory of the Peasant Economy*, Homewood, IL: The American Economic Association.

Visaria, L., and Visaria, P., 1982, 'Population, 1757–1947', in D. Kumar and M. Desai (eds.), pp. 463–532.

Watt, G., 1966 (Reprint edn.), *The Commercial Products of India*, New Delhi: Today and Tommorow's Printers and Publishers.

Wrigley, E.A., 1969a, 'Family Limitation in pre-Industrial England', in M. Drake (ed.), pp. 157–94.

—— 1969b, *Population and History*, London: Weidenfeld and Nicolson.

Youngson, A.J. (ed.), 1972, *Economic Development in the Long Run*, Allen and Unwin.

Chapter Three

Canal Irrigation and Agrarian Change: The Experience of the Ganges Canal Tract, Muzaffarnagar District (U.P.), 1840–1900*

IAN STONE**

I

In most areas of India the availability of water determines, more than any other input, the nature of agricultural production. Cultivators in regions and localities without sources of water supply to supplement rainfall are constrained in their choice of cropping patterns and have to suffer enforced adjustments when the rains are abnormal, either in their timing or in their quantity. Such insecurity can never be the basis for agricultural development, particularly because potential surpluses, and the investment and expansion opportunities these imply, are normally relinquished in favour of risk-minimization and survival needs. In most areas, therefore, the state of agriculture depends vitally upon the security and flexibility provided by irrigation

* Taken from *Economy and Society*, ed. K.N. Chaudhuri and Clive Dewey (New Delhi: Oxford University Press, 1978).
** The research for this paper was financed by the Social Science Research Council, in the form of a doctoral studentship. It has benefited from the helpful comments of Dr Clive Dewey and Dr W.J. Macpherson, neither of whom is responsible for the views advanced.

facilities—a point long recognized in India. It follows, then, that the substantial British activity in the provision of large-scale irrigation systems—begun in the last decades of the Company's rule, and continued, more vigorously, under the Crown—should have been conducive to agricultural progress in the recipient areas.

A substantial amount of this activity was centred in the Ganges-Jumna *Doab* of what is now Uttar Pradesh. Located on the vast alluvial plain of northern India, it was a comparatively simple engineering task to divert the water from the perennial snow-fed rivers into canals which watered the intervening land. The first major work, the Eastern Jumna Canal, was a renovation of an abandoned indigenous work, and was opened in 1830. This was followed, in 1855, by the opening of the Ganges Canal which stretched almost the length of the *Doab*. By the 1870s this system was watering an average of 750,000 crop-acres annually, though 1,078,000 acres had been supplied in the famine year of 1868–9. The Agra Canal and the loan-financed Lower Ganges Canal were opened during the 1870s, and, by the end of the century, the proliferation of channels and distributaries coursing the land between the Ganges and the Jumna annually irrigated around 1,700,000 acres, whilst being able to supply water to 2.5 million in a year of drought. Of the 11.5 million acres normally sown in this tract, 5 million were irrigated in a dry year—a figure shared almost equally by canals and wells.[1]

Though relatively favourable to their construction, the *Doab* in the mid-nineteenth century was not sufficiently protected by wells to render the tract secure from the marked unreliability of the rainfall in the western U.P. Crude estimates suggest that, prior to the opening of the Ganges Canal, only around 15 per cent of the area annually sown in the *Doab* was protected by wells; clearly there was substantial scope for the extension of irrigation facilities in a tract consisting of districts which were 'among the most insecure in Northern India.'[2] It might be ex-

[1] *Report of the Indian Irrigation Commission 1901–3, Parliamentary Papers* [hereafter PP], 1904, LXVI, p. 182.
[2] Ibid. The proportion of well protection is calculated from the statistics of R. Baird Smith, *Report on the Famine of 1860–61 in the NWP of India*, cited in

pected, therefore, that substantial agricultural progress would have resulted from the subsequent expansion of the irrigated area due to the introduction of the canals. A section in a recent book, however, throws doubt on whether this expected result was actually realized. Dr E. Whitcombe, having considered the impact of the canals on the rural economy of the western U.P., seems to suggest that the canal brought disadvantages which outweighed its advantages, ultimately concluding: 'The canals proved a costly experiment.'[3]

That the canals brought with them unfortunate side-effects has never been in dispute. Contemporaries connected with rural matters were well aware of the canal's faults and official reports abound with references to the drawbacks. Dr Whitcombe brings together the criticisms and observations to be found in the reports. Strung together these outwardly present a formidable case against the canals: waterlogging was experienced in low-lying lands, particularly where drainage lines had been obstructed by water channels, and this encouraged the formation of *reh* (saline deposits) on the soil surface, increased the incidence of malaria, and periodically damaged standing crops; unlined wells fell in as the water-table rose, and cultivators were forced to rely on the canals' uncertain supply; the production of the staple food crops was diminished in favour of the more lucrative commercial crops—a situation felt particularly in years of drought when the canal, heavily committed for cash-crop production, could do little to decrease the ravages of scarcity; over-watering and the encouragement to over-cropping reduced soil fertility; the contraction of wastes made grazing land scarce; petty officials exploited their position of influence over the canal water supplies; and pastoral castes were seduced away from cattle thieving and breeding into more agricultural pursuits, thus reducing the supply of cattle to *Doab* cultivators.[4]

The indictment of the *Doab* canal schemes runs counter to

E. Whitcombe, *Agrarian Conditions in North India* (Berkeley, California, 1972), pp. 67–8.

[3] Whitcombe, *Agrarian Conditions*, p. 91.

[4] Ibid., pp. 64–91.

the general view of them, which—with the exception of some rather unconvincing works condemning the canals, published in the 1870s[5]—accepts the existence of side-effects, but would not claim that over-all the disadvantages more than balanced the positive effects. It immediately throws up questions. Why, for example, if the canals were unable to provide effective protection from famine and even increased susceptibility to it through diminishing staple food production, did successive Famine Commissions encourage their construction? Why, when the ill-effects were so well-known, did the construction and expansion activities continue throughout the period? Is it conceivable that the canals could be profitable to the rulers while exhibiting negative welfare effects overall? The basis of this unfavourable view of the canal's impact is really a subjective re-evaluation of the observations of contemporaries which have filtered down to us. Unfortunately the quality of Indian statistics —particularly output data—prohibits the kind of cost-benefit analysis of the direct and indirect effects which would allow quantitative precision in the assessment of the impact of canal irrigation. We are left to make the most of a mass of essentially qualitative data, mainly concerned with revenue matters. The important question is, what was the quantitative significance of these drawbacks in real terms? When arrayed *en masse* the disadvantages appear overwhelming, but were they likely, in fact, to outweigh the benefits? What follows is a closer look at the impact of the canal and a re-examination of some of the benefits and problems generally associated with its introduction into the peasant farmer's environment. If we are forced, as often happens, to rely on subjective evidence in examining historical topics, then it is important to place it into some sort of objective

[5] A.F. Corbett, *Climate and Resources of Upper India* (London, 1874). Corbett believed that all irrigation was unnecessary and urged deeper cultivation as the key to agricultural advance. W.T. Thornton, *Indian Public Works* (London, 1875), commended upon this on p. 128: 'it is all the more to be regretted that Col. Corbett should have damaged a good case by exaggeration, and that he did not content himself with putting forward his proposals as supplementary to, instead of substitutes for, irrigation.' See also A.K. Cormell, *The Economic Revolution and the Public Works Policy* (London, 1883), pp. 121–4 for further criticisms of the canals.

framework in order to obtain a correct perspective. The precise benefits and costs of the canal may always remain out of reach, but a careful examination of its effects may help to reconcile the contradictions noted above, as well as the apparent contradiction between the peasant's observed behaviour and peasant rationality.

II

The area upon which attention will be focused is the Ganges Canal Tract in the eastern part of the Muzaffarnagar district, situated in the upper *Doab*. The tract, 34 miles long and between 12–26 miles wide, consisted of 6 *parganas* bounded on the north by Saharanpur district, with Meerut to the south and the rivers Ganges and Kali to the east and west. It is sufficiently diverse in its characteristics to reflect many of the features found throughout the *Doab*: it possessed both rich loam soils and light, sandy *bhur*; it contained *khadir* (valley) areas as well as upland plain; its cultivators included Rajputs and Gujars as well as the industrious Jats. In natural facilities for irrigation, too, it exhibited contrasts. The extent of irrigation prior to the opening of the canal was both limited and unevenly distributed. While the river levels were too low to provide irrigation for the upland region lying between them, the possibilities for well irrigation were over wide areas restricted by the spring level being generally far below the surface, and by the prevalence of an unstable sandy subsoil, which made the low cost *kachha* (earthen) wells impractical. *Pakka* (masonry) wells were both expensive to construct and to work at such depths. E. Thornton's 1841 settlement report revealed a general pattern of pre-canal irrigation: only the south-western portion of the tract was irrigated, and of that an even smaller portion had any real security. To the north and east, where the sub-soil water was from 60 to 110 feet below the surface, irrigation was 'practically un-known.'[6] The situation was clearly not conducive to careful cultivation, as Thornton

[6] A. Cadell, *Muzaffarnagar Settlement Report (Ganges Canal Tract)* [hereafter SR], (1878), p. 67.

noted: 'The scarcity of wells leads to there being very few villages, the cultivators being obliged to locate themselves at such a distance from the fields as necessarily has an injurious effect upon the state of agriculture.'[7] This regional variation in access to sub-soil water resources is reflected in Table 1, with the well-irrigated areas being largely confined to Khatauli and parts of Jansath and Muzaffarnagar. The irrigated areas contained in the table refer to the land irrigated in the normal course of cultivation, not that irrigated in any one year. Theoretically, land only irrigated in abnormal seasons, such as famine years, was excluded.[8] The general trend, however, is clear: total irrigation between 1841 and 1872 expanded six-fold, to some extent at the expense of well irrigation, which continued to decline throughout the period. The spread of irrigation facilities embraced the whole tract, as A. Cadell, the settlement officer, noted in 1872:

TABLE 1

COMPARISON OF 'IRRIGATED' AREAS OVER TIME, GANGES CANAL TRACT (Acres)

Paragana	1841	1872		1892	
	Well	Canal	Well	Canal	Well
Purchhapar	53	19,997	48	21,192	58
Muzaffarnagar	3,133	19,140	1,522	25,262	306
Bhukarheri	162	23,878	39	25,530	29
Jansath	6,273	21,725	2,520	21,951	2,433
Khatauli	10,812	20,846	3,438	23,077	2,565
Bhuma Sambalhera	1,200	9,077	422	11,816	216
Total	21,633	114,663	7,989	128,828	5,607

SOURCE: Compiled from settlement report statistics.

[7] Quoted in ibid., p. 67.
[8] The average area actually watered by the canal between 1863–4 and 1870–1 was 77,530 crop-acres. In the drought year, 1868–9, the figure rose to 128,203—around half the total cultivated area.
(Throughout, unless otherwise indicated, all statistics are taken from settlement reports.)

'Throughout the whole tract in which thirty years ago wholly dry villages were the rule, there are now hardly any left.'[9]

Innovation and Peasant Agriculture

What were the repercussions of technical change on peasant production systems? The agricultural setting is probably best described by T.W. Schultz's phrase, 'the equilibrium of traditional agriculture', and is characterized by static technology and by unchanging preferences and motives of the cultivators themselves.[10] In the absence of technical advance, the marginal productivity of investment in additional agricultural factors declines to the point where there is no incentive to save for investment in them. The marginal rate of return to investment in known, and has 'existed long enough for an equilibrium to have become established between savings and investment and between the demand for and supply of agricultural factors as a source of income.'[11] The experience of generations enables the farmers to gravitate towards this kind of optimum solution to resource allocation. It is a state of agriculture typified by a low rate of capital accumulation and a preference for low risk, low return production rather than potentially larger—though less certain—returns to production involving greater production outlays. With little incentive to add to the stock of traditional capital either because returns are relatively low, or, obversely, where the cost of traditional inputs is relatively high—cultivation is likely to stagnate: a state of affairs stemming from the constancy of the 'state of arts' causing the price of permanent income streams to be high.[12]

Into this stable and familiar setting came the canal—an innovation which, in many ways, was to revolutionize the

---

[9] SR (1878), p. 67.

[10] T.W. Schultz, *Transforming Traditional Agriculture* (New Haven, 1864), pp. 30–2.

[11] See H. Singh, 'Resource Allocation and Enterprise Combination—A Case Study of UP Farms', *Agricultural Situation in India*, December 1967, pp. 987–96.

[12] See P. Panikar, 'Capital Formation in Indian Agriculture', *Indian Journal of Agricultural Economics*, 24:4 (1969), 31–44.

agriculture of the tract. Its importance, in both the dry and well-irrigated areas, stemmed largely from the fact that it represented a significant departure from the traditional water supply technologies; it marked a break with the constancy in the 'state of arts.' Over large areas, traditional technology was unable to provide an opportunity for the bulk of cultivators economically to avail themselves of the sub-soil water resources. Not only did the canal make available a supply of water, but it required little investment in the actual water-providing process on the part of the recipient; funds could thus be channelled into investments which complemented the new input and thus increased its productivity. In areas which already had an indigenous water supply, resources tied up in well maintenance and operation could be redirected. A completely new cost structure was imposed by the canal, and its impact on perceptions of risk, on yield expectations, and the availability and productivity of factors of production, transformed the production possibilities in the canal tracts. Inasmuch as the canal used fewer resources in the actual provision of water than did wells, it could simply have been substituted for labour, bullocks and capital equipment with no effect on output. Indeed, Dr Whitcombe suggests that this was likely to have been the result.[13] The evidence of both the switch to more intensive cultivation patterns and the response of the Gujar communities suggests otherwise.[14] The evidence strongly

---

[13] Whitcombe, *Agrarian Conditions*, p. 80. She writes: 'Far from firing [the cultivator] with the much-heralded spirit of industriousness which increase was assumed to bring, canal irrigation required less by way of labour than his well had demanded.'

[14] 'With the less industrious castes . . . increased certainty of the result, gives the required incentive to industry, and both on the east and west side of this district there are many Rajput and Gujar communities which have been, comparatively speaking, reformed by what appears to be the most effectual civilizing agent at our disposal—canal water given flush. It is at all events curious to notice the comparative oblivion into which once notorious communities have passed since their estates came under irrigation from the canal, while their neighbours of the same clan, and the same old habits, but without any fresh inducement to adopt an honest life, have more than upheld their ancient evil reputation.' E.T. Atkinson, ed., *Statistical, Descriptive and Historical Account of the NW Provinces of India*, volume III (Allhabad, 1876)

supports the contention that the technical advance represented by the canal shifted upwards the labour-input/crop-output schedule sufficiently to raise the level of returns to the requisite additional labour input to attract it into use—although leisure preference may have led family farms to share the gains with hired labour.

As a result of the new production conditions, the long-standing 'automatic' nature of peasant decision-making had to be substantially revised. Although production adjustments reflected the change in the expected rates of return under the new conditions, the changes represented a complex process. No alterations were made in isolation, the repercussions involved the whole range of economic activity on the farm. One of the questions implicitly raised by the critics of the canal is whether this disturbance of equilibrium was not running out of the control of the individual cultivator.

(a) The more intensive utilization of land: cropping pattern adjustments

The cash crop *par excellence* in the Upper *Doab* was sugarcane, and the provision of canal facilities was followed by an expansion in the production of this thirsty crop. In the canal tract agricultural operations were then, as now, to a great extent regulated by the arrangements for its cultivation. Wrote Miller: 'It is regarded above all others as the rent-paying crop; and where the tenant has a fixed rent, and is not liable to a high crop-rate, he puts under cane as much land as the available supply of manure and a due regard for the rotation of crops allow.'[15] In terms of percentages of 1872 cultivated areas, settlement statistics show that from around 3–3.5 per cent (4.5 in Khatauli) in 1841, the *pargana* areas devoted to cane had risen, by 1892, to between just under 8 per cent in Purchhapar and upland Bhuma and 12.7 per cent in Khatauli. In all cases the

---

[hereafter *Muzaffarnagar DG*], p. 708. Even the weavers, smiths, carpenters, barbers, *purohits* and *fakirs* substituted agriculture for their former pursuits, finding it 'more paying than hereditary callings.' J. Miller, *Muzaffarnagar SR* (1892), p. 48.

[15] SR (1892), p. 27.

largest increases had been registered before 1872, at which time Cadell noted the existence of canal estates in Khatauli which could boast 20 to 35 per cent of their cultivated area under cane.[16] That this was primarily the result of the expansion in canal irrigation rather than the opening of railway communications is shown by the fact that the great bulk of the sugar trade still followed the old route over unmetalled roads to Panipat and beyond.[17]

The statistics, however, do not adequately reflect the profound effect the canal had on the cultivation of sugarcane in the tract. Prior to the canal, the great bulk of the area had been sown on unirrigated land. This necessitated elaborate and factor-intensive preparation of the soil in order to maximize the absorption and retention of rain water. The field would be ploughed between fifteen and twenty times over the cold season, each ploughing followed by the use of the *merah* to pulverize the soil and protect the concealed moisture from the sun. The canal revolutionized the whole process: the large amounts of water made available by the canal reduced the need for such elaborate moisture-retaining activities, almost halving the number of preliminary ploughings required.[18] The labour and bullock power set free could be devoted to the larger areas under cane. But more than this, the canal had an important impact on yields in both the quality and quantity dimensions, and much of this derived from the adoption of higher-yielding varieties. Soon after the canal had opened, *agaul* cane was imported for the first time into Meerut from Bareilly and acclimatized in Gagual, a village near Meerut city. From there this thick cane spread into Muzaffarnagar where it quickly ousted the thin strain called *daulhu*, which had been 'all but universal' in 1841.[19] The spectacular

---

[16] A. Cadell, 'Rent Rate Report, Pergunnah Khataoli' (1871), p. 65. Uttar Pradesh State Archives, Lucknow [hereafter UPSA], Board of Revenue Records [hereafter BR], 'Different Districts', Muzaffarnagar, File 35, Box 8.

[17] A. Cadell, 'Rent Rate Report, Pergunnah Jansath', (1872), p. 9. UPSA, BR, 'Different Districts', Muzaffarnagar, File 29, Box 5.

[18] See SR (1878), pp. 19–20 for additional details.

[19] S.M. Hadi, *The Sugar Industry in the UP* (Allahabad, 1902), p. 32. The same widespread adoption of superior varieties was noted in neighbouring

success of *agaul* was later to be dampened by the repeated appearance of rind fungus, which caused such damage to the crop during the 1890s that cultivators replaced it with local forms of *ukh* cane; even so, the Upper *Doab* was still noted for 'the superior quality of the canes grown . . . . especially the *dhaur* cane, which', noted one investigator, 'is famous for its high outturn.'[20] Larger areas sown with cane along with heavier outturns put pressure—at times severe—on the cane-pressing capacity of the district, and this region was always the first to take advantage of the several developments in cane-crushing technology, which not only increased the extraction and quality of juice, but reduced both labour and bullock inputs.[21]

TABLE 2

AVERAGE AREAS OF CROPS IN THE TRACT IRRIGATED
BY THE CANAL 1869–70-1871–72 (Acres)

Sugarcane	Wheat	Rice	Cotton	Other Grains	Total Area Irrigated
14,291	32,017	15,885	814	6,278	70,436

SOURCE: *Muzaffarnagar DG* (1876): compiled from information on p. 483.

Along with sugarcane, rice dominated *kharif* irrigation, as Table 2 indicates, and this crop, too, underwent significant changes following the introduction of the canal. Encouraged by favourable canal rates, the area devoted to rice increased throughout the tract, particularly in Purchhapar where, between 1841 and 1872, it rose from 3.6 to around 10 per cent of the 1872 cultivated area. Like sugarcane, though, the extent of the changes is not fully reflected in the area statistics. Important improvements occurred in both the quality of the product and its

---

Meerut where, by the early 1870s, it was recorded that 'the cane that was most generally grown in 1807 is now everywhere considered the most inferior of all the four sorts cultivated in this district.' *Meerut DG* (1876), p. 226.

[20] S.M. Hadi, op. cit., p. 110.

[21] See J.F. Duthie and J.B. Fuller, *Field and Garden Crops of the NW Provinces and Oudh*, Part I (Roorkee, 1882), p. 58.

outturn. Fine rice, *munji*, was virtually unknown in the upland region before the opening of the canal; in 1841 rice was normally a dry crop and, except in the river valleys, was a less valuable coarse variety known as *dhan*. As Table 3 shows *munji* was a valuable commercial crop and—unlike *dhan*—was grown on the best land, always with a full supply of canal water. By 1872, rice grown on irrigated land was almost always *munji*[22] Its profitability rivalled that of sugarcane, particularly at times of high grain prices when the quickly-ripening *munji* crop could be followed by gram in the cold season.

TABLE 3

PRICES AND YIELDS OF MAJOR CROPS[23]

	Sugarcane (Gur)	Cotton	Munji	Dhan	Bajra	Wheat	Gram
Produce/Acre (Maunds)	33.9	2.8	21.9	18.2	7.0	16.9	18.7
Average Prices 1861–73 (Rupees per Maund)	2.6	16.0	1.3	1.0	1.7	2.0	1.6

SOURCES: Yield statistics are selected village estimates, A. Cadell, *Muzaffarnagar SR* (1878), p. 21. Market price data are converted figures from ibid., p. xlii and *Muzaffarnagar SR* (1892), p. 43.

The most valuable *rabi* crop, wheat, showed relatively slight progress in area extension—rather surprisingly, when the impact of the canal in other tracts is considered. For the five *parganas* for which data is available,[24] the proportion of area under wheat remains fairly constant, though in fact, when the increase

---

[22] SR (1878), p. 20.
[23] These figures are subject to all the reservations concerning agricultural statistics. They do, however, reflect in a general sense the pattern of yield-price relationships.
[24] Crop figures for Bhuma were not included in the 1841 settlement report.

in cultivated area is taken into account, the period 1841–72 saw the area recorded under wheat increase from 50,729 to 63,146 acres; in addition 7,326 acres are recorded in 1872 as under 'wheat and barley', at least some of which must have been subsumed under 'wheat.' Even so, the expansion of wheat was clearly inhibited by the fact that the expansion of cane, rice, and fodder crops limited the availability of good land. Indeed, the 1903 Gazetteer records that it was a sign of 'careful cultivation' that wheat was often found on poorer soils.[25] On such land it was likely to be sown as a mixed crop, whereas one of the features of wheat sown in good soil with canal irrigation was the high proportion sown alone.

A less clear pattern emerges for other crops over this period. Cotton, the irrigation of which was both irregular and relatively unimportant until the early years of this century, remained overall a marginal crop; it was unable to compete with the other major crops for good land, labour and manure. Acreage movements in other crops, the cheaper *kharif* millets, for example, are particularly difficult to gauge, since they were to a great extent sown according to relatively short-term requirements and sowing conditions. Single-season comparisons have to be viewed sceptically at all times, but especially so when dealing with the cheap and flexible crops such as *jowar* and *bajra*. In a general sense, though, it is possible to identify from the information available, movements which are meaningful in the context of the introduction of a canal-orientated system of agricultural production, and which gain substantiation from trends noticeable elsewhere.

The poor *kharif* crops, such as *bajra*, clearly declined—particularly in those areas with larger proportions of good soil and access to canal irrigation. The more developed the *pargana* the smaller the area occupied by rain-crops such as *bajra*. Although the canal rarely watered the poorer crops—irrigation made little difference to their outturn during normal seasons—its effects extended to them nevertheless. Not only were resources

[25] H.R. Nevill, *District Gazetteers of the United Provinces of Agra and Oudh*, vol. III, Muzaffarnagar District (Allahabad, 1903), p. 40.

diverted to the production of more cash crops, but the area under fodder crops, too, expanded.[26] The more suitable the soil, when cultivated in conjunction with the canal, the more pressure was exerted by these crops on the coarse *kharif* food crops. Their share of the harvest fell, it is true, but in effect the production of coarse food crops was simply switched to that of *rabi*. While the average *bajra* acreage in Khatauli had fallen to 630 between 1877 and 1888, the area under gram and peas had more than trebled, to stand at 5,794 acres, around 11 per cent.[27] As Miller noted, referring to the tract as a whole: 'there has been a decided increase in the cultivation of gram and peas and a decline in that of the less important kharif staples.'[28] The shift indicated is important in itself, since poor *kharif* grains signified careless cultivation and a shift to the *rabi* was indicative of the existence of irrigation facilities and of increased levels of input in the production of a more valuable output. Compare, for example, the price-yield relationship between gram and *bajra* in Table 3. The most likely explanation for this shift is to be found in the fact that the commercial crops required substantial factor inputs,

---

[26] *Jowar*, normally planted thick as *chart* for fodder, also competed for inputs in the *kharif* season. Its area increased significantly between the reports of Martin and Cadell—up to a level of 10 per cent of the cultivated area in Khatauli—and continued afterwards. Dr Whitcombe is critical of the expansion of cultivation into the waste lands reducing fodder supplies: this was just one aspect of the ecological revolution which the canals helped to being about and which 'most farmers' techniques were not adapted to deal with (op. cit., p. 90). The evidence, however, suggests otherwise: 'The spread of cultivation . . . rather increases than reduces the fodder supply', observed the Director of Agriculture and Commerce. Fodder derived from crops was more valuable than the grass of uncultivated lands, and the *Doab* in particular was not favourable for the growth of grass. Moreover, this grass tended to fail everywhere in times of drought. Far from being forced into stall-feeling, the cultivator benefited from a change whereby the new possibilities for irrigated cultivation on the good soils made the marginal land, on which useful food and fodder crops could be raised, far too valuable, through their ability to release better land from food production, to be left underutilized. See North Western Provinces & Oudh [hereafter NWP&O] Revenue Department Proceedings, Par A [hereafter Rev Progs.], October 1885, p. 15.

[27] L.P. Varma, 'Handbook of Pargana Khatauli', (1892), pp. 22–3, U.P. BR Library, Lucknow, Class IVc(7).

[28] SR (1892), p. 42.

and that the switching of the less valuable crops to *rabi* freed factors for the cultivation and processing of the more valuable ones. In this way the demand for factors would be balanced between the seasons; family labour capacity and the seasonal price of hired labour in such tracts would also encourage this balance in decision-making. Moreover, the presence of the canal as a safeguard should the late autumn rains fail, meant that the switch was not a threat to the security of food supplies that the other system may have provided.

These, then, are the essential features of the tract's cropping patterns over the period. The movements occurred within the framework of an increase in the upland's total cultivated area from 221,423 acres in 1841 to 263,997 thirty years later, a growth of 19 per cent. Only 7 per cent of this came from extension into wasteland.[29] 'Cultivable wastes'—land which had either never come under the plough, or had not been cultivated for the previous five years—fell by around 10,000 acres (35 per cent) in this period. The largest contribution came from the reduction in 'recent fallow'—'that lying uncultivated for three years'— which fell from 32,000 acres to 6,000.[30] Marginal land this may have been, but its value must be seen in that it released a portion of the better land from staple food production to the cultivation of market crops. In addition to the expansion in the area cultivated, further output was obtained through an expansion in *dofasli* (double cropped) area. Figures are unreliable on this as far as the early part of the period is concerned, but Cadell estimated that the *dofasli* area had 'at least doubled' since Thornton's time, and that 10 per cent of the upland area, and 20 per cent of the irrigated area, was cultivated the whole year round.[31]

(b) Re-allocation of resources resulting from risk-reduction

The aggregate shifts in the cropping patterns obviously conceal a vast array of reactions by cultivators producing under a variety

---

[29] The increase thereafter was slight: 269,738 acres was the figure for 1892.
[30] SR (1878), pp. 63–4.
[31] Ibid., p. 21.

of conditions. What was the mechanism by which individual farmers adjusted to the new conditions? Where all, or a substantial part of, a holding was under dry cultivation, the introduction of canal water would have the effect of reducing yield variability, raising average yields and extending the range of production. Cultivation under the uncertainty of dry conditions is severely constrained by the large proportion of resources required to secure subsistence, and by the risk of losses associated with costly valuable crops which, because of their relatively high moisture requirements, are especially susceptible to rainfall shortages. Under these conditions the full potential of available resources is not realized: production is wasteful because of the overriding need to reduce risk. The main feature of the system is one where future risk losses are converted where possible into currently budgeted fixed costs—in other words, into a variety of indigenous insurance programmes, such as crop-sharing rent systems, substantial reserve maintenance, a high degree of product diversification, and an overwhelming concentration on food crops of low market value. The main cost of uncertainty is, therefore, output foregone.[32]

Broadly speaking, the canal tended to vastly increase the security of investment in the production of valuable crops: an adequate water supply not only increased their yields, in many cases they simply could not be grown without an artificial supply of water. Less benefit accrued to the subsistence crops whose yields benefited from irrigation only in an unusually dry season. Nevertheless, as will be shown, the canal did provide a degree of security to such staple food supplies in times of rain shortage. Given the relative effects of watering the different classes of crops, it is clear that the new production possibility curve facing the cultivator who has just obtained canal irrigation would show a substantial shift in favour of commercial rather than subsistence production. This differential effect was indeed likely to encourage commercial rather than staple crop production, but was it likely to be at the expense of such foodstuffs? The evidence suggests not.

[32] See R. Schickele, 'Farmers' Adaptations to Income Uncertainty', *Journal of Farm Economics*, 32:3 (1950), pp. 356–74.

For the individual cultivator, it was not only the enhanced possibilities for the cultivation of commercial crops, but the protective properties of the canal as well, which exerted an important influence over the choice of output mix. Peasant-cultivator utility functions tend to place a high value on selfsufficiency; it constitutes a vital part of their policy of risk-minimization, since insufficient provision of food can be disastrous when—as is likely—food crop shortfalls on the farm coincide with general harvest failure and high grain prices. Generally speaking, therefore, the area devoted to commercial crops is thus limited by the need to ensure subsistence; the farmer who allocates away from subsistence production runs the risk of having to purchase consumption requirements at an unfavourable price ratio. He is thus likely to discount price-relationships between subsistence and commercial crops quite substantially and allocate sufficient acreage to food crops for full farm requirements in a year of below normal yields. The relative unprofitability of his surplus of most years—its market price *vis-a-vis* that of the foregone valuable crop—in effect represents the size of the risk premium for his protection.[33]

The increased security the canal imparted to food supplies through reducing the likelihood of serious shortfalls, meant that fewer resources were required to ensure subsistence output levels. 'Even in a year of plentiful rain, when the canal is not much used, it still performs its role as an insurer', observed H.A. Brownlow, the Chief Engineer of Irrigation, in 1872.[34] The canal's influence, therefore, extended beyond the area normally irrigated, and when F. Henvey observed that 'farmers will only take canal water to save, not to improve, the coarser crops', he was, in fact, observing this mechanism of insurance.[35] More important, though, the canal safeguarded *rabi* sowings. *Kharif* coarse crops were seldom crucial to a cultivator able to make up for a deficiency caused by a dry summer by putting down a

[33] See J. Mellor, *The Economics of Agricultural Development* (Ithaca, N.Y., 1970), p. 201.

[34] *NWP Irrigation Revenue Report* [hereafter IRR] 1871–2 (Allahabad, 1872), p. ii.

[35] F. Henvey, *Narrative of Drought and Famine* (Allahabad, 1671), pp. 113–14.

large *rabi* area with the aid of canal water. Canal irrigation was important for this crop, to counter the effects of an autumn too dry to permit any sowings, or too scanty to permit a full area to be prepared quickly enough. The cultivator operating in a canal village could therefore generally place more reliance on his food supplies and thus release some of his resources—primarily land—to extend his production of market crops.

How does this square with Dr Whitcombe's pessimism regarding the impact of the canal, particularly in drought years, on staple food supplies? She writes: 'Generally speaking, canal irrigation did, and could do, little to decrease the ravages of scarcity by expanding the sources of staple food supply; indeed its effect tended to be the reverse, to contract them.'[36] For some producers, it is perfectly possible that the high returns to commercial production led them to discount the importance of subsistence outturn: where, for example, a cash crop was rice or wheat, it could be converted to home consumption should normal subsistence crops prove insufficient.[37] On the whole however, the evidence does not support the contention that the staple food supply was prejudiced by the existence of the canal. Detailed analysis of village crop statistics from Aligarh district, derived from the *pargana* handbook, points to the same conclusion: that the decline in the area under *kharif* coarse food crops in misleading as an indication of the real food situation, since a portion of that category of crops was switched to the *rabi* season where, though more expensive to produce, the yields of gram and barley were larger and their market value greater.

Since it is a feature of peasant farming that priority is given to meeting subsistence requirements, and since the canal enabled commercial crop production to expand substantially without contracting the production of food crops, it is reasonable to contend that the majority of cultivators in canal villages were

---

[36] Whitcombe, *Agrarian Conditions*, p. 75.

[37] Cultivators on very small holdings, who survive by hiring out their labour in addition to farming their own plot, often exhibit a high proportion of cash crops in their cropping patterns, due to their dependence on the market to provide circulating capital. See K. Bharadwaj, *Production Conditions in Indian Agriculture* (Cambridge, 1974), pp. 62–3.

able to supply their own food requirements, or at least make secure provision for them. The landless labourers and cultivators with inadequate holdings, however, had less control over their food sources, and were the ones most likely to suffer due to a contraction in the supply of foodstuffs. Is there any indication here that the new conditions in the canal tract affected them adversely? It would appear that this was not the case. The spread of irrigation facilities meant that water was no longer the major constraint to reaping a greater return from cultivation; the most important limiting factor was now, in most areas, labour. The resulting high productivity of this resource tended to be reflected in a buoyant demand for the labourer's services, as Miller noted: 'the difficulty and expense of procuring labourers is a frequent subject of complaint amongst well-to-do cultivators.'[38] The increased demand for labour was also more sustained throughout the year. The canal made cultivation less dependent on the dictates of nature and the expansion of labour-intensive crops, particularly sugarcane, gave rise to a large demand for labour, not only for cultivation, but for processing and for transportation. It was, indeed, the western part of the U.P. which was the high real-wage region, as Cadell noticed in 1872: 'The labourer in Muzaffarnagar dresses better than the average petty proprietor of the eastern districts, and wheat now forms a much larger proportion than of old of the food of the poorest classes.'[39] The quotation suggests another reason why the decline in *kharif* food staples was a less disturbing phenomenon than it outwardly appears. The increasing incomes of this period were reflected in changing consumption patterns. W.H. Moreland enquired into this matter in 1901 and concluded that the Meerut division was 'as a whole a wheat-eating tract.'[40] There are very strong

---

[38] SR (1892), p. 23.

[39] SR (1878), pp. 17–18. This was noticeable in other canal tracts; see, for example, H. Le Poer Wynne, *Saharanpur SR* (1870), p. 138: 'Such is the call for labour, that I have met instances of a high-caste proprietor—even a Rajpoot, associating with a chumar in partnership with himself in his seer holding, on the condition that the latter should enjoy an adequate portion of the profits.'

[40] Commissioners Office, Meerut, Scarcity Department, File No. 2 of 1900–1, p. 1.

reasons for questioning the existence of a contraction in food supplies available either to the cultivating units or their hired labour, while the urban wage levels predictably responded to those of the villages.

A development of this issue concerns the role of the canal in years of drought. Dr Whitcombe is sceptical as to the canal's usefulness in decreasing the 'ravages of scarcity.' Certainly the canal was not employed to raise extensive crops of *kharif* millets when the monsoons failed. Although the canal had advantages over the well in such conditions—wells tended to dry up and bullocks lifting the water suffered severely in the heat—the efficiency of both was limited by the shrivelling effect on the crops of the scorching winds accompanying dry summers, and the cultivators preferred to use canal water on their commercial crops. Canal facilities were, however, invaluable for the autumn sowings, which were of crucial importance when the summer crops failed. If the autumn months were dry, it became impossible, even with wells, to put down a substantial *rabi* area. With canal water at his disposal the cultivator had the capacity to prepare, sow and ensure the germination of a substantial cold-season crop. The drought of 1868–9, when scarcely any rain fell in the district from late July to February, demonstrated this: the rain crops failed in the unirrigated portion of the upland, and the *rabi* sowings were virtually confined to the irrigable area. Whereas in a typical year around 70,500 acres were watered by the canal, in the famine year 91,500 acres were sown with the aid of the canal in the *rabi* season alone.[41] Assisted by rains falling in February, the yields were described as good.

With the development of trade and communications, the nature of 'famine' in simple terms had changed from being a condition whereby the food supply was insufficient to feed the people, to one primarily of acute unemployment due to the cessation of fieldwork which resulted in a large part of the

---

[41] *Muzaffarnagar DG* (1903), p. 56. See IRR 1896–7, pp. 12–13 for a brief description of the role of the canal when the rains ceased in late August in 1896. With regard to these irrigated areas, it must be remembered that *rabi* crops were often sown in land still moist from *kharif* irrigation, and their areas are not included in the statistics of irrigation.

population being unable to purchase food.[42] It was in this respect that the canal's *kharif* performance was important; it gave substantial protection to the labour-intensive cash crops. While the *kharif* crops in dry areas withered or remained unsown, and while the exhaustion of his cattle led many a cultivator to give up using his well and look hopefully to the *rabi*, the canal villages were tolerably secure against the loss of their commercial crops and made large profits out of their surplus food crops.[43] 'The labouring class of this tract', wrote Cadell, 'has the advantage of continuous employment throughout the year . . . ; and they are not exposed to the loss, distress, and danger caused by droughts.'[44] Throughout the *Doab*, in fact, far from constituting a weakness in the face of drought, the canal villages were the focus of immigration from unirrigated villages in dry seasons; as evidence for 1868–9, and 1896–7 shows, these villages contained considerable stocks of food.[45] Further testimony to this is provided by an observation by Cadell in *pargana* Muzaffarnagar: 'with one trifling exception the population has fallen off in every village in the pergunnah which is not watered from the canal. The single exception is the little village of Sitherali, situated between two sand hills but surrounded by canal irrigated estates, which in years of drought afford employment and subsistence.'[46]

(c) The interaction of irrigation technologies: the implications for factor supplies

The impact of the canal was no less dramatic in areas previously protected by wells than it was in dry areas. This point is important in a wider context, since it was an observed tendency—

[42] W.H. Moreland, *The Agriculture of the UP* (Allahabad, 1910), p. 25.

[43] For example, in *pargana* Muzaffarnagar in the famine of 1860–1, 'those proprietors whose estates border on the canal have been making, during the last year of distress, . . . large but temporary profits', reported the settlement officer S.N. Martin in *Muzaffarnagar SR* (1866), p. 60.

[44] SR (1878), p. 18.

[45] See F. Henvey, op. cit., p. 15, and *Further Papers Regarding the Famine and Relief Operations*. PP 1898, LXII, No. V, pp. 120–1.

[46] A. Cadell, 'Rent Rate Report, Pergunnah Moozuffurnugger', (1872), para. 36. UPSA, BR, 'Different Districts', Muzaffarnagar, File 29, Box 5.

particularly in the earlier years—for the canals to encroach upon well areas. The general effect of the introduction of the innovation into well tracts was the substitution of canal for well irrigation. Much criticism of the *Doab* canals revolved around the tendency of the enhanced water-tables, which occurred in the vicinity of the canal and its distributaries, to result in the destruction of *kachha* wells.[47] This widespread destruction was less noticeable in the Muzaffarnagar Canal Tract, since not only was water comparatively far below the surface, but the sandy subsoil meant that there was a high proportion of masonry wells. If there are doubts about the exact comparability of the available well irrigation statistics, the dramatic decline is unmistakable: from over 21,000 acres in 1841, the area watered by wells fell to 9,620 in 1860–1, 8,000 a decade later, and by 1901–2 reached 5,476. The substitution was not confined to earthen wells, but the masonry and half-masonry wells fell into disuse, even though the rise in water-level actually added to their working power. The main cause, according to Bulandshahr's settlement officer, of the switch from wells to canals was 'undoubtedly the preference most tenants show for the latter.'[48]

The general preference shown by Muzaffarnagar cultivators for canal water reflected a theme which ran the length of the *Doab*. This reaction was perfectly understandable where, for example, the water supply and hence irrigating capacity was limited to that which percolated slowly into the bottom of the well, or where the well water was brackish and hence unsuitable for the germination of crops, as in Muttra.[49] However for two reasons, the abandonment of good wells in the south-western part of the Canal Tract was outwardly surprising. First, whereas costs associated with well irrigation were largely of the non-monetary kind—involving the use of inputs employed in the normal operations of cultivation and available on the farm—the cost of canal water was a cash payment and it is therefore likely that the cultivator would discount the relative costs in favour

---

[47] See, for example, R.S. Whiteway, *Muttra SR* (1879).
[48] T. Stoker, *Bulandshahr SR* (1891), p. 47.
[49] *Muttra SR* (1879), p. 18.

of the non-cash technology. Second, evidence suggests that for certain crops the yields were higher with well than canal irrigation—as S.N. Martin observed, 'well water gives much heavier crops than canal water. All the zemindars agree on this point.'[50] Clearly, for some reason the new form of water technology was preferable to the cultivator, sufficiently so to outweigh these disadvantages. Dr Whitcombe seems to imply at times that leisure preference was the force behind the switch from the toil of well irrigation to the canal. There is, however, another interpretation—important from the point of view of our understanding of peasant motivation—which holds that the switch to the canal was a perfectly rational adjustment within the framework of production processes, and independent of leisure preference. It can be shown, in fact, that the substitution had a positive effect on production.

Assuming no effective limitation on the water supply in the well itself, the limited capacity of this form of irrigation stemmed from the inability of the cultivator to lift more water: 'if a farmer has to work his well he cannot sow more sugar and wheat than he has time to irrigate', observed C.H. Crosthwaite.[51] It is from this point of view that it is possible to appreciate why the wholesale switch away from perfectly good wells occurred. The very factors required in the actual raising of water were those which could be otherwise utilized in cultivation operations associated with an expansion in production. Well irrigation, as a form of water technology, competed for the productive resources which could only be used to raise water at the expense either of the quality of, or the area under, cultivation.[52] It was not, in fact, a very efficient way of irrigating: the labour of three men and a pair of bullocks for five or six days being necessary to irrigate an acre, while 'flow' irrigation required only two labourers and

---

[50] SR (1866), p. 59.

[51] *Etawah SR* (1875), p. 16.

[52] See K. Singh and J. Hrabovszky, 'An Economic Analysis of Bullock Labour Use on Delhi Farms', *Indian Journal of Agricultural Economics*, 20:4 (1965), 17–28. While bullock-use on the farms was nearly identical in January and February, the study showed that bullocks were watering wheat on well-irrigated land and preparing the fields for sugar on the canal-irrigated farms.

'almost any quantity of land' could be irrigated in a day.[53] Moreover, the factors released by the switch to the more efficient source of irrigation were highly complementary to the new water input. In Khatauli, for instance, the bullock-power released from the treadmill of water-raising on nearly 7,000 acres was 'sufficient for all the work connected with the increased cane cultivation, and for the ploughing of 10,000 acres besides.'[54] The saving in labour—especially important because of the flexibility of its use and its relation to direct capital formation—was unquestionably the reason behind the more largely increased area, and the more generally maintained quality of the cane crop.[55] Cadell's Rent-Rate Report for Khatauli provides a good example: adjoining first class villages of Bhainsi, Sontah and Dudaheri—all quite closely cultivated under well irrigation—there were, prior to the canal, the estates of Raipur Nagli with a small hamlet, and Sherpur and Jahangirpur entirely uninhabited. All three were cultivated extensively, inferior crops being grown. They were carelessly cultivated by tenants who 'however industrious, had plenty to do in their own villages.'[56] When canal irrigation came to be substituted for that of wells in Bhainsi, Sontah and Dudaheri, 'the men and cattle working at the wells were set free, and at once began to improve the cultivation in their own and adjoining townships.'[57] The introduction of the canal raised the irrigated area in the adjoining estates from 10 to 80 per cent of the cultivated area, and within a short time their crop patterns corresponded to those of the hitherto far advanced villages.

Those *parganas* formerly dry responded more slowly to the introduction of the canal; complementary inputs moved equally slowly into line with the new production conditions since there was comparatively little labour displaced from well duty. The population only gradually expanded to meet the increased demand, it was—as Cadell observed in Purchhapar in 1869—'not

[53] IRR 1873–4, pp. cvi-cvii.
[54] SR (1878), p. 71.
[55] Ibid.
[56] A. Cadell, 'Rent Rate Report, Pergunnah Khataoli', (1872), pp. 35–6.
[57] Ibid.

in a position to make such inroads on the *bhoor* as were effected elsewhere. The Jats and Tagas, notwithstanding the progress they have made, are still merely working on the edges of the sand waste, while the people of Khutowlee have caused *bhoor* to disappear from estates in which it once formed a large area.'[58]

The value of the released productive factors was clearly an important reason—along with his increased involvement with the market and monetization—why the cultivator overcame his aversion to cash outlays for water. In addition to this, once the canal's impact on cropping patterns is considered, it becomes clear why canal water was preferred to that of the well—even though some crop yields could be lower on canal watered land. The high unit costs of water obtained from wells ensured that the supply was carefully applied to the limited area it would cover, and that such land was meticulously cultivated. This meant that this land would have priority over the manure supply; it grew almost invariably the most valuable crops, and large quantities of labour were lavished upon it. Canal water was not paid for according to volume. Its cheapness in terms of quantity encouraged the cultivator both to dispense with the high degree of input intensity he had to lavish upon his well-irrigated plot in order to obtain favourable returns on the expense—including opportunity cost—of lifting well water, and also to spread inputs over a larger area. With the increased supply of water which was cheaper in terms of resources, the cultivator could expect an acceptable rate of return on land, the intensive cultivation of which would not have been possible when he relied on his well. The cultivator therefore irrigated a larger area than before, though on average he cultivated it less intensively than he had his fields near his well. On balance, he forfeited some of the output he could have expected from his well-watered fields, but gained from the cultivation of better quality crops on land previously only infrequently irrigated, if at all. In other words, the canal enabled the farmer to increase the total product of his holding, and this seems to have been the general response—as reflected in the observation that 'the whole tendency of Jat and

[58] NWP Rev Progs, June 1870, p. 47.

Rawa cultivators is to secure a large average produce, rather than the excessive large average of a few fields.'[59] It is to this tendency that the existence of the yield differentials—where they did exist—is to be largely attributed.

III

External Diseconomies—A Closer Look at the Waterlogging Problem

In the area under discussion the problem of excess water breaks down into two parts: (1) the waterlogging afflicting the *khadir* (valley) areas of Bhuma, Bhukarheri and Purchhapar; (2) saturation problems in the upland plain resulting from the percolation from water channels, the tendency of cultivators to make lavish use of a water-supply charged for by crop-area rather than by volume, and insufficient attention to drainage facilities on the part of the canal engineers.

Within five years of the opening of the canal, *khadir* revenue assessments were reduced, and the pattern established for the rest of the century. The effects upon the *khadir* of canal irrigation of the uplands represents, in fact, a classic case of external diseconomies: the level of consumption of canal water by the upland farmers directly harming the interests of the cultivators in the valleys—mainly the Ganges *khadir*. Water lavished on the uplands percolated down into the valley where it exacerbated a natural tendency to saturation in the soil—particularly below the cliffs where *reh* and swamp-land increased. Nearer to the Ganges, where the land was slightly higher, less damage occurred to the soil, the quality of which was similar to that of the plain. Early official attempts to attribute deterioration in the *khadir* to causes other than the canal soon gave way to the obvious reality: 'It seems that as long as the canal is running, two-thirds of the *khadir* can never form other than a precarious, fever-stricken tract where cultivation is not only financially

---

[59] *Muzaffarnagar DG* (1876), p. 545.

insecure, but is only possible at risk of health.'[60] The *khadir* mirrored in reverse the effects of the new technology on the upland, and the considerable expenditure by the authorities on drainage did little more than slightly slow the decline. All but the highest and sandiest fields below the upland deteriorated and the area under cultivation declined. It is not possible to give an overall statistical impression of the decline, but the fate of a group of villages in the Solani *khadir*, which adjoins that of the Ganges in the north, seems to reflect the general trend. Here the cultivation of wheat, cane and cotton all contracted as the cultivation declined from almost 1,500 acres before the canal to 840 by the 1870s. Cane was by then only to be found in one village, while even rice had fallen—both in area and in quality of output.[61] Over a matter of several thousand acres of low-land in the *khadir*, rents fell as the population thinned and the fertility of the land declined.

It is important, however, that this essentially unfavourable view is not left unqualified, since the canal's impact in this respect was more complex than it at first appears: the tract's deterioration was not entirely due to the canal's encouragement to saturation; nor do the cultivation trends reflect fully the *khadir's* relative importance in the agriculture of the area. Prior to the introduction of the canal, the adjacent upland areas were often extremely insecure and parts of the *khadir* were, for some villages, the only areas in which 'produce was tolerably assured.'[62] Generally, though, it was 'unfit for residence' and the cultivators often lived in the 'strong populous villages' on the uplands from which they descended into the *khadir* to grow some of their crops—the output here being more certain than on the upland. If the expected output of these *khadir* crops fell with the introduction of the canal, so the yields of their upland crops increased. A perfectly natural reallocation of resources

---

[60] The discussion over the cause of the *khadir's* deterioration is summarized in SR (1878), Appendix XII. The quotation is that of the settlement officer of neighbouring Gordhanpur *pargana*, made in 1899 and cited in *Muzaffarnagar DG* (1903), p. 5.

[61] *Muzaffarnagar DG* (1876), p. 477–8.

[62] Ibid., p. 478.

towards the general area of greater returns meant that production in the valleys would inevitably decline—just as the population in the dry villages was observed to decline in favour of the canal villages.[63] At the same time, if the *khadir* became less important for its agricultural produce, its importance as a grazing ground and fuel source was enhanced both by the contraction of waste on the upland, and by the increased demand for bullocks. Through releasing factors to upland cultivation, the *khadir* contributed to the progress of the canal villages at the expense of its own cultivation. The *khadir* had always been considered an unhealthy area in which to live, and that reason alone was sufficient to limit its development potential.

Although *reh* was seldom encountered in the upland tract, certain localities were prone to periodic waterlogging, particularly during the early stages of canal irrigation, before the importance of the crucial technical requirement—that the alignment of irrigation channels should follow the watersheds—became fully recognized. The extensive remodelling of the *rajbaha* (water channel) system during the 1870s and after had as its aim the correction of a situation whereby, at first, 'distributary after distributary was run out without regard for the drainage of country, and . . . there was no practical admission of the necessity of allowing waterway under the irrigation channels.'[64] The abandonment of the original Jansath *rajbaha* occurred as part of the attempt to correct some of the original design defects of the system. Funds gradually filtered through, too, to enable drainage channels to be dug, and the Kali river to be straightened and deepened. These measures greatly improved the situation, though they cannot be said to have removed all the waterlogging problems which accompanied the canal. The engineers 'learnt by doing', even if the improvements were only slowly and imperfectly put into practice.

Indications as to the acreages affected by waterlogging in the tract are not available, and would be of limited use if they were, since it is the *degree* of the problem which is the most important from the agricultural viewpoint. For example, abnormal

[63] SR (1878), p. 78.
[64] *Muzaffarnagar DG* (1876), p. 476.

rains in the 1880s caused a great deal of waterlogging, but although this was felt by Miller to have affected the productiveness of the soil, it was only in low-lying places that this had occurred to a 'serious extent.'[65] Elsewhere waterlogging was observed to prevent *kharif* cultivation, but leave the soil sufficiently moist to allow *rabi* sowings without the aid of irrigation.[66] Clearly the effects of waterlogging varied considerably; they were not always negative. It was quite usual for saturated fields to be devoted to cattle, since they provided water and forage of some kind;[67] furthermore, where *kharif* fields under sugarcane became waterlogged for a time, it was likely that the crop would not suffer serious injury.[68] It seems that, as with other disadvantages and changes, the cultivator adapted to the particular characteristics of the saturation tendencies in his land in accordance with the rates of return and risk associated with its various uses. This adaptation is probably reflected in Miller's observation that the evidence of rent movements was 'against the view that there has been any serious deterioration', and in the remark contained in the 1903 gazetteer that a full supply of canal flow-irrigation was the most valuable quality land could have, and that it commanded a high rent 'even where there is danger of saturation.'[69] The resilience of sugarcane, the suitability of rice cultivation, and the ability to sow *rabi* crops without irrigation are three important reasons why the 'test of competition rents [was] against the theory of deterioration' in the tracts prone to saturation.[70]

In part, too, the problem of saturation was one brought on more by prosperity than by the design faults of the canal system. Increases in product prices and the expansion of population induced farmers to cultivate land which, at the time of the canal's construction, had been much less in demand. Some of the land brought into cultivation during the three decades following 1841

[65] SR (1892), p. 10.
[66] IRR 1873–4, pp. IA–2A.
[67] IRR 1880–1, p. 16A.
[68] IRR 1873–4, p. 20A.
[69] SR (1892), p. 9 and P. 53.
[70] Ibid., p. 9.

was land which made up the wide margins left around ponds and depressions in Thornton's time; only in seasons of exceptional rain did the flooding of the cultivated areas cause much damage. The encroachment of such lands made flooding likely with less than exceptional levels of rainfall, and Cadell noticed a tendency to blame the canal for what was only '... a natural result of the increased value of land which thirty years ago was allowed to its waste, while it now swells the profits of the landlord in a dry year and intensifies the outcry against the [canal] department in a year of plentiful rain.'[71]

IV

The introduction of canal irrigation into eastern Muzaffarnagar set off a complex series of repercussions which left virtually none of the region's inhabitants untouched. Both dry and canal villages were affected, as all agricultural groups, from the landless labourers to the landowning *zamindars*, from the small cultivating tenants to the sturdy peasant-proprietors, and from the industrious Jats to the law-abiding Gujars and the lowly Chamars, adjusted their activities to suit the rates of return associated with the new production conditions. The overall impression is of a substantial contribution to wealth and security, and the larger returns to, and demand for, factors of production—particularly that most vulnerable section, farm labourers—meant that it was unequal gains, rather than actual gains and losses, which accrued to rural society's component groups. As in all periods of economic change, some groups benefited more than others from the adjustments, and our knowledge of the institutional framework leads us to expect that not only would rich peasants be able to benefit most by directly employing the new technique, but that it would also provide them with further opportunities to extend their influence and control within the village community. It was to labour's advantage, though, that an innovation which had the effect of

[71] SR (1878), p. 5.

making labour a comparatively scarce input was not accompanied by the kind of labour-saving complementary technology which has been available in the era of the Green Revolution.

What emerges from this brief glimpse of these six *parganas* is an impression of an energetic response to the new production conditions. Far from being helplessly buffeted by the uncontrollable force of change applied indiscriminately from without —as the deterioration of the *khadir* and the destruction of wells would tend to suggest—cultivators had more control over circumstances than many critics recognized. A close examination of the observed adjustments suggests that they conform to a pattern which our knowledge of peasant motivation and rationality would predict. Peasants, in fact, did not suddenly forget the importance of securing their food supplies—and there is ample evidence that their reluctance to use canal water on coarse autumn food crops during drought years was entirely consistent with the provision of the family's subsistence. Nor did they allow simple leisure preferences to dictate their choice of irrigation technology. Good cultivators, I would argue, did not suddenly become careless ones when they brought canal water to their fields, as many contemporaries implied. It is entirely rational for a cultivator to 'over-water' his crops, for example, if there exists uncertainly over the timing of his next supply of water; perhaps a definition of what 'over-cropping' actually entailed should precede any indictment of the cultivator for following such a practice. It is vital in examining such issues to understand the conditions under which production decisions were made by the various classes of cultivators. It has been the aim of this paper to point out that contemporary comment, when used outside a basic theoretical framework, can result in a misleading impression. Once the qualitative evidence is set into perspective, a more favourable view of the canal's impact— and of peasant adaptability—emerges.

Chapter Four

Growth of Commercial Agriculture in Bengal—1859–1885*

BINAY BHUSHAN CHAUDHURI

Though agriculture in Bengal remained predominantly subsistence agriculture, a striking phenomenon in the period under review was the growth of the cultivation of cash crops—a development which deeply affected the peasant economy of the regions where such crops were produced.

1. SOME GENERAL CIRCUMSTANCES AFFECTING
   THE GROWTH OF COMMERCIAL AGRICULTURE

Apart from the particular circumstances contributing to the growth of the cultivation of different cash crops, which we shall analyse when we deal with individual crops, several developments in the second half of the 19th century stimulated India's foreign and internal trade in general—trade largely consisting of raw materials and agricultural produce.[1] The pace of industrial growth in some nations in the continent of Europe was far quicker in this period than at any time before, resulting in the increased demand for raw materials. In the world economy itself, as Knowles[2] has pointed out, new trends were visible

---

* *Indian Economic and Social History Review*, 7.1 (1970).
[1] For details see Parimal Ray, *India's Foreign Trade since 1870* (London, 1934), Ch. II.
[2] *The Industrial and Commercial Revolutions in Great Britain During the Nineteenth Century* (London, 1930), p. 182.

about the year 1870. The 'period of world economy which means world production, world interdependence and world rivalry may be held to date from 1870, by which time railways and steamships were developed in England, France, Germany, and the U.S.A. to a point where their means of communication were revolutionised.' New developments tended to make quicker, easier and broader the commercial contacts of the industrial west with the sources of raw materials, including India: The opening of the Suez Canal (1869), synchronising with the fast growth of steam navigation, revolutionised the east-west trade. Telegraphic communications between England and India since 1855 further broadened the contact by making possible a more accurate and quicker study of the demand and supply position and of other related phenomena. The liberalisation of tariff policy by the Government of India, particularly after 1867, by abolishing or reducing export duties on very many commodities, and the gradual fall in ocean freight also contributed to the expansion of India's foreign trade. These developments affected not only the volume, but also the commodity composition of the trade. It was no longer practically confined to 'drugs, dyes and luxuries', and now included in increasing quantities foodgrains, fibres and other great staples of universal consumption.

The internal trade and commerce was much stimulated by a gradual development of communications. The most remarkable development at the time resulted from the growth of railways.[3] There was not much appreciable improvement in river communications. On the contrary, the gradual silting up of some rivers, which has been analysed in Chapter I, had undoubtedly adversely affected their efficiency as means of communications. The introduction of steam ships, some of them in fact employed by railway companies themselves as subsidiary means of communication between the main railway lines, made river

[3] For a history of the development of railways in India see Nalinaksha Sanyal, *Development of Indian Railways* (Calcutta, 1930). For the reasons behind the investment of capital by British Companies in India guaranteed railways between 1845 and 1875 and the motives of the different groups interested in railway development in India, see W.J. Macpherson, 'Investment in Indian Railways, 1845–75', in *The Economic History Review*, Dec., 1955.

communications much swifter, but their number was yet too small to perceptibly affect the existing trade. Some improvement occurred in road communications. The introduction of the road cess (1871) considerably stimulated road-building. Railways themselves had brought into existence a number of feeder roads.

The role of railways as an 'economic force', as a 'pace setter', involving innovation in some production function, has been emphasised.[4] It has been shown how 'the initial impetus of investment in railway construction led in widening arcs to increments of economic activity . . . far exceeding in their total volume the original inputs of investment capital'—a feature of modern capitalism called 'multiplier' by Keynes and others. Railways, it has been analysed, created a demand for various factors of production, which directly or indirectly, stimulated a market economy. It has not been possible for me, with the kind of source materials I have mainly used, to attempt such a study. It is also probably justifiable to say that, in the framework of the predominantly subsistence-oriented economy, apart from the fact that the size of the area penetrated by railways in the period under review was still small, this effect was not large. I have mainly emphasised the role of railways as a means of carrying agricultural produce from villages to distant markets.

In the period under review railways had grown to a considerable size.[5] In 1854 when they were first introduced in Bengal their length was about 23 miles, and scarcely any agricultural produce was carried by them. In 1889, at the end of three decades and a half, the mileage was approximately 1927, and they carried a sizeable part of the staple produce. In 1854 their impact was mainly felt in a small industrial zone—from Howrah to Ranigaj. In 1889 most districts could pride themselves on having railways.

Railways linked the metropolis not only with most districts, near and remote, of the Bengal Presidency, but also with numer-

---

[4] L.H. Jenks, 'Railroads as an Economic Force in American Development', in the *Journal of Economic History*, IV, 1944; it is reprinted in E.M. Carus-Wilson (ed.), *Essays in Economic History* (London, 1966), vol. 3, pp. 222–36.

[5] *Administration Report on the Railways for 1888–89*, Part II, (Calcutta, 1890), Appendix D; Also *Report Int Trade, 1833–84*, pp. 26–7.

ous agricultural regions and trade centres in other parts of India —a process to which the numerous feeder roads brought into existence by railways contributed not a little. The internal as well as the external market of the agricultural produce of Bengal thus tended to widen.

The pre-railways means of transport were incompatible with rapid development of commerce. The route of part of the trade of agricultural produce was undoubtedly overland. More important, however, were the river routes, and contemporary reports deal with them in far greater details than with the overland routes. Our study is mainly confined to an analysis of the shortcomings of the river routes, particularly of the regions through which railways passed.[6]

In Bihar the rivers other than the Ganges, i.e. the Gandak, Gogra, Bagmati, Kosi etc., were scarcely navigable except during July-October when they were full of rain water. Even during the rainy season, the rapidity of the current in some rivers rendered boat traffic virtually impossible at times.

In Northern Bengal, between the Mahananda on the west to the Brahmaputra on the east, the Tista on the north-east to the Ganges on the south-west and comprising a part of Purnea and Jalpaiguri, and the whole of Rangpur, Dinajpur, Bogura, Rajshahi, Pabna and Maldah, the river routes had two main directions. The produce sent along the Purnabhava and all the rivers to the west flowing into the Mahananda was partly exported to Bihar and mostly to parts of northern India through Bihar, while the produce passing along the Atreyi and the rivers to the east flowing into the Brahmaputra found its way into Calcutta. Since, as in Bihar, the rivers here were navigable for only three or four months, during the rest of the year the produce was either carted all the way to some point where there was direct water communication with the Ganges or the Brahmaputra, or its transport was postponed till the rivers became navigable at the end of June. Where traders decided to avoid the more expensive overland route, such a delay was unavoidable since the harvest time of the main agricultural produce and the period of navigability

[6] *Rept. Int. Trade*, 1876–77, Chapter II.

of the rivers did not coincide. The rice harvest was not over till January when the water level in the rivers was too low for the movement of boats laden with a bulky and heavy commodity like rice. Jute was grown during the rains, but by the time it was harvested the rivers had already begun to fall, so that boats could carry only a small quantity. Tobacco was the only exception. It was harvested in April, and traders had to wait for only two months to carry it along the river route. But the cultivation of tobacco was purely local, mostly confined to Rangpur. So, 'as a general rule', an official report concludes 'for some eight months of the year, there is no natural outlet for the main productions of the country.'[7]

The rivers of Central Bengal, known as the Nadia rivers—the Bhagirathi, the Jelanghi and the Mathabhangha—were worse in point of navigability. They had, in fact, been worsening over the years. A big sum had to be spent to keep them, navigable. 'If left to themselves, they form sandy shoals, and navigation becomes impossible soon after the . . . cold weather.' The rivers of western Bengal—the Damodar, the Rupnarain and the Kansai—were isolated rivers springing from the Chotanagpur plateau, unconnected, or having little connection, with the main river system of the country. 'They are very unfavourable for navigation, being shallow in the cold weather, and violent during the rains.' The rivers of the delta—mainly of the districts of Jessore, Khulna, Faridpur and Bakarganj—and of some other districts of eastern Bengal, were, however, quite efficient as trade routes.

As an alternative means of transport railways increasingly consolidated their position, mainly by providing easier and quicker transport throughout the year, and where the previous trade route was mainly overland, by reducing their costs. But railways could not make their influence perceptibly felt till the later part of the period under review, say from 1870 onwards. Before that time the only completed lines were the East India Railway and 107 miles of the Eastern Bengal Railway. Almost all other lines came later.[8]

---

[7] Ibid., para 19.
[8] The year of the beginning of construction of some of these is as follows:

150 / AGRICULTURAL PRODUCTION AND INDIAN HISTORY

Even when the size of the railways had become fairly considerable, their performance in carrying commodities was not up to general expectations. This was due to a number of circumstances. The number of the feeder roads was still too small to connect remote areas with the main railway lines, and where railways were far away from the river routes, they naturally failed to divert any part of the river-borne trade. Moreover, contrary to expectations, the rail-transport was not cheaper in the beginning than the river-transport. In some cases inefficient administration alienated prospective traders. 'The weak point of the railway', according to the Commissioner of Rajshahi', 'was the very great delay in the transit of small consignment of goods, owing apparently to the want of trust-worthy and intelligent agents.'

Nevertheless, railways succeeded in considerably replacing the river-transport, particularly where the navigability of rivers was poor. According to an official report of 1877, in western Bengal (excluding Midnapur) rivers carried 'but little trade, and the surplus produce ... so far as it is not tapped by the Hoogly river, is borne away by the roads and the East India Railways to Calcutta.'[9] In central Bengal the loop line of the East India Railway had a big role in the fall by 60 per cent in the traffic along the Bhagirathi river route in the period between 1842–43 and 1876–77.[10] 'The whole of the cotton, the whole of the indigo, and more than half the saltpetre and sugar, as well as a large proportion of other staples, have been attracted from the river by the railway.' The traffic along the other two river routes of central Bengal, Matabhanga and Jelangi, was also considerably reduced by the Eastern Bengal line. The railway carried 'nearly the whole of the cotton piece-goods, indigo, chillies, turmeric and sugar.'

As regards the transport of the two most important com-

---

the North Bengal State Railway and its extension—1878; the Darjeeling-Himalayan line—1880; the Bengal Central line—1882; the Dacca-Mymensingh line—1885; the Assam-Bihar line—1887; the Tirhut State line—1875; the Patna-Gaya line—1879 etc.

[9] *Rept. Int. Trade, 1876–77*, Ch. II, para 56.
[10] Ibid., para 34.

modities, rice and jute, the success of railways was far more striking with jute than with rice.[11] In two successive years, 1881–82 and 1882–83, railways carried a larger quantity of jute than country boats. In the three following years they carried 45.8 per cent and 37.47 per cent respectively of the entire quantity of jute imported into Calcutta. In carrying rice, however, country boats, had the predominant role. In 1877–78, 1878–79 and 1879–80, for instance, railways carried only about one-third of what country boats did.

Contemporary reports tell us little about whether railways stimulated agricultural production and thus added to the volume of trade by providing transport where river communications did not exist, or quicker transport where they did. In view of the poor navigability of most rivers for six to eight months a year, it is justifiable to conclude that such a result was very likely. In 1883 the Collector of Dinajpur thus wrote of the implication for the district's trade of the extension of the North Bengal State Railway to Dinajpur and of a similar extension to the Ganges *via* Purnea:

'Hitherto in this district a bumper crop was looked upon by the tenants as only a lesser evil than the total failure, as prices were entirely ruled by the *mahajans* and there was no means of getting rid of the surplus

11

Year	Jute carried by country boats to Calcutta	Jute carried by railways
	(in maunds)	
1876–77	3,839,404	3,382,406
1877–78	4,784,000	3,978,000
1878–79	5,803,000	3,008,000
1879–80	4,456,000	4,331,000
1880–81	4,086,302	3,701,097
1881–82	4,569,560	5,785,457
1882–83	5,973,703	7,001,950
1883–84	4,908,379	3,252,194
1884–85	4,910,687	4,879,448

SOURCE: *Rept. Int. Trade* (for respective years).

produce. The new extension will tap a rich and prosperous country, and will enable the jute sown in that area to find a ready market.'[12]

The extension of the North Bengal State Railway to Siliguri (1865–76) resulted in the fast growth of jute in the *Tarai*, which had scarcely produced any jute before.

2. Particular Circumstances Affecting the Growth of the Market of Some Agricultural Produce of BEngal

(A) Rice

Several circumstances tended to stimulate the internal as well as the external market of rice itself—the main subsistence crop. The rising internal demand was due to a number of developments. One was the overall increase in population, despite a decline in some regions.[13] In the caste-ridden society of Bengal the size of the non-producing group must have been considerable, and it is reasonable to assume that with the general population growth the group had grown too, though with the available statistics we are unable to precisely measure the growth. Such a growth resulted in increasing the demand for food. The increasing urban and industrial population had a similar consequence. The population growth in Calcutta was particularly rapid after 1881–11.4 per cent in 1881–91 and 24.2 per cent in 1891–1901. In the Serampur subdivision of Hugli, the most important industrial centre of the district, the population rose by 40 per cent in 1881–91 and 24 per cent in the following decade. In the industrial belt of Howrah there was a spectacular growth

[12] Beng. Gen. Misc. Progs., Oct. 1883, File 18–3/25, Annual General Report. Rajshahi Division, 1882–83 by the Commissioner, 27 June 1883, para 53.

[13] Percentage of Increase in the Population:

	1872–81	1881–91
Bengal	5.46	7.4
Bihar	13.30	4.7

Source: *Report on the Census of Bengal, 1881*, vol. I, p. 42; *Census of India. 1891*, vol. III, pp. 46–48.

of nearly 88 per cent in 1872–1901. Similarly large was the increase in the Raniganj subdivision of Burdwan. In the Asansol *thana* the population grew by 130 per cent in 1872–1901. The expanding tea industry in Jalpaiguri, Darjeeling, Chittagong and Assam employed an increasing number of immigrant labourers. A large grain supply was thus necessary to feed this growing urban and industrial population. Of the tea districts, Assam, having the largest number of tea gardens, imported the largest quantity or rice. This was partly due to the insufficiency of the local production, but mainly to 'a prejudice or a well-grounded objection derived from experience, against the use of Assamese rice by any but native-born Assamese.'[14] The role of the growth of the metropolitan demand in the expansion of the rice trade is illustrated by the fact that of the total import of 20 million maunds of rice into Calcutta, according to an official estimate of 1874, 7 million were consumed by the metropolitan population.[15] The growth of commercial crops, particularly of jute, resulting in the reduction of the aggregate acreage of rice cultivation, had probably a role in widening the local market for rice, particularly when under the stimulus of rising prices of jute, the peasants cultivated a larger portion than usual of their holdings with jute, relying for the purchase of food on the cash money derived from its sale.

Part of the demand for Bengal rice in some other Indian provinces arose from a similar development there—growing specialisation in the production of cash crops like cotton.[16] The demand was further stimulated by the expansion of industries outside India but employing emigrant labourers from India. The preference of such labourers for Indian rice accounted for a considerable export of Bengal rice to these regions. We can take, for instance, the growth of the tea industry in Ceylon.[17] Its usual

[14] *Rept. Int. Trade, 1876–77*, p. 53, para 161.
[15] Beng. Irr. Progs; Jan. 1875, Appendix: 'Note on the rice statistics' by J.W. Ottley, Assistant to Chief Engineer, Irrigation Branch, Bengal, 12 Oct. 1874.
[16] *Report of the Orissa Famine Commission (Calcutta, 1867)*, para 19.
[17] The large increase in the export of tea from 162,575 lbs. in 1880 to 7,849,888 lbs. in 1886 is evidence of this growth. (*Rept. Cust. Adm. Dept., 1884–85*, para 65; and Ibid., 1886–87, para 65).

dependence on import of Indian rice considerably increased with the flow of Indian labourers employed in the tea industry. The emigration of Indian labourers in larger numbers, particularly after the abolition of slavery in 1833, to sugar plantations in Africa and the West Indies, had a similar consequence.

The increased demand did not necessarily lead to an increased supply. The real determinant factor was the state of the crops. Hence the wide fluctuations in the volume of the export trade without any appreciable change in the market demand. Consequently, the size of exports cannot be taken as an indication of the size of the market, except in the years of normal production.

The absence of adequate statistics of the intra-district and inter-district movement of rice, which undoubtedly formed a sizeable part of the total rice market, makes inconclusive any study of the actual dimension of the rice trade. In view of this we concentrate on a part of the trade—the movement of rice to the metropolis and its subsequent export from there to various places within and outside India.

Systematic statistics of the movement of rice to the metropolis from different parts of Bengal along different routes are available only from 1876–77 onwards.[18] Government, however, admitted the probable inaccuracy of some of these statistics, in view of the complexity and multiplicity of the private organisations responsible for the movement of rice and the difficulty of registering it at its various phases and levels.[19] The statistics of exports from the Calcutta port were far more reliable.

The rice trade had been expanding for quite some time preceding the beginning of the period under review. Its value rose from £ 200,000 in the mid-1830s to more than £ 4 million in 1864–65[20]—a twenty-fold increase in the course of nearly three decades. The increase in the quantity of rice exported in the period cannot be precisely measured. We have no statistics of

[18] *Rept. Int. Trade, 1876–77*, paras 1–2.
[19] Ibid., p. 27, para 40.
[20] *Rept. Cust. Adm. Dept. 1874–75*, 'Resolution of the Govt. of Bengal, 9 Nov. 1876', para 16.

this, nor of the variation in the price-level. A twenty-fold increase in the price of rice would, however, be a preposterous assumption, in view of the dimension of price rise in our period[21] when circumstances were far more conducive to a wider growth of the rice market. A considerable increase in the export of rice in the 30-year period is thus a legitimate inference.

The statistics of the movement of rice to the metropolis and of its exports therefrom show considerable fluctuations in both.[22] The fall in the exports in 1873 was due to the widespread failure of crops, the purchase of an enormous quantity of rice by Government to relieve the distressed and the consequent price rise depressing the export trade. The same reason caused a large increase in the export of rice from the eastern Bengal districts, where crops did not fail. The rumour that Government would buy as much grain as was procurable caused an unprecedented drain on the local grain-reserve, resulting in an upward

[21] *Supra*, p. 102.
[22]

Year	Exports from Calcutta (in Cwts)	Imports into Calcutta (in maunds)
(1)	(2)	(3)
1873–74	3,637,611	–
1874–75	2,724,232	–
1875–76	3,873,854	–
1876–77	5,243,704	18,465,400
1877–78	5,526,053	25,172,000
1878–79	5,701,035	14,733,000
1879–80	3,831,083	11,848,000
1880–81	5,948,208	15,529,540
1881–82	6,350,092	17,839,279
1882–83	6,607,497	17,944,723
1883–84	6,0820,23	14,384,518
1884–85	4,953,065	13,042,765

SOURCE: For column 2, *Rept. Cust. Adm. Dept.*, for the respective years; for column 3, see *Rept. Int. Trade.* for the respective years.

swing of prices.[23] Throughout the Dacca Division, the Commissioner reported, 'the absorbing topic of public feeling and conversation' was the rising cost of food, and there was a public outcry against the continuing export of rice. In Mymensingh 'the feeling culminated in an attempt . . . to stop exportation by force.'[24] Popular anger boiled over in parts of the Chittagong Division, too. 'Anonymous letters', the Chittagong district Magistrate wrote, 'were . . . struck up in the town, threatening to burn the grain-dealers' *golahs*, if prices were not lowered.'[25] Godowns of some rice merchants were actually burnt down.

With the disappearance of famine conditions the languishing export trade revived. Soon followed one of the biggest booms in the rice trade, when the extensive failure of crops in Bombay and Madras in 1876–77 necessitated import of grain from Bengal. The boom, lasting for three years, was possible because of the coincidence of the Deccan famine with a bumper crop in many parts of Bengal in two successive seasons—1874–75. In Bakarganj, having in normal years the largest surplus of grain, 'the yield is admitted to have been greater this year (1874–75) than for the last 10 or 12 years.'[26] This was more or less true of most districts of Bengal, particularly of the eastern ones. Official reports suggest that the extent of cultivation had also increased— in 1874–75 as a result of the natural reaction of the peasants to the precarious food-supply of the previous year, and in 1876–78, as a result of the stimulus of high prices. The boom collapsed with the cessation of the demand from Bombay and Madras. The large harvest in most districts of Bengal in 1880–81, 1881–82 and 1882–83 and the consequent lowness of price again stimulated the trade, but the shortfall in production in the next two years again reduced its volume.

[23] In the Dacca Division, the prices in 1874, were higher by about 70 per cent than those in 1873. (*Prices and Wages in India*, Eleventh Issue. Calcutta, 1884, p. 2).

[24] Beng. Gen. Misc. Progs., Nov. 1874, File 5–32/33, Annual General Report, Dacca Division, 1873–74, by the Commissioner, 12 Sept. 1874, para 17.

[25] Ibid., File 5–23/27, Annual General Report, Chittagong Division, 1873–74, by the Commissioner, 4 Sept. 1874, para 40.

[26] Beng. Gen. Misc. Progs., Sept. 1876, File 132–1, Annual General Report, Dacca Division, 1874–75, by the Commissioner, 29 July 1875.

A distinct feature of the market demand for Bengal rice was its relative inelasticity, unless exceptional circumstances (like the Mutiny, the Orissa famine of 1866, the Bihar famine of 1873–74 and the Deccan famine of 1876–77) created an extraordinary demand. Again, despite a considerable growth of the normal market demand over the years, its extent and its impact on the fluctuations of prices as reflecting the interaction of supply and demand should not be exaggerated. The market for rice did not develop to the extent that the Commissioner of Dacca assumed when he wrote, 'Now-a-days a larger harvest is followed by an increased exportation, and prices remain much the same.'[27] In normal years, i.e., in the absence of an extraordinary demand, fluctuations in the prices of rice were the result of fluctuations in the level of production—which in Bengal mostly depended on rainfall. From his analysis of the statistics of rainfall and of the prices of rice in some parts of Bihar in the period 1872–73 to 1882–83, the Patna Commissioner thus concluded: 'the prices follow the total rainfall with singular fidelity.'[28]

In the absence of extraordinary demand again, a bumper crop immediately resulted in depressing the prices, as in 1880–82. The harvest was then abundant everywhere, particularly in 1880 and 1881. In the Dacca Division, 'the early and abundant rains resulted in a significant rice crop.'[29] In the 24-Parganas 'the harvest in nearly every part was the most plentiful that has been seen for many years past.'[30] In Jessore, 'the rice harvest was the best known for years.' In Patna 'the general yield was abundant.' In Sahabad 'the cold weather crop is said to have been the best known for 20 years.'[31] Local reports were unanimous on the point.

[27] Beng. Gen. Misc. Progs., Sept. 1872, No. 12, Annual General Report, Dacca Division, 1871–72, by the Commissioner, 3 Aug. 1872.

[28] Beng. Gen. Misc. Progs., Oct. 1883, File 5–4/5, Annual General Report, Patna Division, 1882–83, by the Commissioner, 7 July 1883, paras 7–14.

[29] Ibid., Aug. 1881, File 120–1/2, Annual General Report, Dacca Division, 1880–81, 12 July 1881, para 20.

[30] Ibid., File 29–3, Annual General Report, Presidency Division, 1880–81, 12 July 1881, para 25.

[31] Ibid., File 119–3, Resolution on Patna Divisional Report by Government, 3 Aug. 1881.

Not that the absolute demand for Bengal rice diminished. On the contrary, as the statistics we have quoted show, Calcutta imported and exported a larger quantity of rice in 1880–81 than in 1879–80, the reason having been the lowness of its prices which enlarged the margin of profit of the rice merchants. Yet the unsaleability of rice, the slump in its prices and the consequent sharp fall in the purchasing power of the peasants were universal complaints. The official report on the internal trade of Bengal for 1880–81 referred to the 'accumulation of stocks in the interior . . . throughout the country' as 'a marked feature of interest in connection with the trade and commerce of Bengal.'[32] The situation scarcely improved in 1881–82.[33] In Dinajpur, as the Collector reported, 'sellers had stocks already in hand, which added to this year's produce made an accumulation they were anxious largely to get rid of.' The Collector of Tripura wrote of 'the continued stagnation of the rice export trade.' In Mymensingh 'the rice market was as bad, if not worse than the last year . . . there is a perfect glut in the market.' The fact was that exports, though larger, were still a very small part of the total production.

The inevitable consequence was a slump in the prices.[34]

[32] *Rept. Int. Trade, 1880–81*, p. 55, para 76.
[33] Ibid., 1881–82, p. 63, para 76.
[34] QUANTITY OF COMMON RICE SOLD PER RUPEE IN DECIMALS OF A SEER OF 80 TOLAS

	1879	1880	1881
Eastern Bengal (Bakaraganj, Noakhall, Chittagong, Dacca, Mymensingh)	12.36	20.24	26.78
Deltaic (24-Parganas, Midnapur, Hugli, Nadia, Jessore, Faridpur)	11.80	17.37	22.90
Central (Bankura, Burdwan, Birbhum, Murshidabad, Bogra, Rajshahi, Maldah)	13.08	21.39	29.85
Northern (Rangpur, Dinajpur)	12.98	21.47	27.41
South Bihar (Mongyr, Gaya, Patna, Sahabad)	14.47	19.15	22.70
North Bihar (Purnea, Darbhanga, Muzaffarpur, Saran, Champaran)	14.68	18.87	24.16

Compared with the average prices of 1878–80 the average price of the best rice in 1881 fell by 62.69 per cent and that of the common rice by 58.24 per cent.[35] It is paradoxical that an abundant harvest caused an acute economic distress among the producers. In Mymensingh, the perfect glut in the market made 'things almost as bad for the agriculturists as if there had been a total failure of the crop, as at present they can get little or no sale for their stocks.'[36] In Bakarganj, as a consequence of this, 'the want of money is now so much felt that it has developed the same effects as a scarcity elsewhere.'[37] In parts of Tripura, 'a considerable quantity of rice land has been left fallow, as it did not pay the cost of its cultivation with rice.'[38] In Nadia the peasants complained that 'it was worse to be a cultivator than a labourer, for it was cheaper to buy a maund of rice than to cultivate it.'[39]

(B) Silk

Of the non-food crops, the only one whose cultivation had been continuously declining was mulberry. The Bengal silk trade, on the state of which its fortunes entirely depended, was passing through an acute depression. The statistics of the market prices[40]

A slight improvement occurred in 1882. The price in these zones was 26.20, 21.95, 26.36, 25.07, 20.99, and 21.85 respectively.

SOURCE: *Prices and Wages in India*, published by the Statistical Department, (Government of India).

[35] *Rept. Int. Trade, 1881–82*, p. 64, para 82.

[36] Ibid., p. 63.

[37] Beng. Gen. Misc. Progs., Aug. 1881, File 120–1/2, Annual General Report, Dacca Division, 1880–81, 12 July 1881.

[38] Beng. Gen. Misc. Progs., Oct. 1882, File 113–1/2, Annual General Report, Chittagong Division, 1881–82, 13 July 1882, para 29.

[39] Ibid., File 8–3/4, General Annual Report, Presidency Division, 1881–82, 30 June 1882, para 24.

[40]

Year	Price per seer Rs. As.P.	Year	Price per seer Rs. As.P.	Year	Price per seer Rs. As.P.	Year	Price per seer Rs. As.P.
1855-56	22-15-0	1856-57	20-4-11	1857-58	19-8-9	1858-59	21-1-9

of Bengal silk show that the crisis had set in 1872–73. Excepting a short-lived spurt of prosperity in 1876–77, the descending spiral was not reversed thereafter, Bengal was increasingly losing its market to Italy, France, China and Japan. The Bengal trade recovered from the depression only when the mulberry crop failed in these countries. Its prosperity over a fairly long period, 1864–65 to 1871–72, was entirely due to this.

Bengal was beaten undoubtedly by the superior technology of its rivals—by the application of a superior skill to the process of rearing cocoons and reeling. In 1883 an official report concluded: 'there can be little doubt that the competition of other silks in the European market is too strong for the Bengal article.'[41] Bengal silk, as the Collector Maldah wrote, 'fed on immature shrub mulberry leaf, from the nature of the cocoons when reeled, is wanting in wiriness and is endy, and is hard to throw into organzine', and consequently, an 'abomination to the European silk-thrower-preparer of organzine.'[42] The competition from China and Japan became all the more deadly after the opening of the Suez canal made possible a far larger importation of silk from there.

Year	Price per seer Rs. As.P.	Year	Price per seer Rs. As.P.	Year	Price per seer Rs. As.P.	Year	Price per seer Rs. As.P
1859-60	20-13- 0	1866-67	23- 6- 2	1873-74	15-9-4	1880-81	17- 6- 0
1860-61	18- 1-11	1867-68	24-14- 7	1874-75	12-8-4	1881-82	16-15-10
1861-62	18-15-11	1868-69	23- 6- 9	1875-76	16-4-4	1882-83	15-10- 6
1862-63	18- 3- 9	1869-70	23- 8- 6	1876-77	20-2-3	1883-84	13-12- 7
1863-64	16- 9- 8	1870-71	20-12-10	1877-78	14-8-9	1884-85	12-11- 5
1864-65	21- 0- 9	1871-72	23-10- 7	1878-79	16-1-0	1885-86	17- 6- 9
1865-66	24- 0- 4	1872-73	18- 2- 8	1879-80	17-8-8		

SOURCE: H. Maxwell-Lefroy. Imperial Silk Specialist, *Report on an Enquiry into the Silk Industry in India* (Calcutta, 1917), vol. 1. The table was contained in a letter from Messrs. Jardine, Skinner & Co., to Govt of Bengal.

[41] *Rept. Int. Trade, 1882–83*, p. 84, para 154.
[42] Beng. Agr. (Head: Produce and Cultivation) Progs., March 1883, Colln 5–43/44, Collector of Maldah to Bhagalpur Commissioner, 2 Feb. 1883, para 9.

The cause of the failure of the Bengal silk-worm rearers to adopt better techniques was a debatable point. The Government of Bengal disagreed with the view which attributed the failure to the 'ignorance' of the rearers. Their technical ability, Government argued, was beyond doubt, but tended to be depressed by some recent economic changes, particularly the rising cost of rearing cocoons and their falling market price. The 'failure to realise a profit', as a result of this, Government argued, made rearers indifferent to the 'proper rules' regarding rearing of worms. The result was a gradual deterioration in the breed of worms.[43]

The argument is faulty in that it completely overlooked the failure of Bengal silk to compete with other varieties, which caused the depression in the Bengal silk trade and, consequently, slashed its prices. Thus, the entire silk trade had occasionally suffered from such depressions, but even when it had revived, Bengal silk could not regain its old position. This is to be related to some special circumstances of Bengal silk production. Here the more or less agreed view was that Bengal silk failed to compete because of its inferior quality.

It is difficult to account for this inferiority and the failure to adopt measures to remove it. A partial explanation is that the people responsible for the rearing of worms, on the quality of which the quality of raw silk largely depended, were invariably people of small financial resources. The high cost of direct mulberry cultivation and of rearing worms on the basis of wage labour dissuaded European manufacturers from rearing worms themselves. Worm-rearing thus remained a specialised function, in some cases of mulberry growers themselves and in others, of a distinct group having no connection with mulberry cultivation.

Even small reverses of fortunes were, naturally, enough to dislocate the production organisation of these small producers, whereas a silk production on a capitalistic basis under the management of wealthy manufacturers could have withstood much

---

[43] Ind. Silk and Fibre Progs., Jan 1882, Govt. Of Bengal to Govt. of India, 6 Oct. 1881.

severer jolts. Such reverses were not infrequent, particularly after 1874. It was not unnatural for people thus preoccupied with survival not to have been eager for innovations in the technique of production.

Some local officers emphasised the role of higher differential rent on the mulberry land in depressing its cultivation. Such rent in Maldah, Rajshahi and Murshidabad was as high as Rs 12 to Rs 16, while the rent on the ordinary rice land was only Rs 1-As. 8 to Rs 2. In 1873 Skrine, Assistant Magistrate of Rajshahi wrote of having seen 'acres of mulberry field, prepared at great expense, covered with jungle or maintaining less valuable crops', because ryots, pressed for higher rent 'threw up the land in disgust.'[44] In Rajshahi it was a 'common feeling', as the Rajshahi Commissioner reported in 1881, that 'the high rates for mulberry lands . . . have affected the cultivation to a large degree . . . [and] that unless there be a general fall in the rates for such lands, the manufacture of silk will altogether disappear from the district in the course of a few years.'[45] A similar feeling led Thomas Wardle, an authority on sericulture, to propose in 1886 to the Government of India to investigate 'whether mulberry land rentals are not acting as a barrier against the extension and development of sericulture in Bengal.'[46] At a conference held in Calcutta in 1886 'there was evidently a decided feeling that the zamindars were killing the industry by exorbitant rents.'[47] To the extent that the high rate of rent increased the price of mulberry leaves, and eventually led to the deterioration in the breed of worms from insufficient food, i.e. mulberry leaves, it adversely affected the silk industry. But to say that it alone killed the industry is exaggeration.

Other important cash crops prospered. Our study is confined to some major crops.

[44] Beng. Agr. Progs., April 1873, File 17–2, Skrine's letter enclosed in the Rajshahi Commissioner's letter to the Board of Revenue, 23 Jan. 1873.
[45] Beng. Gen. Misc. Progs., July 1881, Annual General Report, Rajshahi Division, 1880–81, 2 July 1881, para 21.
[46] Beng. Agr. (Head No. 2) Progs., April 1887.
[47] H. Maxwell-Lefroy, *Report on an Enquiry into the Silk Industry in India* (Calcutta, 1917), vol. I, p. 11, para 10.

(C) Sugar

The cultivation of sugarcane remarkably increased particularly in the last three decades of the 19th century. Between 1884 and 1899 it trebled.[48] It was mostly concentrated in the Bihar districts, of which Sahabad had the largest area under it.

The history of the growth of sugarcane cultivation in Bengal since the last decade of the 18th century had a striking feature. The growth in the first phase was entirely due to foreign demand, while in the last three or four decades for the 19th century it was entirely due to internal demand. By then Bengal was an importer of foreign sugar rather than an exporter.[49] Bengal started exporting sugar to England on a large scale when the revolt in the sugar-producing colonies of France during the French Revolution practically destroyed the sugar industry there. This happened at a time when the consumption of sugar had been fast increasing consumption of tea, particularly after the reduction of duty on imported tea by the Commutation Act of 1784. An alternative source of supply was this necessary. Bengal provided a part of the supply. The Napoleonic wars considerably dislocated the Bengal sugar trade. With their end, however, it regained its former position. The partial opening of the India trade (1813) and the wild speculation that followed on the part of British private traders further stimulated it. The greatest obstacle to its expansion at the time was the high duty on Bengal sugar in the British market, while the West Indian variety, its chief rival in this market, was admitted, till 1836, on a much lower duty. The reason why in spite of high duties Bengal could send any sugar at all to the British market was the smaller cost of manufacturing sugar in Bengal than in the West Indies and the increasing importance of the Bengal sugar trade

---

[48]

Year	Acre
1884	282,000
1899–1900	862,200

SOURCE: Ind. Agr. Hort. Progs., May 1884, No. 57. *Area and Yield of certain Crops, 1891–92 to 1899–1900*, p. 27.

[49] In 1884–85 Bengal 'had almost ceased to export, and has become an importing country instead' (*Rept. Cus. Dept. Adm., 1884–85*, para 64).

as a medium of remittance for India to England in the context of a sharp decline in India's export trade in cotton goods.

The equalisation of duties in 1836 and some other circumstances resulted in a nearly nine-fold increase in about a decade in the exports of Bengal sugar.[50] The dislocation in the West Indian sugar industry caused by the emancipation of slaves (1833), who provided the necessary labour for the industry, and the equalisation of duties resulted in initiating a process of change which, if successfully completed, would have created a new era in the history of the Bengal sugar industry. Adversely affected by these measures some West Indian planters contemplated transferring their capital to India to build up a sugar industry there. Capital continued to flow in. This led in 1846 to the sugar craze in Bihar which Minden Wilson, a sugar planter of Mauritius leaving his country to grow sugar in Bihar, called 'that golden dream that swamped so many good men in 1847–48.'[51] Much capital was frittered away on large buildings, on expensive but unnecessary machinery and on attempts at acclimatising West Indian canes, which invariably failed since the Bihar soil was not the right kind to grow them on. It would have been cheaper to purchase canes than to grow them. Even where they succeeded in growing some canes, their refined sugar failed to compete in quality and price with the sugar which, despite the disorder in the West Indian labour market, continued to pour into Europe from the English, French and other colonies. No wonder that the Bihar sugar craze soon petered out, and, as a

[50] Parl. Papers, 1847–48, xxiii, Pt iv (361 11), Select Committee on Sugar and Coffee Planting, 1848 supplement, 1 to 8th Report.

[51] Minden Wilson, *History of Bihar, Part: Tirhoot and its inhabitants*, (Calcutta, 1908), p. 247. As Wilson tells us, the miscarriage of the sugar project provided the theme of a poem by George Williamson:

'The Lion King streched out his hand,
Talked of the cheapness of labour and richness of land,
Of twenty maunds a *begah*
Take the cypher from the aught, divide the ten by two,
The result will be the produce exceeded but by few,
Then things went on right jolly,
Till the district was dotted o'er with monuments of folly'

pp. 222–33.

consequence, the West Indian enterprise could hardly leave any mark on the Bengal sugar industry.

The temporary chaos in the West Indian sugar industry encouraged new experiments in cultivating sugarcane also in other parts of Bengal. In Hugli the proprietor of the Kishoriganj Indigo concern took the initiative in introducing the two kinds of exotic plants. With the success of the initial experiments 'brick-built houses sprang in every direction.' But the attempt ultimately failed. A sugarcane blight (*dhusa*), the first symptoms of which appeared in 1854–55, completely destroyed the plants. As a result, 'after struggle for more than three years the cultivation of both [kinds] . . . was given up in despair.'[52] A similar attempt in Bogra by the uncle of Payter, farmer of the principal Government estates in the district, miscarried. The story was the same—initial success followed by subsequent failure.[53]

It is likely that despite the revival of the West Indian sugar industry (partly due to the emigration of indentured labour from India) and the absence of any technical innovation in the Bengal sugar production, Bengal would have continued to export to Europe a certain quantity of sugar. But the Bengal sugar was priced out of the European market by beet sugar. The beginning of the beet sugar dates from 1801 when the first beet sugar factory was established in Silesia. By 1840 it 'had grown to a national enterprise, especially in Germany.' The sugar craze in Bihar in the late 1840s is evidence that beet sugar had not succeeded till then in dislodging other varieties from the market. But its impact made itself increasingly felt. 'In 1850 beet fields and sugar factories were spread from northern France to Russia and may have contributed a seventh of the world's sugar production. By 1870 the output had quadrupled, the fraction was one-third, and both figures were rising rapidly.'[54] At the cheap price which this enormous production resulted in, Bengal sugar was no match for beet sugar.

[52] Beng. Agr. Progs., April 1873, File 17–2, Report of Jai Krishna Mukerji enclosed in the Burdwan Commissioner's letter to the Board of Revenue, 19 March 1873.
[53] W.W. Hunter, *Statistical Account of Bogra* (London, 1876), pp. 218–219.
[54] *The New Cambridge Modern History* (Cambridge, 1964), vol. X, Ch. 2, pp. 27.

Bengal thus ceased to produce for Europe. But the internal market tended to widen. Bengal peasants responded by increasing the cultivation of sugarcane in which they were helped by the invention of the Behai iron mill and the new canal irrigation. The use of new tools in manufacturing sugar—a hand-turbine, for instance, was purchased from England in 1883—increased its output.[55] But for the loss of the foreign market the extent of sugarcane cultivation would undoubtedly have been far larger.

(D) Indigo

The history of the indigo cultivation in the Bengal Presidency in the second half of the 19th century had three main features: its gradual decline in Bengal proper; a continuous expansion in Bihar till the mid-1890s and a rapid decline there too thereafter. Despite occasional reverses, particularly those caused by the failure of the Agency Houses (1803–34) and of the Union Bank (1848),[56] which provided capital to the indigo planters, indigo cultivation, mainly as a European enterprise, continued to expand in Bengal till 1860—the year of the indigo peasants' revolt. The immediate impact of the revolt was seen in the sudden fall in the indigo exports in the following year.[57] Contemporary reports do not provide us with precise statistics of the declining cultivation that resulted, but they all give an impression that the decline had irretrievably set in. In Nadia 'since the disturbances in 1860, indigo planting has never regained its former position; and although by the help of European capital, which flowed in plentifully, a large number of factories were reopened, the uncertain nature of the crops rendered the business less profitable, and in some years, absolutely losing.'[58] In Jessore indigo cultiva-

---

[55] *Rept. Int. Trade, 1882–83*, p. 103, para 166.

[56] For details see Benoy Chowdhury, *Growth of Commercial Agriculture in Bengal, 1757–1900* (Calcutta, 1964), vol. 1, pp. 90–108 and 111–120.

[57] The quantity exported fell from 1,23,552 factory maunds in 1855 to 68,710 in 1861–62. According to an official report the exports in 1861 were the 'smallest.' (*Rept. Cus. Dept. Adm., 1874–75*, Resolution on the Report by the Govt. of Bengal, 9 Nov. 1875, para 12.)

[58] Beng. Gen. Misc. Progs., Aug. 1879, File 132–1, Annual General Report, Presidency Division, 1878–79, 7 July 1879, para 48.

tion fell by nearly half in 1868–1874.[59] In Rangpur 'the western capitalists who had laid large sums of money on indigo factories have all been ruined', and in 1873 not a single European-owned factory existed.[60] In Rajshahi most of the factories other than those with a plentiful water supply were closed.[61] In Bogra the cultivation was 'given up entirely.' The indigo cultivation in Midnapur was entirely confined to a frontier estate—the Jungle Mehals. Hugli was 'at one time studded with indigo factories, the property of Europeans . . . All these have now changed hands. The majority of them have been closed, and stand as relics of a former age.'[62] In Mymensingh, the Dacca Commissioner reported in 1876, the cultivation 'has been entirely discontinued for the past four years.'[63] In Dacca it had 'almost disappeared. Only one factory was working on . . . a small scale.' The cultivation in Faridpur had shrunk to 'a fiftieth part' of what it had been in 1861. 'European capital has within the last few years almost ceased to be employed in the cultivation.' The withdrawal of European capital from indigo was encouraged by investment opportunities of other kinds. 'Capitalists prefer investing their money in speculation like tea, which involves less risk and gives larger profits over a term of years than indigo.'[64]

Indigo cultivation in Bengal thus tended to decline but did not entirely disappear. Bengal continued to produce nearly one-third of what Bihar produced.[65] A new feature of the cultivation

---

[59] Ram Sunker Sen, *op. cit.*

[60] Gopal Chandra Das, *Report on the Agricultural Statistics of Rangpur* (Calcutta, 1874), para 19.

[61] Beng. Gen. Misc. Progs., July 1881, File 67-3, Annual General Report, Rajshahi Division, 18801–81, 2 July 1881.

[62] Beng. Agr. Progs., (Head No. 2), Jan. 1876; Colln 5–7/8, Burdwan Commissioner to Govt. of Bengal, 12 Jan 1876.

[63] Ibid., Colln 5–6, Dacca Commissioner to Govt. of Bengal, 8 Jan. 1876.

[64] Beng. Gen. Misc. Progs., Aug. 1876, File 122–1, Annual General Report, Rajshahi Division, 1875–76, 24 July 1876, para 24.

[65] STATISTICS OF PRODUCTION IN FACTORY MAUNDS IN 1879–84:

Year	Bengal	Bihar	Year	Bengal	Bihar
1879–80	14,400	28,400	1882–83	18,000	59,000

in Bengal was the increasing indigenous enterprise. The Rajshahi Commissioner found that 'except in . . . Murshidabad and Rajshahi, the business is now wholly in the hands of the natives, and carried on apparently with but little capital.'[66] In 1873 the Deputy Collector of Rangpur saw 'several persons, with a capital of Rs 300 or Rs 400, carrying on indigo cultivation.' Some 'influential zamindars', however, had 'good concerns on larger scales.'[67] This rare business enthusiasm of Bengal zamindars did not last long. In Rajshahi many of them 'deriving little or no profit, have entirely given up the manufacture.' In times of depression in the indigo trade the marginal concerns with small resources naturally went down first. Where indigenous factories survived, as in Rangpur, the quantity of indigo produced was small, the quality poor, and 'almost wholly used locally.'

In Bihar, on the other hand, the cultivation of indigo was fast increasing.[68] Indigo had already a sound base there, particularly

Year	Bengal	Bihar	Year	Bengal	Bihar
1880–81	23,000	66,000	1883–84	17,000	58,500
1881–82	17,200	58,000			

SOURCE: *Rept. Cus. Dept. Adm.*, for the respective years.

[66] Beng. Gen. Misc. Progs., Oct. 1875, File 149–1, General Annual Report, Rajshahi Division, 31 Aug. 1875, para 37.

[67] Gopal Chandra Das, *op. cit.*

[68] Complete statistics for the whole of Bihar are not available. Indigo cultivation under the factories belonging to the Indigo Planters Association increased from 70,000 *bigahs* in 1878 to 85,000 *bigahs* at about the end of the century—an increase of 21.4 per cent. (Stevenson-Moore, *Muzaffarpur Settlement Report*, para 899). The following are the statistics of indigo cultivation in Muzaffarpur, one of the most important indigo districts of Bihar:

Year	Bigahs	Year	Bigahs	Year	Bigahs
1882–83	65,000	1884–85	76,960	1897	99,959
1883–84	69,700	1885–86	80,000		

In the last two decades of the 19th century the cultivation had thus increased by nearly 53 per cent. (Ibid., para 875). In Saran the cultivation increased from 22,000 *bigahs* in 1860 to 65,000 *bigahs* in 1877–78 (Beng. Gen. Misc. Progs., Oct. 1878, File 153–1, Annual General Report, Patna Division, 1877–78), and the number of factories increased from 25 in 1860 to 69 in 1876.

in Tirhut. The convulsion which shook the indigo industry in Bengal proper was entirely absent there at the time. Other circumstances were stimulating its growth since the 1840s. The disappointment over the sugar craze in Bihar led the frustrated sugar entrepreneurs to transfer their capital from sugar to indigo—a decision considerably influenced by the bright market prospects for indigo at the time. It was about 1850 that 'sugar was finally superseded by indigo as the European industry of ... Tirhoot.'[69] Of the 86 indigo factories found by the Revenue Surveyor in 1850, several were originally meant for sugar and afterwards converted into indigo factories. The basis of the Bihar indigo industry was further strengthened by the transference of a considerable amount of capital from Bengal to Bihar after the indigo revolt of 1860.

The Bihar indigo planters could confidently continue increasing cultivation, since Bengal indigo held in fact a near-monopoly position. Attempts at growing indigo in other parts of the world (for instance, in Central America and Java) were not as great a success as to perceptibly affect it. The chemical dye began to tell on it only since the mid-1890s. As late as 1888 the Collector of Customs, confidently said: 'Indigo holds its own against all rival chemical preparations.'[70] But once the dye established its reputation in the market, Bengal indigo was no match for it, and the continuous decline in its cultivation could not be arrested.[71]

(SOURCE: Beng. Agr. (Head No. 2) Progs., Oct. 1876, Colln. 5–29, Collector of Saran to Patna Commissioner, 20 Sept. 1876, para 3).
[69] Stevenson-Moore, *op. cit.*, para 872.
[70] *Rept. Cus. Dept. Adm.*, 1887–88, para 43.
[71]

Year	Acres	Year	Acres	Year	Acres
1892–93	645,950	1897–98	509,500	1902–03	255,500
1893–94	648,928	1898–99	512,100	1903–04	249,700
1894–95	629,100	1899–1900	449,200	1904–05	223,100
1895–96	552,700	1900–01	363,600	1905–06	170,000
1896–97	582,200	1901–02	311,200		

SOURCE: *Area and Yield of Certain Crops, 1891–92 to 1905–06).*

(E) Opium

Despite periodical variations in the extent of poppy cultivation, the long term trend was a considerable growth.[72] In the 40-year

[72] STATISTICS OF POPPY CULTIVATION IN THE BIHAR AND BANARES OPIUM AGENCIES FROM 1860–61 TO 1889–90 (in Bigahs)

Year	Bihar	Banares
1860–61	281,126	154,211
1861–62	398,251	222,914
1862–63	425,353	287,008
1863–64	450,552	358,107
1864–65	417,344	347,840
1865–66	410,505	227,325
1866–67	444,530	257,546
1867–68	461,675	265,572
1868–69	416,554	287,785
1869–70	468,584	307,446
1870–71	487,550	337,812
1871–72	497,801	358,922
1872–73	471,780	342,705
1873–74	410,278	319,432
1874–75	510,313	359,300
1875–76	470,926	378,242
1876–77	517,377	372,245
1877–78	405,622	342,663
1878–79	415,289	302,820
1979–80	461,086	438,531
1880–81	434,786	423,265
1881–82	460,382	389,659
1882–83	394,232	398,952
1883–84	399,518	409,831
1884–85	433,161	471,232
1885–86	453,510	498,360
1886–87	458,266	441,018
1887–88	447,759	410,813
1888–89	405,866	329,917
1889–90	398,230	373,861

SOURCE: *Report on the Administration of the Opium Department* for the respective years.

period, 1845–46 to 1885–86, the cultivation increased by 221 per cent.[73] Since Government had the sole control over poppy cultivation, it was Government decisions to increase or reduce cultivation that were responsible for the changes in its size. The Government policy was to derive the maximum amount of revenue from exercising a monopoly control over the production and sale of opium. The method adopted to that end was to maintain poppy cultivation at the level at which a relatively high price resulting from insufficient production would not encourage the import of opium into China from other countries (Malwa, for instance), or the cultivation of poppy in China herself, or large production resulting in the fall in the market price of opium[74] and at the same time increasing the total cost of opium manufacture[75] would not reduce the revenue of Government. Given such a policy, wide fluctuations in the extent of poppy cultivation were inevitable. Government could control the number of peasants who would be permitted to grow the crop, but not the productivity of the poppy land, which entirely depended, in any particular season, on climatic variations. Other conditions remaining the same, an abundant poppy harvest for two or three successive seasons would persuade Government to reduce the cultivation. The decision of Government on the desirable extent of cultivation was also affected by varying interpretations by Government of the precise role of other variables in the fluctuations of the market prices of opium, i.e. competition with other countries and the size and rate of growth of poppy cultivation in China. Government at times related the upward trend of opium prices to a supposed insufficiency of the Bengal production and believed unless such production was raised the resultant rivalry of other varieties of opium would result in reducing the opium revenue of Government. To counteract this probable menace Government extended poppy cul-

[73] 1845–46-296,282 *bigahs*.
1885–86-951,870 *bighas*.

[74] *Papers Relating to the Opium Question*, Calcutta, 1870, Government of Bengal to the Board of Revenue, Lower Provinces, 22 May 1865, p. 52.

[75] Because Government had to pay the peasants for whatever quantity they produced.

tivation as rapidly as possible, only to find that its apprehension was unfounded.

The growing importance of the trade in opium, a commodity produced mainly for the China market, was associated with a particular phase in the development of trade relations between Great Britain, China and India.[76] As regards the size of poppy cultivation the policy of Government till the 1820s was not to increase it, since Government was against increasing the quantity of opium sold in the Calcutta market.[77] The policy worked well as long as Bengal opium held a virtual monopoly position, since, assuming a rising demand, this enabled Government to derive a larger revenue from the higher sale price of opium in the Calcutta market without any increase in the opium production. Such a policy, however, eventually proved self-defeating. The consequence of a fixed supply of Bengal opium while the demand in the China market was growing was a gradual rise in the price of opium. The obvious reaction of the opium consumers to this was to look for a cheaper variety and they soon found it in Malwa opium—a generic name given to the opium produced in Malwa, large parts of Central India and the native states of Rajputana. This, Government apprehended, would eventually destroy the monopoly position of Bengal opium—with all its sinister implications for the opium revenue of Government. The failure of a number of measures[78] adopted by Government to control the production of Malwa opium and to shackle its trade left Government with no other alternative but to increase the Bengal opium production, so that the threat from Malwa opium could be directly met. As a result, poppy cultivation increased by 122 per cent in the period 1828–29 to 1838–39—from the time the new policy came into force to the outbreak of the Opium war. A further growth occurred subsequently till

---

[76] See M. Greenberg, *British Trade and the Opening of China, 1800–1842* (Cambridge, 1951), Chs. 1 and 5.

[77] The opium thus sold was all smuggled into China. Government had no role in the export trade.

[78] For details see Benoy Chowdhury, *Growth of Commercial Agriculture in Bengal, 1757–1900*, vol. 1, pp. 10–18.

in 1848 Government decided not to increase the cultivation any further.[79]

The period of the next phase of growth was 1859–1864. The main consideration behind the decision of Government to increase poppy cultivation again was the same—to ensure the security of the opium revenue by counteracting the threat to the monopoly position of Bengal opium from a rival variety, but unlike in the 1820s, the threat was produced by a sharp decline of poppy cultivation in Bengal.

The falling poppy cultivation in Bengal was caused by the discontent of the poppy growers over the smallness of the price paid them by Government of crude opium. Since 1848 when Government decided not to further increase the cultivation, the price was cut twice—from Rs 3 As. 10 to Rs 3 As. 8 in 1850 and, again, from Rs 3 As. 8 to Rs. 3 As. 4 in 1855. The first cut by two annas did not appreciably affect the level of cultivation. It was the bigger cut of 1855 that soon began to tell on it. The poppy growers everywhere deeply resented it. The Bihar Opium Agent wrote of an 'attempt that the ryots have made to obtain their object by demonstrating a combined spirit of opposition.'[80] In the Aliganj Sub-Agency the peasants rallied with the cry that they would not 'submit to [the] reduction [of the price] to Rs 3 As. 4 per seer and other absurdities.'

Without the conjunction with the price cut of some contemporary economic changes, the poppy-growers would probably have been gradually reconciled to it, at least with some small concessions. Of these changes the most important was the rising commodity prices and wages of labour, particularly since the Mutiny. In 1859 the Bihar Opium Agent found that 'every sort of country produce is now double what it was three years ago and labour proportionately high.'[81]

---

[79] For details of the different phases of increase in poppy cultivation before 1859, ibid., pp. 7–23.
[80] Board of Revenue (Opium) Progs., 30 July, Nos. 18–19, 1856.
[81] The Agent quoted the following statistics:

| | Quantities sold per rupee: in maunds and seers | |
	1855–56	1858–59
Tobacco	0–20	0–10

The rise in the commodity prices and wages of labour affected the poppy peasants in various ways. Against the fixed price for opium, in itself very low, they now found before them the prospect of a much higher income from cultivating other crops. This was enough to tempt them away from the poppy. Even where they did not grow such other crops, they were hit by the rising commodity prices, since these increased their cost of living, while at the same time the price for crude opium, a main source of their income, was cut. The increased wages of labour made poppy cultivation, in some cases, more expensive. As a result the small income from the poppy became smaller.

The increase in the wages of labour did not equally affect all the sections of the poppy growers. The *koeries*, the traditional poppy growers, were much less affected than the Rajputs, Brahmins and others who were drawn into poppy cultivation since its gradual expansion in the 1830s. Caste prevented the women of a Brahmin or a Rajput family, unlike *koeri* women, from working in the field. Where hired labour was thus indispensable the rise in its wages naturally hit the Rajputs or Brahmins. This persuaded them to change over to cereals. This preference for cereals was thus explained by the local officers: given the need to hire labour, the cultivation of cereals had one clear advantage over that of the poppy, because of the customary payment of wages in kind in the former case. The officers, however, did not explain why it cost the Rajputs less to pay the wages in kind than in cash. The rising commodity prices, however, had the more decisive role, in most cases, in the abandonment of the poppy by the peasants.

	Quantities sold per rupee: in maunds and seers	
	1855–56	1858–59
Mustard	0–30	0–15
Linseed	1–00	0–16
Potato	2–20	1–10
Wheat	0–30	1–16
Barley	1–10	0–20
Gram	1–00	0–22

SOURCE: Beng. Misc. Rev. Progs., 3 Nov. 1859, No. 20).

As a result poppy cultivation had been shrinking everywhere. In Tirhut, one of the best poppy-growing regions, it fell by 32.6 per cent in 1850–51 to 1858–59.[82] The Sub-Deputy Opium Agent of Tirhut made a dismal prophesy, 'if the cultivation goes on decreasing in the ratio it has for several years past, it must come to a total annihilation.'[83] Between 1853–54 and 1858–59 the aggregate cultivation in Bihar fell by about 24 per cent. The quality of the production also deteriorated. The skilled peasants left, and the inefficient ones, attracted to the poppy mainly by the opium advances[84] and without any intention of persisting in the cultivation, took over. A Deputy Agent wrote of 'the increased trouble and anxiety experienced from the bad class of ryots who have substituted themselves for our former good men and the incessant supervision they require to force them to give the least ordinary care to their poppy fields.'

The diminishing cultivation alarmed Government. 'This decrease, if allowed to continue', the Bengal Board of Revenue observed in 1859, 'will inevitably cause a most serious diminution in the opium revenue.'[85] The Government of Bengal too concluded: 'The whole opium revenue is in a precarious state.' The alarm was heightened by the information Government received about the growth of poppy cultivation in China. The new policy now was 'to extend the cultivation, the manufacture remaining still profitable, until the increasing cultivation of opium in China is decisively checked.' The increase of the cultivators' price from Rs 3 As. 8 to Rs 5 within two years, as an incentive to them to increase the cultivation, was unprecedented. Government was so keen on increasing the cultivation that it did not object to the employment of even inferior soils, generally unsuitable for the poppy. The Bihar Opium Agent, who held a different view was reprimanded, 'Certainly the ryot must know that his own land would grow better than the Opium Agent, and as Government pays only for the quantity produced, it is

---

[82] Beng. Opium Progs., 3 Nov. 1859, Nov. 19, Sub-Deputy Opium Agent, Tirhut to the Bihar Opium Agent, June 1859.
[83] Ibid.
[84] *Infra*, pp. 283, 304
[85] Ibid., No. 16.

the affair of the ryot, not of the Government, what sort of land opium is grown upon.'[86] Poppy cultivation increased, as a result, by about 62 per cent between 1860–61 and 1863–64.[87]

A reaction followed soon. In 1864 Government suddenly realised that the extension of cultivation at the current rate would soon result in overstocking the market, which by lowering the prices would eventually reduce the opium revenue. Government promptly acted. The cultivator's price was reduced from Rs 5 to Rs 4 As. 8. The Opium Agents were instructed

> to take no new lands, to make no engagements with any new ryots, to confine those with whom engagements were made last year strictly within the limits of their former cultivation; to give them facilities for diverting land to other purposes . . . and on no account to allow the ryots to cultivate in excess of the area for which they may engage.[88]

This restrictive policy did not change till 1868–69, when a shortfall in the production and the disquieting news about the fast growth of poppy cultivation in China led Government to revise it. Government again wanted the cultivation increased as far as possible, and succeeded in increasing it by about 22 per cent between 1868–69 and 1874–75.[89] In 1875 the policy of unlimited expansion was changed for one of concentration on the best lands without reducing the aggregate production. Even the abundant harvests of 1875–76 and 1876–77 did not induce Government to take 'active' measures for reducing the cultivation. Government thought it 'quite enough for the present to leave the reduction of the price . . . to produce its natural and legitimate effect in the reduction of the area cultivated.'[90] Unfavourable seasons appreciably reducing the production reinforced the need for continuing the policy. The question now in fact was to make up for this short-fall by a fresh initiative for increasing

---

[86] Beng. Opium Progs., June 1861, No. 78, Govt. of Bengal to Board of Revenue, 25 June 1861.

[87] *Vide* the table of statistics on p. 49.

[88] Beng. Opium Progs., Dec. 1864, No. 19, Govt. of Bengal to Board of Revenue, 25 Aug. 1864.

[89] *Vide* the table of statistics on p. 49.

[90] Beng. Opium Progs., June 1877, Colln 1–64, Resolution of the Government of India, Finance Dept., 6 June 1877.

cultivation. Such a policy continued till 1885–86, when Government reverted to the policy of 1870—'neither advancing, nor going back.' A radical change followed in 1887–88. 'Stringent' orders were given to the Opium Agents 'to give up the least remunerative tracts, to refuse advances to unsatisfactory cultivators and bad villages, and to close inferior cultivation in the neighbourhood of large towns.'[91] This was what distinguished 'active' measures from the passive ones for eliminating unwanted cultivation.

(F) Jute

Jute, unlike the crops we have studied so far, was a new crop. Its emergence as an important cash crop dated only from the mid-1850s. The growth rate of jute cultivation was strikingly high.[92] No other cash crop had a comparable record. Moreover, jute occupied a much larger cultivated area than any other cash crop. Sugarcane cultivation at the end of the 19th century was

---

[91] *Report on the Administration of the Opium Dept., 1887–88*, Banares Opium Agent to the Board of Revenue, 30 Nov. 1888, para 3.

[92] STATISTICS OF THE GROWTH OF JUTE CULTIVATION IN SOME IMPORTANT JUTE-GROWING DISTRICTS:

District	Year 1872	Year 1900	Percentage increase (approximately)
Rangpur	100,000 acres	277,000 acres	177
Rajshahi	14,300 acres	107,800 acres	514
Mymensingh	84,000 acres	519,000 acres	517
Faridpur	16,000 acres	100,000 acres	525
Tripura	78,400 acres	219,000 acres	179

Source: N.C. Chaudhuri, *Jute in Bengal*, Calcutta, 1908, pp. 63–66. Chaudhuri belonged to the Agricultural Department, Govt. of Bengal. In Purnea the cultivation had increased in about the same period by 68.6 per cent. (J. Byrne, *Final Report of the Survey and Settlement Operations in Purnea*, Calcutta, 1908, para 317). In the period between 1891–92 and 1905–06 the cultivation had increased by more than 124 per cent, (*Area and Yield of Certain Crops, 1891–92 to 1905–06*).

estimated at 860,200 acres. Poppy cultivation seldom exceeded 9.5 lakhs of *bigahs* (approximately 3.20 lakhs of acres). In the early 1890s, before the effects of the chemical dye made themselves appreciably felt on the indigo trade, the area under indigo was about 6.5 lakhs of acres. Even in Sahabad, which had the largest extent of sugarcane cultivation, it did not occupy more than 3 per cent of the total cultivation in the first decade of the 20th century.[93] Poppy cultivation hardly exceeded 1 per cent to 3 per cent of the cultivated area.[94] This percentage in the case of indigo in the most prominent indigo districts—Saran, Muzaffarpur, Champaran and Darbhanga—was 3 45, 5 62, 6 63 and 3 08 respectively.[95] The average of jute cultivation in the 10-year period 1891–92 to 1900–1901 was 2,030,548 acres.[96] In 1901–02 jute cultivation in the districts of Rangpur, Tripura, Mymensingh and Dacca occupied nearly 30 per cent, 27 per cent, 18 per cent and 13.5 per cent respectively of the net cropped area.[97]

The market for raw jute tended to widen as a result of its increasing consumption in the new jute mills in Calcutta, its neighbourhood and Sirajganj for the manufacture of gunny bags, and of its increasing exports.

The number of jute mills continued to rise since the establishment of the first mill in 1855.[98] In 1873 the number of mills was 5 with 1250 looms. By 1875, 13 new mills came into existence and the number of looms rose to 3500. Then followed a brief period of a very low growth rate, 1875–82, when only one mill was added. Then the growth resumed. The boom in the jute trade in 1882–83 led to an enormous expansion of mill activity.

---

[93] J.A. Hubback, *Final Report on the Survey and Settlement Operations in the District of Sahabad, 1907–1916*, para 310.

[94] *Report on the administration of the Opium Dept., 1887–88*, Bihar Opium Agent, to Board of Revenue, 27 Nov. 1888, para 4.

[95] Kerr, *Saran Settlement Report*, para 640 and Kerr, *Darbhanga Settlement Report*, para 486.

[96] *Area and Yield of certain Crops, 1891–92 to 1905–06*, published by the Commercial Intelligence Dept., Govt. of India.

[97] *District Gazetteer (Statistics)*, Calcutta, 1905) for the respective districts; Table no. VII in each case.

[98] The establishment of the jute mills was made possible by the coming of the coal and railway age.

One thousand new looms were needed in 1882–84.[99] The growth was temporarily checked by the resultant crisis of over-production. The Jute Mills Association, formed in 1884 as a united move of the mill owners to fight the crisis, decided to curtail production by under-employing the mills. This measure, it was believed, by decreasing the demand for raw jute would reduce its price, and consequently, the cost of jute manufacture, and at the same time would tend to raise the price of gunnies by gradually lessening the glut in the market.[100] During the following decade, only one new mill was set up.

Precise statistics of the increasing consumption of raw jute in the mills are wanting. The increase in the number of exported gunny bags may, however, be taken as an indication of the size of the increased consumption. The number increased more than ten-fold between 1874–75 and 1884–85.[101] This resulted from the expanding trade, both within and outside India, in commodities like rice, cotton and wheat.

The export of raw jute, too, fast increased over the years, the increase having been particularly remarkable since the 1850s. In the 1830s and also in the early forties the export was negligible. A strong feeling at the time against the 'adulteration' of hemp and flax by combining them with jute—a feeling best conveyed by the commercial phrase 'warranted free from Indian jute'—retarded the increasing use of Bengal jute. An appreciable progress resulted from the decision of the Netherlands Government, about 1838, to replace flax by jute in the manufacture of coffee bags for the East Indies. But the decisive turn in the tide occurred with the Crimean war (1854–55). The war cut off the supply of flax, the main source of which was Russia. The old prejudice against jute completely disappeared, and the Bengal jute began to flow in, the main importer being the Dundee mills.

The increase in the exports in the 30-year period, 1838–43 to 1868–73, was more than forty-one times.[102]

---

[99] *Rept. Cus. Dept. Adm.*, 1883–84, para 55.
[100] Ibid., 1884–85, para 55; ibid., 1885–86, para 55.
[101] Ibid., 1878–79, para 54, and ibid., 1884–85, para 54.
[102] The exports increased from 585,238 Cwts to 24,290,814 Cwts in the period. (Hem Chandra Kar, *Report on the Cultivation of and Trade in Jute*,

Excepting the years 1866–68 the period was one of unbroken prosperity. A major setback occurred in 1873.[103] The depression was, however, shortlived. The popular confidence in the stability of the jute trade was shown by the eagerness of investors to purchase the shares of the new companies floated during the slump.[104] But a distinct trend for a few years after 1873 was a decline in the exports of raw jute. They fell from 7,061,951 Cwts in 1872–73 to 4,532,148 Cwts in 1876–77. The Collector of Sea Customs, Calcutta related it to the increased consumption in the local mills. The explanation is only partially valid, since the stationary state of jute cultivation is to be accounted for. It was an indication that the price received by the cultivators continued to be low since the depression of 1873. Their decision to gradually substitute rice for jute was reinforced by the rising prices of rice from the beginning of the Bengal famine in 1873–74 to the end of the famines in Madras and Bombay in 1877–78.

The export trade in the subsequent period considerably fluctuated.[105] This was due partly to variations in the size of jute cultivation, and partly to fluctuating demand from the Dundee mills. Apart from natural conditions the cultivation was extremely sensitive to price changes. A fall in the price was

Calcutta, 1877, para 264.)

[103] *Infra*, pp. 297–99.

[104] An official report says: 'The shares of the jute companies . . . are selling at premia considerably in advance of former quotations. Indeed, the trade has laid so deep a hold on the public confidence that, in the case of the Seebpore Jute Company, just formed [1873], double the number of shares to be allotted were subscribed for in less than a week after the scheme was afloat.' (Hem Chandra Kar, *op. cit.*, p. 209).

[105]

Year	Quantity (Cwts)	Year	Quantity (Cwts)
1877–78	5,319,318	1883–84	5,953,147
1878–79	5,701,346	1884–85	7,305,391
1879–80	6,298,693	1885–86	6,803,335
1880–81	5,367,855	1886–87	7,003,740
1881–82	6,896,610	1887–88	8,226,688
1882–83	9,191,884	1888–89	9,269,483

SOURCE: *Rept. Cust. Dept. Adm.*, for the respective years).

GROWTH OF COMMERCIAL AGRICULTURE IN BENGAL / 181

invariably followed by a fall in the extent of cultivation. The foreign market for jute was adversely affected by the worldwide trade depression in the 1880s. Moreover, as the Collector of Customs, Calcutta wrote in 1886, 'Dundee has severe competition to contend against, and has also the misfortune to find, owing to protective legislation, that other markets on the continent of Europe are being closed against her manufactures.'[106]

---

[106] Ibid., 1885–86, para 55.

Chapter Five

Expansion of Commodity Production and Agrarian Market*

A. SATYANANARAYANA

In a general analysis of colonial agriculture one can identify divergent and conflicting notions and interpretations about the impact and significance of the expansion of commercial agriculture and agrarian market. The imperialist historiography[1] and the colonial bureaucracy viewed commercialization of agriculture, the expansion of trade in agricultural products and the rising agricultural prices as an indication of the 'growing prosperity of the peasantry.' On the other hand, anti-imperialist historiography[2] (both nationalist and radical-Marxist) emphasized the negative impact of commercialization of agriculture and the integration implied that agricultural production in India was to be determined by imperial preferences and needs. In other words, production was solely geared to export and the internal requirements were not met. For the anti-imperialist historians, India was only a 'granary' of Europe and its raw material supplier. It has been suggested that 'the subordination

---

* A. Satyananarayana, *Andhra Peasants Under British Rule: Agrarian Relations and Rural Economy* (Delhi: Manohar Publishers, 1991).

[1] For a general statement of this view see J. Strachey, *Finance and Public Works of India*, (London, 1882); J. Strachey, *India: Its Administration and Progress*, (London, 1911); V. Anstey, *The Economic Development of India*, (London, 1952).

[2] B.M. Bhatia, *Famines in India, 1860–1965*, New York, 1967; R.C. Dutt, *The Economic History of India*, Vol. II, (Delhi, 1963); D.R. Gadgil, *The Industrial Evolution of India in Recent Times, 1860–1939*, (Bombay, 1971); R.P. Dutt, *India Today*, (Delhi, 1948); A.R. Desai, *Social Background of Indian Nationalism*, (Bombay, 1949).

of the agrarian economies of the countries of the Third World to the needs of imperialism, as suppliers of food and raw materials, is a common place.'[3] Therefore, commercialization meant 'impoverishment of the producers who became the victims of the forces of fluctuating market.'

Samir Amin and Hamza Alavi have characterized the commodity production and exchange under colonialism as an 'internally disarticulated system.' There was 'external integration' with the metropolis without an accompanying growth of an 'integrated' internal market ('internal disarticulation'). Hamza Alavi observed thus:

in the colonies the pattern of production was progressively lop-sided, geared to the requirements of metropolitan economy (i.e., exports), and also providing a market for the products of metropolitan industry (i.e., imports). Thus the circuit of generalized commodity production was not completed within an integrated and internally balanced economy but only by way of the linkage with the metropolitan economy, through dependence on exports and imports. It was a disarticulated generalized commodity production, characteristic of a colonial economy and not an integrated generalized commodity production characteristic of the metropolitan economy.[4]

For Utsa Patnaik, 'India . . . never saw an integrated development of . . . generalised commodity production . . . whatever the possibility which might have existed for such an independent integrated development, it was made historically irrelevant by imperialism . . . generalised commodity production was *im-*

---

[3] Hamza Alavi, 'India and the Colonial Mode of Production' in *Economic and Political Weekly*, Vol. X, August 1975, pp. 1235–62. 'Generalised commodity production in the colony did not have the same character as that in the imperialist centre itself, because of . . . disarticulation. It was a disarticulated, generalised commodity production precisely a colonial form of deformed, generalized commodity production' (p. 1257). Also see, Bipan Chandra, 'Reinterpretation of Nineteenth Century Indian Economic History' in Morris D. Morris et al., *Indian Economy in the Nineteenth Century: A Symposium*, (Delhi, 1969), p. 31.

[4] Hamza Alavi, 'India: Transition from Feudalism to Colonial Capitalism in *Journal of Contemporary Asia*, Vol. X, No. 4, 1980. See Samir Amin, *Accumulation on a World Scale: A Critique of the Theory of Underdevelopment*, 2 vols., (London, 1974).

posed from outside . . . '[5] (Emphasis in original). Therefore, she suggested that commodity production under colonialism remained peripheral and did not alter the existing relations of production. In criticizing Patnaik's view, Paresh Chattopadhyay has maintained that 'agricultural production, though it did not assume the form of commodity production in general by the close of the British rule, did increasingly show its commodity character at least since the last quarter of the nineteenth century.'[6] Some others, like Sunil Sen have argued that ' . . . commercial agriculture ushered in new trends in production relations and laid the basis of capitalist mode of production in some regions.'[7]

In the light of the above interpretations, this chapter will attempt to analyse the nature of commodity production in colonial Andhra. Though it is not intended to put forward a detailed critique of the above arguments, it proposes to examine, in concrete historical terms, the process of the expansion of commodity production[8] and its long-term stability and sustenance. It also analyses the concrete modalities through which the market expansion and integration unfolded over time.

CROP PATTERN

The cropping pattern in colonial Andhra was complex and a tendency towards crop-diversification was noticeable. A look at the crop statistics indicates that during the period under study

[5] U. Patnaik, 'On the Mode of Production in Indian Agriculture, A Reply' in A. Rudra et al., *Studies in the Development of Capitalism in India*, (Lahore, 1975), p. 213.

[6] P.C. Chattopadhyay, 'Mode of Production in Indian Agriculture', An Anti-kritik', Ibid., p. 246.

[7] S.K. Sen, *Agrarian Relations in India, 1793–1947*, (Delhi, 1979), p. vi.

[8] By commodity production we mean 'an organisation of social economy in which goods are produced by separate, isolated producers, each specialising in the making of some one product, so that to satisfy the needs of society it is necessary to buy and sell products (which, therefore, become commodities) in the market', V.I. Lenin, *On the So-Called Market Question*, (Moscow, 1976), p. 20.

more and more crops were coming under the sway of market. In a sense, a definite trend towards commercialization was visible, as the new 'industrial' crops eroded the traditional crop structure which was dominated by the subsistence crops, i.e., foodgrains. In other words the 'industrial' crops swallowed up the area under foodgrains.

TABLE 5.1

AREA UNDER PADDY CULTIVATION
(PERCENTAGE OF AREA TO TOTAL CROPPED AREA: 4 YEARLY AVERAGES)

Years	Ganjam	Visagapatam	E. Godavari	W. Godavari	Kistna	Guntur	Nellore
1	2	3	4	5	6	7	8
1901–4	57.4	31.0	49.4	–	40.1	16.4	23.4
1905–8	57.1	34.0	50.2	–	51.1	14.5	26.1
1909–12	59.7	31.6	48.7	–	55.8	14.7	31.4
1913–16	59.3	33.2	48.3	–	57.7	15.3	30.0
1917–20	64.5	27.2	48.8	–	56.2	15.4	34.9
1921–24	60.0	32.7	47.0	65.0	51.0	19.5	29.2
1925–28	63.0	38.0	49.0	69.2	44.7	16.1	27.2
1929–32	59.6	33.5	47.9	67.9	46.5	16.3	28.5
1933–36	64.0	35.6	49.0	68.0	46.3	16.3	26.0
1937–40	–	33.2	50.7	68.7	47.0	17.0	27.7

SOURCE: *Season and Crop Reports of the Madras Presidency.*
NOTE: West Godavari district was formed in 1924–25.

During the late 19th century and the beginning of the 20th century paddy dominated the crop structure of Coastal Andhra. It occupied around 50 per cent of the total cropped area in Ganjam, Godavari (East and West) and Kistna districts.

The delta districts of Godavari and Kistna, with their rich alluvial soils primarily produced paddy under canal irrigation. It was estimated that between 1895–99 and 1919–24 the total irrigated area in the deltas was increased by 36 per cent. Consequently, the total cultivated area also went up by 22 per cent in Kistna district and West Godavari district and 21 per cent in East Godavari district.[9] In Kistna the new irrigation systems (Divi and Munniycru projects) have contributed to a large increase in the area under paddy. It was also estimated that by the end of the 1920s 'nearly one lakh acres of land were brought under wet cultivation for the first time.'[10] However, during the same period the expansion of the area (of 26,041 acres) under paddy in East Godavari district was rather low, as compared with the Kistna figures. It was because 'there was not . . . the same possibility of extension of delta irrigation in this district as in Kistna and the second crop was more developed' by the beginning of the 20th century.[11] The 'very marked increase' in the area under coconuts in the delta taluks of Amalapuram and Razole was also responsible for the relatively slight increase under paddy in this district. In Kistna and West Godavari districts the fall from 9 to 5 per cent of the total sown area between 1899 and 1923 under oilseeds, which were grown as a second crop on the delta lands, was mainly due to the increase in the area under second crop paddy. Likewise the shrinkage (from 11 to 7 per cent during the same period) of gingelly area in the East Godavari district was also due to increase in second crop paddy cultivation.[12] In fact, other crops such as *jowar, bajra, ragi*, etc., which occupied around one-third of the total cropped area in the 1890s declined to around one-fifth in the 1940s. Therefore, the 'immense' area irrigated from Godavari anicut had naturally resulted in paddy being the most important crop in these districts. 'The delta present a vast expanse of rice fields dotted with gardens of plantains and coconut and with innumerable palmyras', thus remarked a government official.[13] Dharm Narain has also noted that the extension of irrigation had contributed to an

[9] Resettlement Report of the districts of Kistna, West Godavari and East Godavari, p. 36, G.O. BOR(LRS), No. 29, 18 May 1927.
[10] Ibid., p. 37.
[11] Ibid., p. 38.
[12] Ibid., p. 37.
[13] F.R. Hemingway, *Godavari District Gazetteer*, (Madras, 1907), p. 3.

expansion in the rice acreage.[14] The expansion of irrigation also caused upward shift in the level of rice acreage in Coastal Andhra Districts.[15]

In fact, 'rice specialisation had come to stay in Coastal Andhra' since the late 19th century. A recent study on 'Irrigation and Agrarian Change in Andhra' revealed that

> due to the provision of anicut-irrigation large portions of Coastal Andhra increasingly came to specialise in rice-cultivation from mid-19th century onwards. In the Southern half of the Coastal Andhra the process of commoditisation of paddy—a subsistence crop—had started as early as eighties of the 19th century . . . In 1891–92 Total Sown Area (TSA) in the region was a little over 45 lakh acres. But in a short period of 20 years i.e., 1910–11 onwards it had got more than doubled, standing at a little over 97 lakh acres. The interesting point to note is that this increase in TSA was primarily due to an increase in the area under food crops, for during these two decades area under food crops rose from 37.6 lakh acres in 1891–92 to 82.4 lakh acres in 1910–11. A closer look at the data further reveals that during this period area under rice increased by 11.4 lakh acres i.e. from 12.5 lakh acres in 1891–92 to 33.9 lakh acres in 1910–11 . . . considering the fact that in Andhra in general and Coastal Andhra in particular rice has been raised primarily with irrigation, the role that irrigation played in the agrarian expansion and shift in cropping pattern in favour of rice-specialisation needs hardly being emphasised.[16]

---

[14] Dharam Narain, *Impact of Price Movements on Areas under Selected Crops in India, 1900–1939* (Cambridge, 1965), p. 127.

[15] G.N. Rao and D. Rajashekhar, 'Irrigation and Agrarian Expansion in Andhra—A Long-term Overview' (Mimeographed), pp. 4–5. 'In the Godavari district total cultivated area (Ryotwari) 'in 1859–60, i.e., four years after the anicut irrigation commenced was nearly 3,34,000 acres. By the early 1880s it rose to 5,97,000 acres. During the same period "wet" area increased from 1,13,300 acres to 2,91,00 acres . . . As for cropping pattern . . . total area under all crops increased from 9,71,000 acres in 1878–79 to 11,14,000 acres in 1890–91, over the same period proportion of rice in the total cropped area varied between 54 per cent to 59 per cent . . . However, a significant feature is that the area under rice as a proportion of area under all foodgrains increased from 75 per cent in 1878–79 to 80.6 per cent in 1890–91. The above underline the fact that rice increasingly came to occupy an important place not only in total area under food crops but also in total cropped area. Thus, irrigation gave significant boost to the paddy cultivation . . . '

[16] Ibid., pp. 6–7.

In Nellore district the improvement of irrigation during the early decades of the 20th century considerably increased the area under paddy. The increasing paddy cultivation 'took the place of cholam' and other 'inferior' crops.[17] As in the Kistna-Godavari delta region, the area under garden crops in Nellore also increased over time. The Rushikulya irrigation system in Ganjam district, besides increasing the area under paddy also brought large area under coconuts, mangoes and plantains.[17a]

Thus, the Coastal Andhra region as a whole presents a similar crop structure, i.e., paddy remained the dominant crop. As we shall see later, it was not simply a subsistence crop but the single most predominant commercialized product. It was increasingly grown for the market. Needless to say, that the uniform crop pattern in the Coastal Districts was mainly due to the expansion of canal irrigation. In fact, the completion of the Godavari Kistna anicut during the second half of the 19th century improved the agricultural situation in delta region. The effects of irrigation on the rural economy of Andhra during the late 19th century had already been analysed by scholars, and need no repetition.[18] However, it may be pointed out that the advantage of irrigation in Coastal Andhra was that the danger from the fluctuations of season was certainly minimized. And when the immense effects of famine in the dry districts are taken into account the importance of irrigation in this region can be better understood. Irrigation prevented almost all the worst effects of famine, which, in turn, led to agricultural 'improve-

---

[17] Resettlement Report of Nellore District, G.O. Revenue No. 1171, 26 November 1906, p. 24. 'The crops . . . have changed considerably, there is an increase . . . in the area under paddy, but this of course was only to be expected from the improvement in irrigation: more curious are the entire disappearance of aruga . . . and the reduction in . . . cholam.'

[17a] Resettlement Report of Chicacole and Berhampur Taluks in the Ganjam District, BOR(RS, SUR, LR & AGRI.), No. 46, 6 February 1909.

[18] A.V. Raman Rao, *Economic Development of Andhra Pradesh, 1765–1957* (Bombay, 1958), p. 283: V.V. Sayana. *The Agrarian Problems of Madras Province* (Madras, 1949), pp. 3–5: G. Niranjan Rao, 'Changing Conditions and Growth of Agricultural Economy in the Kistna and Godavari Districts 1840–90', unpublished Ph.D. thesis, (Waltair, 1975). Also see Bibliography, Ian Stone, E. Whitcombe.

ment' and expansion. The protected irrigation system also induced the cultivators to grow remunerative and other 'special' garden crops. It also encouraged the 'well-off' ryots to sink capital in land, as there was no fear of uncertainty.

TABLE 5.2

AREA UNDER SUGARCANE IN ANDHRA (IN LAKH ACRES), 1901–1940

Year	
1901	0.11
1904	0.13
1910	0.29
1921	0.83
1925	0.94
1929	0.92
1933	1.04
1937	0.90
1940	1.08

SOURCE: *Season and Crop Reports of the Madras Presidency.*

Next to paddy the irrigated crops chiefly grown were sugarcane, tobacco, turmeric, chillies and plantains. They were more local in expansion, generally required irrigation and relatively large outlays of capital. In general, there were 'immense' potentialities in the climate, soil varieties and irrigational facilities favourable for their cultivation in Andhra districts. In the districts of Vizagapatam, East and West Godavari, Ganjam and Bellary sugarcane cultivation had begun to expand since the beginning of the 20th century. Vizagapatam, 'the district with largest area of sugarcane in the province', accounted for one-fourth (25 per cent) of the provincial total before the onset of the World Economic Depression.[19] Vizagapatam, Bellary and Godavari districts exported considerable quantities of *gur* (unrefined sugar) to other places. However, as we shall see later, the area under sugarcane did not increase until the early 1930s when (in 1932) protection was granted to the domestic sugar industry. In

[19] Dharm Narain, *op. cit.*, pp. 103–4.

the mid-1930s Vizagapatam district accounted for about 33 per cent of the total sugarcane area in the presidency. Following the Sugar Committee recommendations, the Madras Government, in the mid-1920s, explored the possibilities of extension of cane cultivation and the development of sugar industry. The Government expected the expansion of sugarcane cultivation along the lines of developing a factory industry. They also considered the possibilities of making agency (waste) lands available for cane cultivation and/or compelling the ryots to lease out lands to a sugar factory. But they found that the Coastal Andhra districts offered 'very limited prospects' for cane, owing chiefly to the widespread preference for rice and the scattered area under cane.[20] The 'petty landholders' offered no possibility of leasing out their lands for cane cultivation to a central factory, while the 'large landholders and speculators' favoured paddy, because it meant a 'safe income' from a large area without much trouble.[21] Moreover, the rising prices of paddy did not encourage the ryots to cultivate sugarcane. As a government official pointed out:

The area cropped with sugarcane is influenced not only by the price obtained for jaggery in the local market, but by the possibilities that exist of growing other remunerative crops, the cultivation of which is not attended by the same risk of loss as is the case with sugarcane. In years in which the prices of cereals rise the area under cane has a tendency to fall.[22]

---

[20] G.O. Development No. 1691, 27 November 1931. 'The limiting factors for an expansion of sugarcane cultivation . . . are the ryot's ingrained preference for paddy. Cane is an expensive crop to grow as it is on the ground for 11–12 months whereas other equally valuable crops which can be grown . . . more easily do not have to remain on the ground for such a long time . . . This is the reason why ryots generally cannot be induced to take to cane growing to a great extent.' Letter from Director of Industries (in charge) to the Secretary to Govt., 21 August 1931: 'In many places the ryot is afraid to touch the crop, especially in the more intensively farmed parts . . . there is no need for him to do so as he has plenty of other valuable crops which do not have to remain on the ground for so long a period.' Letter from the officiating Director of Agriculture to the Secretary to the Govt., 20 June 1922, G.O. Development No. 1119, 6 August 1923.

[21] G.O. Development No. 942, 14 July 1922 (Notes, p. 12).

[22] G.O. Development No. 495, 29 March 1926, p. 11.

However, the commencement of World Economic Depression and government protection provided increasing possibilities for sugarcane cultivation in Coastal Andhra during the mid-1930s.

Tobacco, 'the crop of overwhelming importance', was grown in Guntur district from the fifties of the last century. Later the delta region also cultivated this crop. Thus, by 1940 Madras Presidency produced one-fifth of India's total production and accounted for 70 per cent of the total Indian exports.[23] However, Guntur had been the largest tobacco growing district in the Madras Presidency.

TABLE 5.3

AREA UNDER TOBACCO IN ANDHRA (IN LAKH ACRES), 1901–1940

Year	
1901	0.47
1904	0.50
1910	0.99
1921	1.08
1925	1.75
1929	1.92
1933	1.91
1937	2.34
1940	2.53

SOURCE: *Season and Crop Reports of the Madras Presidency.*

Although a number of varieties of tobacco such as snuff tobacco, chewing tobacco, *beedi* tobacco, *hukka* tobacco, and cheroot and cigar tobacco were grown, the prominent type was the Virginia variety (flue-cured Virginia). With the increasing consumption of cigarettes more and more cigarette tobacco had been grown in Guntur. The growth of cigarette tobacco con-

---

[23] A.V. Raman Rao, *op. cit.*, pp. 280–81. Also see Nata Duvvury, 'Tobacco Trading and Forms of Market Organization: A Case Study of Guntur District' in K.N. Raj, *et al.*, *Essays on the Commercialization of Indian Agriculture* (Delhi, 1985), pp. 273–305.

tinued to draw the attention of the government throughout the 1930s and 1940s. 'The cultivation of cigarette tobacco is pursued with interest in Guntur. There is awakening among the ryots to cultivate cigarette tobacco',[24] remarked an official of the Agricultural Department of the Madras Government. As a result, there was a 'great demand' for the seed of the 'superior' varieties. The cultivation of Virginia tobacco 'diverted a considerable foodgrains area' in Guntur. Guntur district was also the largest tobacco export centre in the Madras Presidency. From there tobacco was exported to different places within the Presidency. Nevertheless, the largest buyer was the United Kingdom. Other countries like China and Belgium also imported Guntur tobacco.

In addition to tobacco, Guntur district was also known for its cultivation of turmeric and chillies. They were grown under wells 'with very good results.' In some parts of Guntur, soils were far superior to the Godavari delta soils and were 'ideal' for growing such 'excellent' crops. The Director of Agriculture noted in 1922: 'During the last 20 years the number of wells increased . . . a large increase . . . is expected. The most important crops grown now are turmeric . . . whereas twenty years ago this cultivation was non-existent.'[25] Though it was an expensive crop, it had brought higher returns. The net income from it was more than what could be expected from 'good single crop rice land.' Therefore, the people in this area had 'invested a large capital in sinking wells.' Cuddapah district was second only to Guntur in the cultivation of turmeric which was basically confined to the taluks of Proddatur, Cuddapah, Rajampet, Sidhout and Badvel. It was grown as a pure crop under the Kurnool-Cuddapah canal, river channels, tanks and wells. Chillies, though of minor importance, were mostly grown under wells in Guntur district.[26]

[24] G.O. Development No. 485, 29 March 1926.
[25] G.O. Development No. 942, 14 July 1922 (Official Memo No. 1332 111/20). Revenue Special, dated 13 September 1920, Notes by Director of Agriculture, 6 May 1922.
[26] Bh. Sivasankaranarayana, *Cuddapah District Gazetteer* (Revised Edition) (Hyderabad, 1974), p. 213.

In the Ceded, Districts cereals like cholam and cumbu and pulses were predominant, while paddy was 'scantily' cultivated. Though cereals and pulses occupied more than 70 per cent of the sown area the area under them diminished over a period of time as can be seen from the table below.

TABLE 5.4

AREA UNDER CEREALS AND PULSES IN RAYALASEEMA, 1901–1940
(PERCENTAGE TO TOTAL CULTIVATED AREA: 4-YEARLY AVERAGE)

Years	Kurnool	Bellary	Anantapur	Cuddapah
1901–4	78.9	77.2	77.1	79.9
1905–8	77.8	78.1	76.8	77.5
1909–12	78.4	75.1	78.4	79.7
1913–16	73.2	72.4	72.4	71.1
1917–20	76.9	72.5	76.4	72.8
1921–24	53.0	60.2	50.0	58.2
1925–28	49.7	54.3	41.7	55.0
1929–32	49.6	48.5	44.0	60.3
1933–36	62.0	56.5	55.7	68.3
1937–40	59.2	54.7	53.2	65.7

SOURCE: *Season and Crop Reports of the Madras Presidency*.
NOTE: The area under cereals and pulses declined marginally between 1901 and 1920, but there was a rapid fall after 1921. It appears that the Ceded Districts as a whole responded favourably to the growth of foreign market for industrial crops, particularly groundnut.

TABLE 5.5

AREA UNDER CASH CROPS 1901–1940
(PERCENTAGE TO TOTAL AREA: 4-YEARLY AVERAGE)

Years	Kurnool	Bellary	Anantapur	Cuddapah
1901–4	17.1	20.6	17.3	13.0
1905–8	17.9	19.6	16.9	14.4

Years	Kurnool	Bellary	Anantapur	Cuddapah
1909–12	18.4	19.3	17.6	14.5
1913–16	22.4	25.5	25.9	21.9
1917–20	21.7	25.6	19.4	22.7
1921–24	28.2	24.0	19.5	22.7
1925–28	32.7	30.2	28.5	27.1
1929–32	30.2	34.2	24.0	23.2
1933–36	30.0	33.2	25.2	20.2
1937–40	32.8	36.4	27.9	23.5

SOURCE: *Season and Crop Reports of the Madras Presidency.*

Two dry crops, viz., cotton and oilseeds (specially groundnut) steadily eroded the traditional crop structure which was dominated by subsistence foodgrains. In other words, a steady process had set in about shifting of land from the cultivated area under foodgrains to the area sown under 'the great industrial crops.' An important development in the agrarian economy of dry districts, therefore, was the shift in cropping pattern. Over a period of time, cash crops claimed an increasing share in the expanded area under cultivation. In terms of proportions the area under foodgrains and millets declined sharply over time. According to the findings of Niranjan Rao and Rajashekhar, 'the decline in the importance of these millets started right in the last decade of the 19th century itself. For instance, by 1895–96 millet area declined to 20 lakh acres forming only 38 per cent of the total sown area. By 1930–31 it went down to 31 per cent.'[27]

The author of the Cuddapah District Gazetteer has also pointed out (1915) that:

... the marked predominance of cholam is at length threatened by the enormous increase in the cultivation of groundnut ... Improved varieties of the cotton plant have been introduced and are making some headway ... By far the most important change which dates from only two years ago is ... the extensive substitution of the groundnut for cholam. It may perhaps be thought strange that the ryot of this

[27] G.N. Rao and D. Rajashekhar, *op. cit.*, pp. 8–9.

part of the country has been so slow to recognise the advantages of cultivating the remunerative crop ... appears to have cast aside all his hesitation in the matter of its adoption ... there is no doubt that thousands of acres, which were grown with cholam 3-4 years ago are under groundnut.[28]

As far back as in 1906–7 the Director of Agriculture observed that ' ... the expansion of groundnut is displacing cumbu and varagu, the comparatively unremunerative crops.'[29] Almost all the settlement officers noted that the 'most striking feature' of the crops structure in dry region was the 'large increase' in the area under industrial crops. The Settlement Officer of the Pulivendla taluk, Cuddapah district, remarked in 1909 thus:

Of all the industrial crops however the most remarkable development is shown in the case of groundnut ... oilseeds ... find a ready market .... the average area devoted to such crops is now double of what it was at the time of the last settlement (1878). The market for these products is so brisk that even the rise in the price of foodgrains has not induced the farmer to devote a larger area to them ... Presumably he finds industrial crops more remunerative ... cotton, the great industrial crop of the black cotton soils maintains its relative position.[30]

Similarly, in the Cuddapah, Jammalamadugu and Proddatur taluks a remarkable feature was the decrease of area under acreage under the two principal dry grains: *korra* and *arika*. The acreage under them decreased from 10 per cent of the total sown area in 1863 to 2.63 per cent in 1902–7,[31] whereas cotton showed a large increase in area. 'Its cultivation is inexpensive and in a good year the ryots ... can pay their whole assessment from cotton alone and have at had some profit ... '[32] In the Pattikonda taluk of Kurnool district, there had been a large displacement

---

[28] Brackenbury, *Cuddapah District Gazetteer* (Madras, 1915), Vol. 1, pp. 79–80.
[29] *Season and Crop Report of the Madras Presidency* (hereafter referred to as SCRMP), 1906–7, p. 3.
[30] Resettlement Report of Pullivendla Taluk of the Cuddapah District, BOR(RS, SUR, LR & AGRI.), No. 428, 29 November 1909, para 18.
[31] Resettlement Report of Cuddapah, Jammalamadugu and Proddatur Taluks of the Cuddapah District, BOR(RS, SUR, LR & AGRI.), No. 294, 26 August 1909.
[32] Ibid., para 21.

of cholam in favour of other crops. However, there was a distinct increase under industrial crops such as cotton and still more in the case of oilseeds.[33]

In the red soil taluks of Anantapur district, groundnut was non-existent in 1894–95, but occupied 10 per cent of the cultivated area in 1924–25. It indicated that 'ryots have begun to appreciate the profitable nature of the cultivation of this crop.'[34] It was supposed to have also 'put (more) money into the pockets of the ryots and enabled them to pull through a season of distress.' In the red soil taluks of Bellary, cotton accounted for 15 per cent of the total sown area. The increase was most marked in some taluks which contained the bulk of the better regar lands. Nevertheless, there was a considerable extension of this crop in 1917, 1918, and 1919 due to the prevailing high prices. Besides, the area under oilseeds was far greater in this tract as elsewhere and a large export trade was carried on in this commodity. Though it was unknown in this district till 1907, it covered 20 per cent of the cultivated area in 1921–22.[35] A more or less similar situation existed in the black soil taluks of Bellary and Anantapur districts. Cotton occupied 22 per cent of the cropped area and groundnut, which was non-existent in 1890–91, accounted for 6 per cent of the total sown area of black soils. However, it was by far the most popular industrial crop in the taluks of Tadipatri and Gooty, where it covered 15 and 13 per cent of the total cultivated area respectively. The increase in the industrial crops naturally caused a decline 41 per cent in 1900–1 to 32 per cent in 1919–20.[36] Cotton also displaced many less important food and industrial crops and promised 'in the fullness of time to become a powerful rival to cholam and Korra.'

Of all the Coastal Districts, Vizagapatam and Guntur

---

[33] Resettlement Report of Pattikonda Taluk of the Kurnool District, G.O. Revenue No. 710, 25 July 1906, p. 735.

[34] Resettlement Report of the Red Soil Taluks of Anantapur District, BOR(LRS), No. 2, 4 January 1926, p. 14.

[35] Resettlement Report of the Red Soil Taluks of Bellary District, G.O. Revenue No. 1392, 13 September 1923, p. 13.

[36] Resettlement Report of the Black Soil Taluks of Bellary and Anantapur Districts, BOR(RS, SUR, LR & AGRI.), No. 170, 29 October 1920.

witnessed moderate expansion under cotton and oilseeds. During the early 1920s many taluks in the Vizagapatam district cultivated groundnut, 'a crop, which fifteen years ago, was practically unknown.' Thus, the settlement officer of Vizagapatam district remarked: 'Before the war, the profits obtained from gogu and groundnut were very large, and the introduction of increased cultivation of these crops had undoubtedly brought a considerable influx of wealth . . .'[37]

The above discussion suggests that there was a significant response of industrial crop acreage to changes in relative prices. In the Deccan Districts, market orientation and responsiveness to price changes was high. In fact, it may be pointed out that it was not the rainfall alone, but also the prices which determined the crop structure in this region. Dharam Narain in his discussion on cotton price and acreage noted that 'the overall fit of the two curves, plotted to scale so chosen that their amplitude of fluctuation about agree is in fact so close that the price factor alone would seem to account for most of the change in area . . . Price bears in an unusually large degree on the variation of Madras cotton area.'[38] Generally speaking, his study shows that no other province which he studied possessed the same degree of price determination. Also in the case of groundnut, 'the directional affinity between the movements of price and area is positive and persistent.'[39] In his study Sayana also observed that 'the growth of groundnut in the province . . . is an instance of the remarkable effect which the stimulus of a commercial demand exercises over the expansion of cultivation of any crop.'[40] The Government of Madras remarked that 'there has been a regrettable tendency to violate the old established rotation by too persistent cropping with commercial crops.'[41] 'Groundnut is making headway. It is certainly replacing the food crops. The

[37] Resettlement Report of Vizagapatam District, BOR(RS, SUR, LR & AGRI.), No. 40, 18 March 1919, p. 12.
[38] Dharam Narain, *op. cit.*, p. 24.
[39] Ibid., p. 71.
[40] V.V. Sayana, *op. cit.*, p. 24.
[41] *Royal Commission on Agriculture in India*, Vol. 14 (Appendix to Report (Madras) (Calcutta, 1928), p. 238.

ryot is getting more money into his pocket', a big cultivator told the Royal Commission on Agriculture.[42]

Therefore, it can be pointed out that the crop pattern in Andhra, during the period under study, underwent a clear-cut reorientation in terms of industrial crops progressively cutting into the traditional crop structure. *The Season and Crop Report for 1924–25* put it:

> This (the cultivation of non-food crops), especially cotton and groundnut at the expense of food crops, regardless of rotation, is increasing rapidly.... The rapid spread of the crop (groundnut) at the expense of cereals, especially in the unirrigated tracts of the Circars and the Deccan which are liable to frequent scarcity or famine, coupled with export of groundnut by sea is reducing the local supplies of fodder.... and increase the prices.[43]

It is also striking to note that the rising agricultural prices encouraged the cultivator of more 'profitable' and 'remunerative' crops. Consequently, a more diversified crop structure made inroads into the Andhra agriculture.

INTERNAL TRADE

Contrary to the assumptions of Hamza Alavi and others, we argue that the rhythms of agrarian production were not determined solely by the 'external articulation' of the market. As a matter of fact, the impact of expanding trade and market differed from region to region. In Andhra some regions like the Deccan districts were directly integrated into the circuit of the world market, producing primarily for foreign exports, while, the coastal districts were linked to the internal market (within the Presidency and in India) to a much greater extent. Thus, Dietmar Rothermund remarked: 'a considerable specialisation in crop production had emerged in the Andhra region ... the areas outside the delta producing certain cash crops and getting their rice supply from the delta districts.'[44] The bulk of the

---

[42] Ibid., Vol. III (Evidence taken in the Madras Presidency), p. 581.
[43] *SCRMP* for 1924–25, p. 24.
[44] D. Rothermund, 'Agrarian Distress in India, 1900–1935' (Mimeo), p. 34.

marketable produce (cotton and oilseeds) from the dry districts was exported to Europe. As we shall see later, the Deccan region responded much more to the fluctuating rhythms of external demand. On the other hand, a large part of the foodgrains from the Coastal Districts was sent to deficit regions in the province (Rayalaseema *districts*, Southern Tamil *districts* and Malabar), and to other neighbouring provinces (Bombay, Bengal and Central Provinces) and native states (Hyderabad and Mysore). There was also a regular import and export trade in *gur*, spices, etc. besides foodgrains, with the neighbouring provinces.

TABLE 5.6

EXPORT OF FOODGRAINS, SUGAR AND PROVISIONS FROM ANDHRA, 1900–1920 (QTY. IN LAKH TONS: 4-YEARLY AVERAGES)

Years	Foodgrain exports by rail to other places within the Presidency	Places outside Madras	Foodgrain exports by sea	By rail	
				Exports of sugar to places outside Madras	Exports of provisions to places outside Madras
1	2	3	4	5	6
1901–4	1.82	0.50	0.37	0.12	0.61
1905–8	1.88	0.74	0.30	0.13	0.54
1909–12	1.56	0.79	0.09	0.17	0.68
1913–16	1.67	1.07	0.06	0.17	0.51
1917–20	1.05	0.72	0.02	0.19	0.67

SOURCE: *Rail-borne Trade Statistics of the Madras Presidency.*
NOTE: The statistics of the rail-borne trade were discontinued after 1920.

IMPORTS FROM NEIGHBOURING PROVINCES INTO ANDHRA BY RAIL
(QTY. IN TONS: 4-YEARLY AVERAGES)

Year	Grains and pulses	Refined sugar	Provisions
1901–4	41,519	2.611	13.129
1905–8	55,000	5,224	14,420

Year	Grains and pulses	Refined sugar	Provisions
1909–12	57,862	7,528	18,252
1913–16	52,872	5,685	14,262
1917–20	67,629	5,269	N.A.

SOURCE: *Rail-borne Trade of the Madras Presidency.*
N.A.: Not available.

It is evident from the available sources that any famine or scarcity within the Presidency or outside led to a diversion of foodgrain export to these areas. For instance, the distress in North India during 1899–1900 led to 'large' exports of grains from Nellore district.[45] Similarly, the demand for foodgrains from Western India resulted in a 'considerable' export from the Coastal Districts. The scarcity in Kurnool district (1900) necessitated imports of grains from the Hyderabad State and Kistna district.[46] High prices and 'dearer' markets outside the Presidency also caused increased exports of paddy and rice from the Northern Circars.[47] The crop failure and famine in Bellary and other Deccan Districts (in 1918) led to the diversion of large amounts of foodgrains not only from the Kistna and Godavari deltas but also from outside.[48] The delta districts also exported to other deficit areas within Coastal Andhra, upland taluks of Guntur and Vizagapatam districts. Considerable quantities of cheap rice from Rajahmundry (East Godavari district) were imported into Guntur district (Tenali). The chief consuming centres of delta rice in Vizagapatam district were the taluks of Vizianagaram and Anakapally. They imported around 50,000 bags annually during the 1930s.[49] Anakapally, the important centre

[45] 'Letter from Collector of Nellore to the Board of Revenue', 15 December 1899, BOR(RS, LR & AGRI.), No. 438, 30 December 1899.

[46] 'Letter from Collector of Kurnool to the Board of Revenue', 3 October 1900, G.O. Revenue No. 1125, 5 November 1900.

[47] *Review and Returns of Sea borne Trade and Navigation of the Madras Presidency for Years 1899–1900 and 1903–04.*

[48] G.O. Revenue No. 2385, 22 June 1918; G.O. Revenue (Special), No. 565, 17 October 1918, Letter from Director of Civil Supplies to the Government, 12 September 1918.

[49] Letter from Director of Agriculture to the Govt., 3 March 1933, G.O.

EXPANSION OF COMMODITY PRODUCTION / 201

of rice trade in Vizagapatam district, also imported grains from the Central Provinces and Orissa. The opening of Vizianagaram-Raipur railway line 'greatly' facilitated imports from the Central Provinces. Nellore district largely exported rice to the Tamil districts, 'where the more well-to-do people' preferred the Nellore variety to others. Nellore district, including the Godavari delta, also exported rice to Coimbatore district, which imported (throughout the 1920s) more than 5–8 times of its local production,[50] whereas the Kistna delta exports were directed mainly to the Nizam's (Hyderabad) territories, the Deccan Districts and partly to the Bombay Presidency. It also exported directly to Madras through the Buckingham canal.

Unlike Burma and Indo-China, Andhra districts exported mainly raw rice. Small quantities of cheap boiled rice, produced in Rajahmundry, Cocanada, Nidadavole, Palakole, Bezwada and Gudivada, were exported largely to Ceylon, West Coast and Pondicherry. Raw rice was mostly exported by rail from Tadepalligudem, Narsapuram, Tanuku, Nidadavole, Ellore, Masulipatam and Bezwada. During the early two decades of the twentieth century, the Masulipatam-Bezwada region, 'a great rice producing tract', exported the 'largest quantity' of foodgrains. Owing to its railway and canal facilities 'its stocks were usually the first to come into the market.'[51] The available data also suggest that the relatively developed transport and communication system in Andhra districts meant an easy, regular and faster mobility of agricultural commodities.

The available evidence suggests that besides foodgrains and pulses, the Andhra districts also exported unrefined sugar (gur or jaggery) and provisions. *Gur* exports were mainly from Vizagapatam and East Godavari districts. Guntur district ex-

---

Development No. 560, 26 April 1934.

[50] Ibid., Letter from Deputy Director of Agriculture VIII Circle, to Director of Agriculture, 16 February 1933, '40 per cent of the rice comes from Nellore, Guntur, Kistna, and Godavari; 15 per cent from South Malabar; 10 per cent from Trichinopally, Tanjore, North Arcot; 10 per cent from South Arcot and 25 per cent from Burma . . . '

[51] G.O. Revenue (Special) No. 158, 9 August 1918; Letter from the Secretary to the Govt. of Madras to the Govt. of India, 24 July 1918.

ported spices (chillies), *ghee* and other dairy products. Dry fruits and other provisions were exported from the Kistna-Godavari delta.

The Deccan Districts also exported 'large quantities' of *Jowar* and *Bajra* (the local staple) to both within and outside the Madras Presidency.

TABLE 5.7

EXPORT OF JOWAR FROM RAYALASEEMA, 1901–1920

Years	Quantity (in tons)
1901	10,367
1901	10,367
1905	52,912
1909	24,559
1913	30,735
1917	23,382
1920	18,860

SOURCE: *Rail-borne Trade Statistics of the Madras Presidency.*

We are told that by the turn of the century in Cuddapah district the export trade in cholam had been 'very active' and in the taluks of Jammalamadugu and Proddatur 'the entire area under dry grain crops has been devoted to meet the increased demand.'[52] Most of the settlement officers of the Deccan Districts, since the early decades of the present century, 'called attention to the large exportation of cholam that was going on' from there. The Collector of Kurnool district reported to the Board of Revenue in 1900 that 'there has been a rise in the price of cholam, which is mainly due to the very large exports to Kathiawar in the Bombay Presidency.'[53] The Koilkuntla taluk, which was able to meet a 'large external demand' also contributed to large ex-

[52] BOR(RS, SUR, LR & AGRI.), No. 294, 26 August 1907, para 21.
[53] Letter from Collector of Kurnool to the Board of Revenue, 5 September 1900, G.O. Revenue No. 1982, 24 September 1900.

portation of *cholam* from Kurnool.[54] Owing to famine in Northern India, Anantapur district exported 'considerable' quantities of foodgrains, which led to rise in prices and a 'diminution of stock in the district.'[55] Evidence of this kind can be multiplied. What comes out clearly is the fact that besides 'industrial' crops, the 'subsistence' foodgrains were also getting commercialized, although to a limited extent. In general, during scarcities, famines and crop failures the Deccan Districts were heavily dependent upon the neighbouring places for foodgrains (mainly *jowar*).[56] Nevertheless, it may be pointed out that the commercialization and expansion of foodgrain exports of this region, in a fundamental sense, was linked with the impoverishment and disintegration of small peasant production. The petty peasants had to alienate, not the surplus over and above the consumption requirements, but the whole produce. Therefore, exchange had become an essential basis for the reconstitution of the production cycle. However, we are not suggesting that it was basically a 'forced commercialization.' We shall argue in a later chapter (chapter IV) that commercialization of agriculture has to be located within the context of class structure.

The fact that a more tightly integrated market was in the process of emerging over time is suggested by the behaviour of (agricultural) commodity prices (Graphs). The available evidence of the 19th century suggests wide amplitude of price fluctuations from one year to the next. It also shows that an integrated structure of market was yet to crystallize and that the climate intervened very crucially as far as the price movement was concerned. The absence of transport facilities precluded the rapid inflow of commodities, which resulted in discrepancies in prices not only at the district level but also at the taluk level. Thus the prices were largely determined by local factors.

We suggest that since the turn of the century, the local factors

---

[54] 'Resettlement Report of the Koilkuntla Taluk of Kurnool District', G.O. Revenue No. 882, 11 September 1900, p. 205.
[55] 'Letter from Collector of Anantapur to the Board of Revenue, 13 February 1900', G.O. Revenue No. 267, 22 March 1900; G.O. Revenue No. 119, January 1900.
[56] See footnote 51 above.

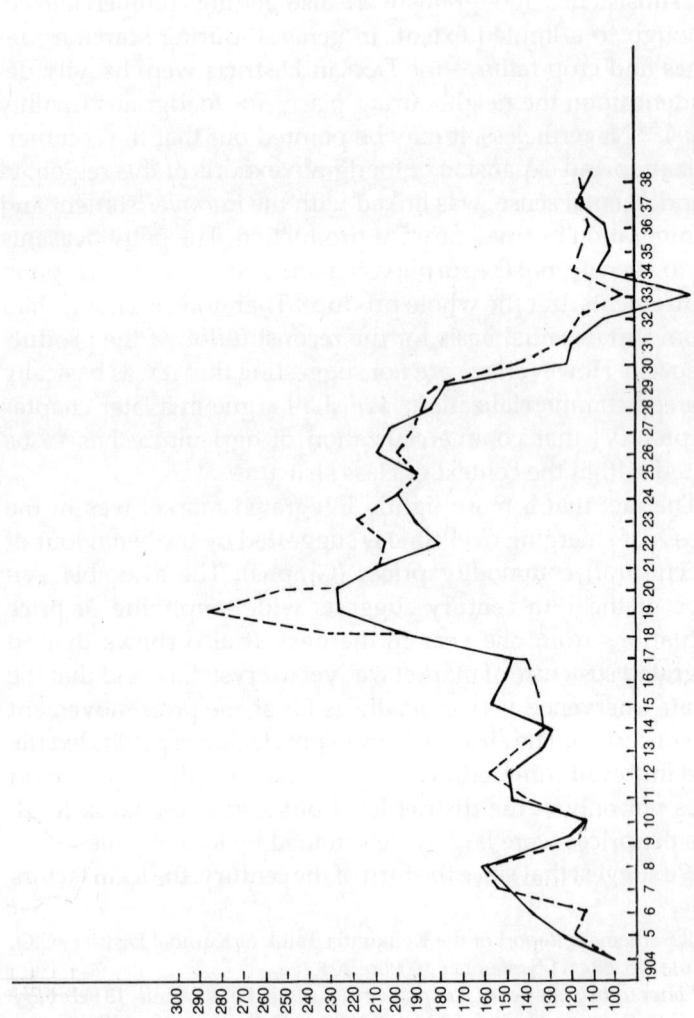

Chart I Price Indices of Rice in Rayalseema and Coastal Andhra

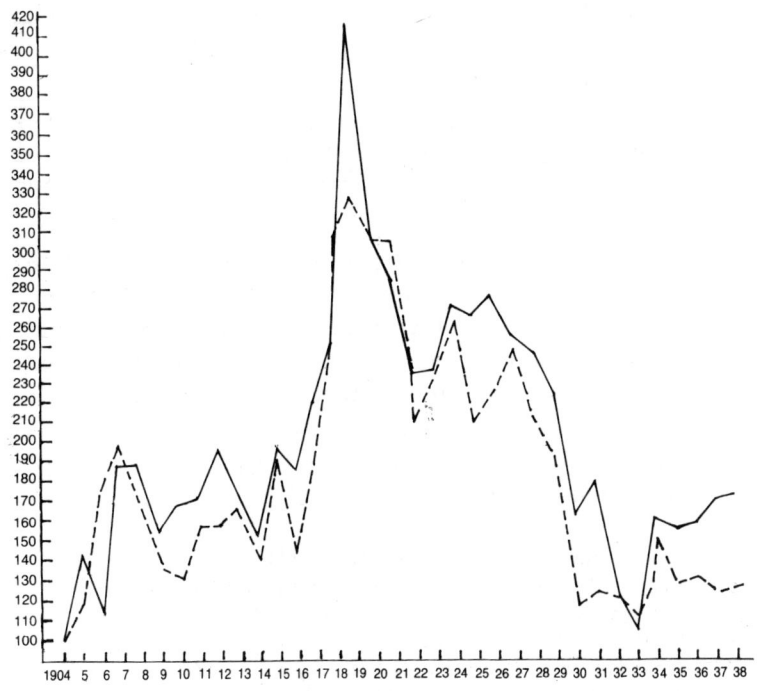

Chart II Price Indices of Cholam in Rayalseema and Coastal Andhra

which regulated price movements during the early years were fast overcome with the region's 'transport revolution' of the late 19th and early 20th centuries. For the country as a whole John Hurd and others have established the effects of railway expansion in bringing about the equalization of market prices.[57] The important factor behind the price rise and relative equalization during the first two decades of the twentieth century, as the local bureaucracy tried to explain it, was the 'large' exports of foodgrains to the national and international markets. The Madras Government remarked (in 1912): 'From a purely economic standpoint the main feature of the decade has been the high range of prices for all foodgrains. A general levelling of rates throughout the Presidency was noticed . . . Prices are indeed governed not only by the local market but by the rates prevailing in other parts of India.'[58] The Settlement Officer of Anantapur district pointed out (1925) that

the one noticeable feature about the prices of the chief foodcrops is that the variations in them were quite in accord with the general trend of prices of these crops in the Presidency as a whole, which indicates that local conditions played only a minor part in their rise or fall . . . the prevalence of high prices . . . indicates that the local plenty or scarcity has the least influence on prices.[59]

During 1907–8 the foodgrain prices in some parts of Andhra reached the 'scarcity' or 'famine' rates, 'although the local seasons were favourable.' In other words, the news of famine and scarcity in other parts led to the rise in prices, despite good harvests.[60] At the turn of the century the increase in exports sometimes 'exhausted' the local markets and price rise was always 'chiefly due to exportation to other places.' The Board of Revenue noted in 1901 that

[57] John Hurd, 'Railways and the Expansion of Markets in India, 1861–1921' in *Explorations in Economic History*, Vol. 12, 1975, pp. 263–88; D. Rothermund, *Government, Landlord, and Peasant in India: Agrarian Relations Under British Rule, 1865–1935* (Wiesbaden, 1978), p. 26 ff.
[58] *Memorandum on the Moral and Material Progress of the Madras Presidency*, p. 3, G.O. Revenue No. 2864, 6 September 1912.
[59] BOR(LRS), No. 2, 4 January 1926, p. 16.
[60] *SCRMP* for years 1906–07 and 1919–20.

during the earlier months of the year of the extreme dearness of foodgrain in western India maintained rates at a high level within the presidency and owing to the depletion of stocks . . . considerable imports were necessary to meet the local requirements . . . in the north of the presidency the activity of the grain trade was great almost throughout the year . . . . [61]

In fact, as early as 1896 the Collector of Godavari district observed:

I found that the condition of the local crop had little to do with the sudden rise in price. The merchants have been acting for most part on orders received from Bombay and Secunderabad (Nizam State) and it appears that the Marwari merchants are speculating on the probability of a famine in the Punjab and in the country between Secunderabad the Raichur where there has been failure of the rains. If there is a good rain here, prices will probably fall slightly, but they depend almost entirely on the condition in other parts of India.[62]

In 1900, 'a considerable demand' for export to Western India led to a very rapid rise in the price of cholam in the Deccan Districts.[63] Similarly, the scarcity of 1918 resulted in large exports, which in turn brought 'a famine price.' In the delta districts prices rose (1918–19) because the merchants purchased 'large quantities' for export to Bombay.[64] The rising prices led to 'rioting' and 'organized looting' of grain shops at various places in the districts of Kistna, Guntur, Kurnool, Bellary and other places.[65]

The role of transport and communications in the development of an integrated internal market structure was also noted by the colonial bureaucracy:

The volume of trade by rail with other parts of India nearly trebled during the decade (1891–1901). In recent years a large part of trade in

---

[61] BOR(RS, LR & AGRI.), No. 278, 29 October 1896, p. 4.
[62] Letter from Collector of Godavari to the Board of Revenue, 6 October 1896, BOR(RS, LR & AGRI.), No. 282, 20 October 1896.
[63] G.O. Revenue No. 11, 9 January 1900.
[64] G.O. Revenue (Special) No. 2806, 27 July 1918.
[65] Letter from the Secretary to the Govt. of Madras to the Govt. of India, 19 June 1918, G.O. Revenue No. 2385, 22 June 1918.

foodgrain was done with those parts .. The volume of goods moved from one section of the presidency to another increased by more than a third . . . chiefly in the traffic of those parts of presidency outside the sea-port blocks. In this branch of the traffic grain accounts for about 1/3 of the whole volume. The expansion of the internal trade is largely attributable to the general extension of railways.[66]

The price levels which varied sharply from region to region earlier, then, tended to project an increasingly distinct tendency towards convergence. In other words, the lowering margin of price fluctuation and the increasing consistency in prices were possible with the faster and regular mobility of commodities. In a general sense, price convergence and equalization necessarily presupposed market integration and a moderately developed system of physical conditions of circulation, which marginalized the transport cost of commodities. It meant a faster mobility of commodities and cheaper rate of circulation.

The other important aspect of price movement was that there was a tendency of closer equalization of prices between the Andhra districts and other parts with which it had an integrated trade. For instance, prices in Tenali (Guntur) district, one of the important rice exporting markets, were ruled by the prices fixed in Bezwada, Masulipatam, Madras and Secunderabad.[67] In another important rice centre of West Godavari (in the 1930s) it was revealed that 'Rangoon appears to be the price fixing market and Tadepalligudam prices closely follow Rangoon prices.'[68] Likewise, the prices of *cholam* in Deccan Districts 'closely followed the Hyderabad prices', with which it had considerable trade.[69] During 1918 when the Hyderabad Government prohibited the export of *cholam* ('export embargo') the prices in Kurnool district immediately shot up . . . ('. . . the embargo . . . did in fact affect very adversely the conditions in

[66] *Memorandum on the Material Progress of the Madras Presidency*, pp. 205–06, G.O. Revenue No. 711, 11 August 1902.

[67] 'Resettlement Report of the Guntur Portion of the Kistna District', G.O. Revenue No. 287, 11 March 1904, p. 281.

[68] G.O. Development No. 60, 4 April 1934; Letter from Deputy Director, VIII Circle, to Director of Agriculture, 16 February 1933.

[69] G.O. Revenue (Special) No. 2385, 22 June 1918.

Madras).'[70] The Economic Depression Enquiry Committee in 1931 also remarked that 'Madras prices are really governed by the prices at which Burma rice could be dumped into the East Coast ports.'[71] 'The market prices for paddy have come to be influenced more and more by the imports from Burma and other places . . .'[72] The Director of Industries wrote in 1937 that '. . . prices of Rangoon rice which when analysed over a period of three years were found to have a high degree of correlation with the prices of Cocanada rice. Further, the trend of prices in Madras in 1936 was found to follow closely the trend observed in Rangoon in the same year . .'[73] A study of the selling prices of rice at Anakapally market (Vizagapatam district) indicated that the price of Godavari rice closely followed that of Central Provinces. Nevertheless, the tendential price equalization was best reflected in periods of scarcity and famine when there was greater velocity of circulation of products.[74] Very often, the process of price convergence and lower amplitude was also at-

---

[70] Ibid., p. 7.
[71] *Report of the Economic Depression Enquiry Committee*, G.O. Revenue No. 948, 1 May 1931, p. 26.
[72] *Report of the Economic Enquiry Committee*, Vol. II (Appendices), p. 23.
[73] *Report on the Administration of the Department of Industries in the Madras Presidency*, 1937–38, pp. 24–25; (hereafter referred to as *RADIMP*).
[74] G.O. Revenue No. 3214, 14 December 1940. 'Owing to the improvement in communications there were ample facilities for imports from unaffected parts and outside the districts. The prices of foodgrains did not rise abnormally as a result of famine'; 'owing to the extent of railway and motor lorry traffic there was prompt supply of grains from outside and there was no rise in prices.' A settlement officer remarked that 'Though some people have deplored that with the increased facilities for the disposal of produce, the tendency to hoard grain had decreased, there is no doubt that the opening of . . . railways made famine a matter of no serious concern now-a-days. The railways have rescued the people from absolute dependence upon the local harvest and have overcome also the immobility of grain. The usefulness of the railway has become more and more apparent in the famines subsequent to 1876. It has enabled foodgrain to be imported from other districts and has in consequence prevented any abnormal rise in prices. In 1876–77 and in 1891–92 the famine resulted in the inflation of prices by 60 to 70 per cent in the former case and 36 to 40 per cent in the latter, but in the famine of 1924 the increase in prices ranged about 11 per cent.' BOR(LRS), No. 2, 4 January 1926, p. 15.

tributed to the 'full operation of the laws of supply and demand.'[75]

A cursory look at the various District Gazetteers, manuals and Settlement Reports reveals the fact that the absence of an integrated market during the first half of the 19th century was due to the lack of an organized transport and communication network. Then some parts of Andhra were still very much on their own; they were 'locked up' from the surrounding provinces. In other words, both internal and external communications were in a state of utter backwardness. However, a major transformation in the transportation of commodities was visible since the late 19th century. By the late 1880s the Andhra districts were linked to Madras, Bengal and Bombay and the neighbouring princely state of Hyderabad, by means of the Madras and Southern Mahratta Railway, Madras Railway, etc. The Madras-Calcutta main (broad-gauge) line entered the Kistna district at Bezwada, 'one of the main junctions', and continued across the East Coast (i.e., Coastal Andhra). A meter-gauge line which connected Bezwada with Masulipatam tapped the 'rich section of the Kistna delta.' Nizam's State Railway also facilitated the export of delta produce to Hyderabad.[76] Another branch line (opened in 1893) connected the two important ports of Andhra, Cocanada and Waltair. The Bengal-Nagpur Line, which traversed from south to north of Vizagapatam and Ganjam districts linked them to Orissa and Bengal. The Deccan Districts (the major cotton and groundnut growing region) were also 'fairly well served with railways.' They were linked to Madras and Bombay. The Bombay-Madras broad-gauge line ran across them. The Bellary-Kistna line linked the districts of Bellary, Kurnool and Anantapur to Coastal Andhra. This line was important, specially in times of scarcity, as it facilitated the import of foodgrains from the deltas to the Deccan region.[77]

[75] 'Letter from Collector of Bellary to the Govt. of Madras, 12 June 1918.' G.O. Revenue No. 2385, 22 June 1918.
[76] 'Resettlement Report of Kistna, West Godavari and East Godavari Districts', p. 30, BOR(LRS), No. 29, 18 May 1927.
[77] This para is based on the District Gazetteers of the dry region. See Bibliography.

EXPANSION OF COMMODITY PRODUCTION / 211

In addition to railways, the delta districts also contained navigable canals and rivers. They provided an 'efficient means of cheap and easy transit. They were, in fact, practically the sole means of through transport' from the delta taluks of Amalapuram and Razole (East Godavari district).[78] Therefore, they were extensively used for transporting produce. The canal and river systems were instrumental in transporting the paddy of the interior of the delta to the milling centres. The transport costs were comparatively cheap. They were said to be 'undergoing reduction on account of competition among the boat owners.'[79] There was also a commercial waterway from Cocanada to Madras. The East Coast was served for a distance of 230 miles by the Buckingham Canal, which runs parallel to the coast through the districts of Nellore and Chingleput and linked the Kistna canal system to Madras. Nevertheless, with the introduction of railways and the improvements of roads for motor transport, navigation along the canals was increasingly restricted over time.[80]

Side by side with the extension of railways in Andhra, went on the construction of (metalled and unmetalled) roads. Writing on the backwardness of roads before the 1880s, the Civil Engineer of Madras remarked that 'roads are none deserving the name, there certainly are tracks through some parts marked out by aloc and milk-bush hedges, but for want of bridges and drains these tracks are divided into isolated portions by the rivers that intersect them . . . without means either for produce . . . or for articles of commerce.'[81] Therefore, specially in the Deccan Districts, roads were largely developed after the 1880s 'with a view to opening up coast to the interior.' Thus the construction of roads received 'a great impetus' during the famine works. The Andhra districts were served by three different kinds of roads—provincial, district and local. The Grand Trunk Road, 'the metalled highway', runs across all the Coastal Dis-

[78] BOR(LRS), No. 28, 18 May 1927, p. 31; A. Vipan, *Scheme of Road Development for the Madras Presidency* (Madras, 1935), p. 7.
[79] G.O. Development No. 548, 4 April 1932.
[80] A. Vipan, *op. cit.*, p. 7; G.O. Revenue No. 287, 11 March 1904.
[81] Cited in W. Francis, *Bellary District Gazetteer* (Madras 1904), p. 116.

tricts. The delta districts were 'very well served with an elaborate system of roads',[82] while the Bombay-Madras Highway linked the Deccan Districts to both Madras and Bombay. It provided an outlet for the cotton of these districts. The Kurnool-Bangalore road also stimulated trade and commerce between Mysore and the dry region. The provincial roads, connected by numerous 'feeder' roads, played an important role in the export of commodities to other parts of India by land. The regional trade and commerce was mostly mediated by the local roads. However, by the turn of the century most of the Andhra districts were 'well connected internally.' The Vizagapatam District Gazetteer remarked that 'the condition of the existing roads is usually excellent . . . and the market for produce had been immensely widened by new roads.'[83] The development of district and local roads, though important, had regional variations. It may be said that the extension of roads was as rapid as that of railway lines and the road construction affected the rural economy rather more directly, because they increased the importance of the village weekly markets and other local fairs.

By and large, an expansion of transport and communications widened the sphere of commodity circulation. A village survey report of the upland villages of West Godavari district pointed out: 'The physical facilities for marketing have considerably improved in almost all the villages . . . They are better connected with centres of wholesale trade markets.'[84] Thus with the expansion of transport network, commodity circulation acquired greater velocity and thereby involved lesser costs. Railways also contributed to the development of market towns like Bezwada, Gudivada, Rajahmundry, Adoni, Nandyal, etc., as important commercial centres.

The process of price convergence and lower amplitude of fluctuations and greater mobility of commodities due to improvement in the means of transport and communication also

[82] A. Vipan, *op. cit.*, p. 95.
[83] W. Francis, *Vizagapatam District Gazetteer*, Madras, 1905, p. 134.
[84] P.J. Thomas and K.C. Ramakrishnan, *Some South Indian Villages: A Resurvey*, (Madras, 1940), pp. 377–78.

implied the emergence of a whole number of market centres in various parts of Andhra. In fact, the late 19th and the early 20th centuries witnessed a growingly integrated structure of the domestic market. Innumerable markets sprang up where most of the internal trade was carried on. 'The weekly markets and fairs played on important role' in so far as the local trade was concerned. The delta districts could 'boast of more than their share of important commercial centres.' The small local markets were mainly used for the purchase and disposal of sundry staples. In Vizagapatam district 'the numerous markets' took a 'prominent part in collecting produce for export and in distributing imports to the villages.'[85] Also in Anantapur district the rural markets played a 'very important part in the collection and distribution of local produce and bringing within the reach of the rural consumer necessaries . . . otherwise procurable only in towns.'[86] At various places in Andhra the local markets were held at convenient distances to serve a village or a group of villages. Retail sale was the rule but in the larger collecting centres merchants purchased articles wholesale. Therefore, in the general process of commodity circulation the function of such local markets was to connect the larger market towns and ports with the rice and cotton producing interior and to collect the scattered commodities for sale and to centralize the small agrarian market at local level. However, the integrated structure of internal market was far from being uniform, since there were district-wise and region-wise variations, while the deltas with relatively better roads and water communication had more market centres than the Deccan Districts. Thus, the former districts stood in a more advantageous position, both in relation to internal and external markets, than the latter. The integration of regional market extended beyond the Coastal Districts to the Deccan and the neighbouring provinces and princely states, where since the early decades of the twentieth century the sphere of exchange was enlarged.

Over and above the local markets, there were *'Mandis'* con-

---

[85] W. Francis, *Vizagapatam, op. cit.*, p. 130.
[86] BOR(LRS), No. 2, 4 January 1927.

stituted the wholesale markets in the district.[87] They were important 'major centres' for the marketing of agricultural produce. All kinds of export and import commodities were dealt with at these markets. In these market towns lived the local merchants (*Bania* and *Marwari*) and agents of big European exporting merchants, who collected foodgrains and other exportable commodities. The integrated formation of the wholesale market was most pronounced in the delta districts. They had important 'agro-market towns' like Bezwada, 'the nerve centre of delta trade', and 'the teeming emporium',[88] where products of upland and delta region were brought to markets. The other 'thriving' towns were: Rajahmundry, Ellore, Gudivada, Guntur, Palakole, Tadepalligudem, Nidadavole, etc. They were predominantly rice-exporting centres. In the dry region the major wholesale business centres were: Adoni, Guntakal, Tadipatri, Nandyal, Gooty and Proddatur. In addition to *Mandis*, the port towns (Cocanada, Masulipatam, Vizagapatam, Bimlipatam) were also the most important terminal marketing centres. Most of the leading mercantile firms (both Indian and European) stationed their representatives in these busy and flourishing centres.

From the above discussion it is clear that during the first half of the twentieth century, the region under study had different levels of market structures and the expanding transport and communication systems uniformly integrated the regional market. The integration not only functioned as the medium of quick commodity circulation at the local and regional levels but also contributed a necessary momentum to the regions' widening integration with the external (distant) markets. Consequently, the development of the regional market increasingly drew the commodities from the interior to national and international exports.

---

[87] For a description of the wholesale market structure in Andhra see the District Gazetteers (Bibliography).
[88] BOR(LRS), No. 29, 18 May 1927, p. 31.

FOREIGN TRADE

In the numerous reports of the contemporary colonial bureaucracy, references to the enormous expansion of commodity export to foreign countries are found very frequently. In general, the sea-borne trade of Andhra districts, since the late 19th century, was marked by a secular expansion. The ports were the primary outlets for the agricultural produce from the hinterland. The available trade statistics indicate that besides a linear growth in the coastal trade export trade also witnessed a significant growth during the period under study.

TABLE 5.8

IMPORT AND EXPORT AND TRADE OF ANDHRA PORTS 1901–1940
(VALUE IN LAKHS: 4-YEARLY AVERAGES)

Years	Rs
1901–04	204.12
1905–08	398.14
1909–12	321.10
1913–16	N.A.
1917–20	107.28
1921–24	387.81
1925–28	668.57
1929–32	537.04
1933–36	362.41*
1937–40	273.61*

SOURCE: *Sea-borne Trade Statistics of the Madras Presidency.*
NOTE: Andhra ports include Vizagapatam, Cocanada, Bimlipatam and Masulipatam.
* The figure relates to Cocanada port only.

TABLE 5.9

Share of Raw Cotton, Oilseeds and Foodgrains, etc. in the Total Exports from Andhra Ports, 1898–1932
(5–Yearly Averages)

COCANADA:

Years	Value of total exports (Rs in lakhs)	Raw cotton (Rs)	Oilseeds (Rs)	Grains (Rs)	Tobacco (Rs)
1898–1902	109.45	17.80 (16.27)	14.86 (13.5)	27.79 (25.3)	16.33 (14.9)
1908–1912	241.56	76.74 (31.7)	35.72 (14.7)	73.48 (30.4)	14.26 (5.9)
1921–1925	244.80	14.05 (5.7)	119.10 (48.6)	53.84 (22.0)	12.05 (4.9)
1926–1930	400.24	4.78 (1.9)	230.67 (57.6)	106.32 (26.5)	7.93 (1.9)

VIZAGAPATAM

Years	Value of total (Rs in lakhs)	Sugar	Oilseeds	Jute
1898–1902	16.95	7.08 (41.8)	6.34 (37.4)	Nil
1908–1012	19.81	7.86 (40.6)	2.96 (14.9)	Nil
1921–1925	85.00	2.50 (2.95)	67.94 (79.9)	8.34 (9.8)
1928–1932	72.86	Nil	63.19 (86.9)	4.00 (5.5)

MASULIPATAM:

Years	Value of total exports (Rs in lakhs)	Oilseeds	Grains
1898–1902	11.99	0.65 (5.4)	8.65 (72.1)
1908–1912	12.88	1.06 (8.2)	10.61 (82.3)
1928–1932	10.51	5.73 (54.5)	4.58 (43.5)

BIMLIPATAM:

Years	Value of total exports (Rs in lakhs)	Oilseeds	Grains
1898–1902	32.94	14.56 (44.3)	1.85 (5.6)
1908–1912	24.09	11.68 (48.6)	4.70 (19.5)
1921–1925	27.84	14.69 (52.7)	7.95 (28.5)
1928–1932	36.99	31.65 (85.3)	4.08 (11.0)

SOURCE: (1) *Sea-borne Trade Statistics of the Madras Presidency.*
(2) *District Gazetteers of Godavari, Kistna and Vizagapatam,* Vol. II (Statistics)

NOTE: Figures in brackets indicate percentages to total.

The predominant export products were: foodgrains (rice), cotton and oilseeds which accounted for more than 75 per cent (in value) of the total export trade (see Table 5.9), whereas the import trade consisted of oils (kerosene), cotton piece-goods, cotton twist and yarn, and gunny bags.[88a] Oils were imported from foreign countries, viz., the USSR and the USA, while gunny

[88a] IMPORTS OF CHIEF ARTICLES INTO COCANADA, 1898–1932
(5–YEARLY AVERAGES)

	1900 Rs	1902 Rs	1910 Rs	1912 Rs	1921 Rs	1923 Rs	1925 Rs	1928 Rs
Total value of Imports (Rs in lakhs)	35.81	–	52.61	–	64.50	–	93.94	–
Cotton goods	7.17	(20.0)	3.61	(6.9)	2.10	(3.2)	2.34	(2.4)
Grain and pulses	6.29	(17.5)	5.15	(9.7)	4.33	(6.7)	3.00	(3.1)
Gunny bags	3.09	(11.1)	4.92	(9.3)	6.47	(10.0)	4.14	(4.4)
Oils (kerosene)	4.13	(11.5)	15.74	(29.3)	32.28	(50.0)	48.41	(51.5)

SOURCE: Compiled from *District Gazetteer of Godavari,* Vol. II.
NOTE: Figure in brackets indicate percentage to total.

bags and cotton twist and yarn were imported from Calcutta and Bombay respectively

From the Andhra region cotton, oilseeds and tobacco were exported mainly to Europe: Britain, France and Netherlands. However, till about 1930 Cocanada and Masulipatam ports also exported paddy and rice to Ceylon, Mauritius, Reunion and Straits Settlements. During the World Economic Depression, exports to foreign markets were sharply reduced owing to the competition from Burma rice. Between 1926 and 1935 exports of paddy and rice from Cocanada fell by 62 per cent.[89] Imports of Burma rice into the Madras Presidency also threatened the Andhra rice market in the West Coast (Malabar, Goa and Cochin) and the southern Tamil districts. Although the imports from Burma into the Presidency increased over time, it did not affect the Andhra rice markets very much. Contrary to contemporary belief, Andhra rice, by and large, retained its markets. The reasons for this are given below:

(i) As we have already suggested, unlike Burma the delta districts exported predominantly raw rice. Only a small percentage of parboiled rice was exported to the West Coast and other places. Thus the Burma rice had to compete with a smaller proportion of the delta rice. There again, imports of Burma rice did not mean total loss to Andhra rice market. Moreover, the 'Nellore rice' preferred by the well-to-do people continued to be popular in some of the southern (Tamil) districts, particularly in Salem and Coimbatore. It is true that owing to fall in prices and competition from Burma rice, a few mills in Bezwada, Rajahmundry, Cocanada, etc., preparing parboiled rice for export, were closed. But having lost the outside markets, boiled rice found a limited market within Coastal Andhra, viz., in Tenali in Guntur district.[90]

(ii) A large part of the delta rice was exported to the neighbouring princely states and Bombay, where imports from

---

[89] Calculated from The Sea-borne Trade Reports of the Madras Presidency.
[90] G.O. Development No. 548, 21 April 1932. Also from *Ceylon Rice Report* (1940).

Burma were not considerable. In spite of depression in trade and fall in prices, the quantity of rice exports from the Kistna-Godavari districts by rail were doubled in 1930 as compared to 1920.

(iii) The opening of Vizianagaram-Raipur railway line in the 1930s facilitated the import of paddy from the Central Provinces and Berar. Therefore, the increased imports of foodgrains after this date consisted of paddy. The imported paddy was thus milled in the delta districts and rice was exported to other markets. As we shall see shortly, the number of rice mills in Andhra also increased during the thirties. Thus the delta region managed to sustain its exports to other places.

(iv) While it is true that the foreign (Ceylon) rice market of Andhra was, to a large extent but not entirely, replaced by Burma rice, the share of foreign rice in the total rice exports from Andhra (Cocanada) was not very significant.

However, it is not argued that the impact of Burma rice was minimal. It is only to suggest that despite the 'dumping' of Burma rice in the East Coast, Andhra rice retained its main markets. As a matter of fact, Burma rice mainly replaced Tanjore rice, for unlike mills in Coastal Andhra, Tanjore mills only produced and exported boiled rice. Therefore, Burma rice did not dislocate the rice economy of Andhra to any considerable extent. Nevertheless, the Depression certainly reduced the value of foodgrain trade by causing a fall in prices. But it did not dislocate and displace the foodgrain market as such. The available data, which suggested a decrease in the value of foodgrain exports, did not indicate the loss of market. The data also revealed that foodgrain trade sustained throughout the Depression period. In other words, it may be stated that the Depression did not alter the production of Coastal Andhra as adversely as in the Deccan Districts, since they produced mainly for the world market.

The World Economic Depression affected the agricultural prices very severely. During Depression they were almost halved. Consequently, the commodities fetched only about half the price in the mid-thirties as compared to a few years earlier i.e., in the 1920's. The fall in prices, in turn, increased the burden of taxation and indebtedness. There were widespread agitations and revolts in some parts of Andhra, and demands were made for the reduction of rent and revenue. The severity of the Depression also led to agrarian unrest. Coastal Andhra thus became a 'seedbed of peasant unrest.'[91]

The fall in prices led to a gradual decrease in the value of shipments of oilseeds and raw cotton.[92] The falling off in the

---

[91] D. Rothermund, 'Agrarian Distress in India 1900–1935', (Mimeo.) p. 24. Also see Brian Stoddart. 'The Structure of Congress in Coastal Andhra' in D.A. Low, ed. *Congress and the Raj*, London, 1977, pp. 109–30; C.J. Baker, *The Politics of South India, 1920–1937*, Cambridge, 1976.

[92] EXPORTS OF GROUNDNUT FROM ANDHRA PORTS

Years	Quantity in lakh tons	Value Rs in lakhs	Per ton Rs
1923–24	0.24	72.68	296
1924–25	0.74	218.83	294
1925–26	0.81	215.59	266
1926–27	0.49	121.68	249
1927–28	0.82	207.88	254
1929–30	1.34	303.45	226
1930–31	1.37	198.44	144
1931–32	1.26	188.41	149
1932–33	1.14	277.89	155
1933–34	1.34	167.90	125
1934–35	1.61	190.40	118
1935–36	0.97	156.68	161
1936–37	2.88	473.66	164
1937–38	2.15	296.80	137
1938–39	3.47	407.76	117
1939–40	1.61	203.04	125
1940–41	1.09	133.67	122

EXPORTS OF RAW COTTON (VALUE IN LAKH RUPEES)

Year	Value
1900–01	154
1905–06	229

demand, specially for groundnuts, and decline in prices can be attributed not only to the trade depression but also to the overproduction in the Presidency. The increased number of sources of oil of similar type such as palm oil, soyabean oil and whale oil also contributed to the trade depression in groundnuts. During the Depression period there was a rapid increase in the supply of alternative oil sources. It was stated that 'even at the low prices that prevailed during the year (1932–33) . . . the competition of soybean and whale oil on the continent was very severe and Germany, among other countries, increased largely the import of these products at the cost of groundnut kernels.'[93] The position of groundnuts was further worsened because of

Year	Value
1910–11	414
1915–16	174
1920–21	227
1925–26	791
1930–31	164
1935–36	97
1940–41	172

GROUNDNUT PRICE (PER CANDY OF 500 LB.)

Year	Rs
1926	53
1927	56
1928	51
1928	51
1930	43
1931	31
1932	40
1933	26
1934	18
1935	34

SOURCE: *Sea-borne Trade and Navigation Reports of the Madras Presidency.*
[93] *RADIMP*, for 1932–33 p. 14.

the formation and development of a powerful European buying combine.[94] Besides oilseeds, oil milling industry also suffered a setback during the Depression years. The Director of Industries wrote in 1929 thus:

This does not appear to have been a good year for groundnut crushers on the East Coast. There was little or no margin of profit between the price of the raw material and the sale price of the oil, for although the price of groundnut has been ruling at lower level than in the past few years the price of the oil has been and still is relatively lower.[95]

Merchant-moneylenders and other trading classes were severely hit by the Depression. During the pre-Depression years moneylending had been 'the safest and most profitable' outlet for surplus capital in the countryside. Consequently, there was a massive investment in moneylending. In fact, the rising prices and 'trade optimism before the Depression tempted the people with surplus capital to speculate, but when the prices dived down such speculators were washed away.'[95a] During the Depression, 'hundreds of thousands of small traders and speculators have come to grief.'[96] The moneylenders who advanced money on the security of crop were obviously distressed to find that they made a loss. Therefore, many *dalals* and other brokers, were ruined. When the Depression set in, many moneylenders refused to lend and were busy recovering money from debtors. As a result rural credit dried up.[97] Many of the small moneylenders and traders who had borrowed money from the commercial banks and other bankers (*Marwaris*) were pressed by their creditors for repayment.[98] The Imperial Bank also restricted credit by stopping advances to the merchant-moneylenders.

---

[94] G.O. Revenue No. 948, 1 May 1931, p. 24. (Note of Dissent by II. Suryanarayana).

[95] *RADIMP*, for 1928–29, p. 21.

[95a] G.O. Development No. 365, 10 March 1931.

[96] W.R.S. Sathianathan, *Report on Agricultural Indebtedness*, Madras, 1935, p. 28.

[97] N.G. Ranga, *Agricultural Indebtedness and Remedial Measures*, Tenali, 1931. Also see D. Rothermund, 'The Great Depression and British Financial Policy in India, 1929–34' (Mimeo).

[98] G.O. Revenue No. 948, 1 May 1931.

Since a vast amount of agriculture was carried on with the help of loans from the moneylenders, commission agents, etc., demands were made on the government to assist in financing the agricultural operations.[99] Government indifference towards rural credit and money market was criticized. As more and more merchant capital was sunk in the expansion of industrial crop production since the early 20th century, there were also demands that the government should extend financial help to the creditor-traders.[100]

More importantly, in the context of dislocation of markets for industrial crops (specially groundnut), the landlords and the trading community alike demanded that the government should develop an industry for the manufacture of vegetable oil products within the Presidency. Thus the Economic Depression Enquiry Committee in 1931 recommended that government should 'give all possible help to develop the vegetable oil and allied industries.'[101] The Oilseeds Association of Madras suggested, 'the development of an industry to crush groundnut locally, to purify and deodorize the oil and hydrogenate the same.' It argued: 'the development of an industry here to manufacture vegetable ghee by purifying and hydrogenating the oil would meet all the local demand in India, Burma, Ceylon and Malaya.' 'Establish the Margarine industry in India, so that India might eventually supply these products to the European markets instead of selling raw produce.'[102] 'Develop the oilseed crushing industry in the land and stimulate local demand for groundnuts', a spokesman of the peasantry suggested.[103] In order to develop the oil milling industry discouragement or prohibition of the export of oil-cake was also suggested.

In fact, from time to time a section of the Madras bureaucracy had also suggested measures for increasing the domestic consumption of oilseeds by developing the oil milling industry and

[99] *Ibid.*
[100] *Ibid.*, (Dissent Note by S.V. Chetty.)
[101] *Ibid.*, p. 12.
[102] G.O. Development No. 365, 10 March 1931; Letter from Presidency, Oilseeds Association to the Governor of Madras, 28 October 1930.
[103] *The Hindu*, 2–11–30, in G.O. Development No. 365, 10 March 1931.

other allied industries.[104] But the Government of Madras did not pay much attention. However, with the onset of the Depression, pressure was exerted by means of petition, memoranda, deputation, etc., for state intervention in the development of local industries. During the early 1930s the Director of Industries (Madras) argued in favour of the development of the oil crushing industry, for he thought that the 'industrial potential of Madras Presidency is high.' 'If the manufacture of ghee substitutes from vegetable oils were developed it would result in the utilisation in the country itself of an increasing quantity of the oilseeds produced in the Presidency . . . while an increased quantity of oil-cakes would be available for the *ryots* at a cheaper at a cheaper price for use us manure.'[105] Such suggestions were repeatedly made throughout the 1930s. The compulsions of Depression, dislocation of foreign market, and the need for local market for industrial crops necessitated limited state intervention. In the late 1930s Madras Government sanctioned experiments for the deodorization and hydrogenation of oils at the Kerala Soap Institute. They had also begun to grant, though few, loans to the 'enterprising *ryots*', under the State Aid to Industries Act for developing the oil mills. The post-Depression period also experienced the development of agro-industries in the Presidency. As we shall see shortly, with the rise and growth of agro-industries the internal market was further crystallized. In other

---

[104] G.O. Development No. 1691, 27 November 1931; Letter from the Director of Industries to the Government, 25 August 1931. 'All possible help should be given to develop the vegetable oil and allied industries . . . the exploitation of the vast oilseed resources of the Presidency must depend to a greater extent on industrial development and the setting up of factories utilizing vegetable oil as a raw material such as soap factories . . . ' An important outlet for groundnut oil in the future, in addition to its use in the manufacture of soap probably lies in the manufacture in India of vegetable ghee . . . If the oil milling industry is developed in India and the imported vegetable ghee is replaced by a product manufactured in the country itself, groundnuts should command a better price.' Letter from Director of Industries to the Govt., 19 September 1931. Also see 'A Note on Economic Resources of the Province of Madras and Possibilities of their Development in *RADIMP*, for 1938–39.

[105] G.O. Development No. 365, 10 March 1931.

words, while the Depression caused loss of foreign market, its aftermath witnessed the widening of the sphere of local market, although at a slow pace.

AGRO-INDUSTRIES AND THE ARTICULATION OF DOMESTIC MARKET

It can be pointed out that the small-scale agricultural processing and agro-based industries had come into existence when the region was linked to the railway system. Among the agro-industries the following were important: cotton ginning and pressing factories; cotton spinning and weaving mills; groundnut decorticating factories; oil mills; jute mills; rice mills; and sugar mills. The expansion of cotton pressing and ginning factories was rather enormous in the dry districts. In Bellary district, during the 1870s there were only two factories but by 1900 they were increased to eight. The total out-turn was valued at Rs 27 lakhs per year,[106] while in Anantapur district, the presses dealt with cotton worth Rs 10–15 lakhs annually.[107] The number of cotton gins and presses in Cuddapah went up from 6 in 1911 to 11 in 1914. In 1914 nine groundnut factories were established.[108]

The District Gazetteer of Cuddapah remarked that

the district has not been unaffected by the industrial tendency exhibited since the beginning of the century in the direction of the suppressing of hand labour by power driven machinery. The growth in the factory movement during the last few years appears remarkable. The recent boom in the groundnut is responsible for the establishment of husking mills.[109]

Nevertheless, as the market for industrial crops increased since the first two decades of the present century, the agricultural processing industries and others witnessed a rapid growth.

---

[106] W. Francis, *Bellary District Gazetteer*, Madras, 1904, pp. 109–10.
[107] W. Francis, *Anantapur District Gazetteer*, Madras, 1905, p. 70.
[108] Brakenbury, *op. cit.*, p. 113.
[109] *Ibid.*, p. 113.

TABLE 5.10
GROWTH OF AGRO-INDUSTRIES IN ANDHRA 1899–1931

Year	Coastal Andhra	Ceded Districts	Total
1899	3	8	11
1902	8	12	20
1905	6	10	16
1906	10	10	20
1920	174	68	242
1921	179	75	254
1922	246	94	340
1923	298	170	468
1924	309	170	479
1928	403	270	673
1929	448	315	763
1931	432	293	725
1932	431	299	657
1935	429	226	655
1936	412	253	665
1937	453	240	693
1939	470	278	748

SOURCE: *Report on the Working of Factories Act in the Madras Presidency.*

In Coastal Andhra, numerous rice factories using hand-hullers were said to have sprung up along the canals.[110] There were also a fair number of large-scale steam driven rice mills throughout the delta districts. Most of them were situated near railway stations or/and on the banks of canals. The rice milling industry in the delta districts developed differently from that in Tanjore district. It has been said that a considerable number of them were evolved directly from the hand milling stage without the introduction of the small self-contained hullers driven by oil engines as an 'intermediary stage of development.' Although

[110] *SCRMP*, for 1907–8 and 1908–9.

several of the small mills were installed in some parts of the Kistna district (notably at Masulipatam) mills of this type were not adapted as in Tanjore. In other words, large-scale mills operating on the Rangoon system were 'the rule and not the exception.'[111] 'The reason for this direct transition' was, as an official pointed out, 'either the spirit of enterprise is more fully developed in the Telugu districts or that there is a large concentration of capital in the hands of individuals which enables them to carry on industrial operation on a large scale.'[112] With the expansion of the rice trade, there was also a rapid increase in the rice mills, particularly in the delta districts. Consequently, they were able to import paddy regularly from the neighbouring provinces and husk it locally. The imports were increased in the 1930s, when the acreage under paddy declined due to the expansion of other crops. Thus the deficit was, to some extent, met by imports from outside and the delta districts were able to export rice to other markets. Therefore, the growing mill industry sustained the export trade.

Unlike rice, the industrial crops suffered dislocation of markets. As we have noted above, the exports of cotton declined very rapidly between 1929 an 1935. Though foreign exports declined, there was a demand for Madras cotton from the local mills. The fact that Madras mills consumed more local cotton was an important and sufficient reason to distinguish it from the other important cotton producing provinces such as, Bombay, Punjab and the Central Provinces and Berar. It has been said that Bombay, 'the biggest cotton growing' province, was obliged to export outside India a large quantity of cotton owing to its unsuitability for local mills. It imported a fair amount of East African and Egyptian cotton, since a large number of mills in Bombay and Ahmedabad specialized in finer counts and naturally needed finer cotton. The local consumption of cotton in the Punjab and Sind was small when compared to production, hence they had to depend entirely on exports. The same was the case with the Central Provinces and Hyderabad State. In

---

[111] G.O. Revenue (Special) No. 584; 22 October 1918, p. 6.
[112] K. Ramaiah, *Rice in Madras, A Popular Handbook*, Madras, 1937, p. 90.

Madras, it was estimated that the total annual production of cotton (in the 1930s) varied from 5 to 6 lakh bales (or ⅓ of total production) were annually exported, while around one lakh bales were imported from outside the province. As the number of cotton mills increased, for instance, in Coimbatore and in other parts of the Presidency, it is reasonable to assume that there was further reduction in the foreign exports of Madras cotton. Therefore, 'in none of the important cotton growing provinces and states, the production and consumption and export are so well balanced as in the Madras Presidency.'[113]

In fact, till before the commencement of the World Economic Depression when the general conditions of raw cotton exports were normal, the cotton mills also fared well, as the demand for yarn and cloth continued to grow over time. However, with the onset of Depression the difficulties of cotton spinners increased because of the growing Japanese competition. Particularly during the latter part of 1932, the spinning and weaving mills 'without exception' suffered from trade depression which was 'intensified by heavy importations at uneconomic prices of cotton cloth and yarn from Japan resulting in heavy accumulation of stocks . . . '[114] 'It was becoming apparent', remarked the Director of Industries, 'that unless an improvement set in, most of the mills would have either to restrict production or reduce wages.'[115] It was only after 1936–37 that the prices of both raw cotton and cotton yarn began to improve. The preoccupation of Japan with the war in China during 1937–38 afforded an opportunity for local mills to increase their output. The increasing local production of cotton yarn and cloth reduced imports to a considerable extent. The imports of foreign piece-goods were reduced from 19.64 million yards in 1933 to 9.33 million yards in 1937 (a reduction of 50 per cent).[116] The increase in the local production meant increasing consumption of local cotton. For example, raw cotton from the Deccan Districts was exported to

[113] G.O. Development No. 1420, 6 June 1938. Letter from Director of Agriculture to the Govt., 12 March 1938.
[114] *RADIMP*, for 1932–33, p. 14.
[115] *Ibid*.
[116] *RADIMP*, for 1938–39 (A Note, p. 24).

mills in Madras and Coimbatore. Thus the local mills were able to supply large quantities of yarn to local handloom weavers. In order to increase local production some of the Madras Government officials also pleaded with the authorities to set up more spinning and weaving mills, for example, in the Ceded Districts where a considerable quantity of cotton suitable for the spinning of coarse and medium counts was grown. By and large, the loss of external market for Madras cotton was made up by the internal mills.

The case of groundnut oil mill industry was, more or less, the same as that of the cotton industry. After the initial slump, there was an improvement in the demand for groundnut from foreign countries. Thus the market for oilseeds became firm and the prices had begun to rise after 1936. Commenting on the recovery of groundnut trade a government official remarked:

> Dealers . . . seized the opportunity to dispose of most of their stocks in view of the excellent prospects for the new crop. At the close of the year . . . it was estimated that 85 per cent of the total crop had been sold to exporters, and in view of the more remunerative rates . . . it was expected that an increased area would be sown . . . for the next season.[117]

Exports of oil-cake to Britain and Germany also increased after this period.

Simultaneously the export trade in groundnut oil further developed with Burma, Singapore and other places within India—Bengal and Bombay. In fact, since the beginning of the present century, an export trade in oils had been developed with Burma and Singapore. However, by the late 1930s output of oil in the province far exceeded the local demand. The growth of oil milling industry greatly increased the market for groundnut. Thus in the 1930s, 'there was good demand for groundnut oil and fair quantities were railed from the Vizianagaram area to Calcutta and from Kurnool to Delhi and the Punjab. There were regular exports of oil to Rangoon and Malay states . . . '[118] For the first time groundnut oil was also shipped from Madras to the United

[117] *RADIMP*, for 1936–37, p. 35.
[118] *RADIMP*, for 1939–40, p. 58.

Kingdom in 1939–40.[119] The increasing crushing of groundnut locally also made available considerable quantities of cake to be used as manure for sugarcane, tobacco and paddy crops at a relatively cheap rate. As a result, there was an increase in the consumption of fertilizers. During 1939–40, for instance, crops such as paddy and sugarcane consumed larger quantities of manure than ever before. In short, the decline in foreign demand for oilseeds was to a large extent sustained by the domestic demand. The internal demand which had been gradually growing evidenced a more rapid advance by the end of the 1930s. Increasing quantities of Vanaspati which till about the early 1930 was used to be imported from foreign countries came to be manufactured within the country. Consequently, the imports of vegetable products into India recorded a sharp decline over time.[120] Regarding the expansion of internal demand for groundnut Dharm Narain observed:

... influence of domestic demand is faithfully reflected in the difference between the smoothed area curves of Madras on the one hand and Bombay and Sind and Central Provinces and Berar on the other hand ... The depression, which is more marked in the Madras curve, becomes somewhat blurred in the curves of the other two provinces—a feature remarkably consistent with the fact that Madras, far from being the supplier of three-fourths of our total groundnut exports in the early 1930s becomes responsible for over nine-tenths by 1938–1939; this happened in the face of a decline in its relative weight in total groundnut acreage of British India, thus showing beyond doubt that an increasing proportion of the growing domestic market was benefiting the groundnut producers.[121]

Among all the agro-industries in Andhra, the sugar industry registered an impressive growth, though less dramatic as compared to the United Provinces, throughout the 1930s. The impact of protection granted in 1932 to the sugar industries was immediate: 'Consequent on the high degree of protection afforded in industry, a great deal of public interest was concentrated ... on the setting up of factories for the manufacture of sugar in the

[119] *Ibid.*
[120] Dharam Narain, *op. cit.*, p. 82.
[121] *Ibid.*

Madras Presidency', a Madras Government official remarked in 1933.[122] The number of factories increased from two in 1930–31 to eight in 1936–37.[123] In fact, prior to the World Economic Depression the refined sugar industry was not a noticeable feature of the regional economy of Andhra. However, in some of the Andhra districts, like in the western United Provinces, jaggery or *gur* was the preserve of independent peasant commodity production. With the grant of tariff protection there was an increase in the value of cane production and trade. There had been a continuous rise in the area under sugarcane as well as in the refined sugar output (from 4,996 tons in 1930 to 12,479 tons in 1939).[124]

Increasing local production of sugar replaced foreign sugar, particularly from Java. The displacement of foreign sugar by local production, in turn, led to an increase in the demand for sugarcane grown in the region. It may be pointed out that, by and large, the supply of cane to factories was also dependent upon jaggery (*gur*) prices and market. Since some of the cane producing districts (Vizagapatam and East Godavari) had a long

---

[122] *RADIMP*, for 1932–33, p. 19.

[123] The first two sugar factories to the built up in Andhra in 1847 and 1899 at Aska (Ganjam district) and Samalkota (East Godavari), respectively, were set up by Messrs. Parry and Co. Madras. The Samalkota factory refined *gur* only. The Indian Sugar and Refineries, Ltd., Bellary (1934), the Vizagapatam Sugar and Refinery, Ltd., Vizagapatam (1934), and the Sree Ramakrishna Sugar Mills, Kirlampudi, East Godavari (1935), were set up by private capitalists and wealthy peasants. The Sri Rama Sugar Mills, Ltd., Bobbili (1934), was set by the Raja of Bobbili, who paid a *peshkush* of Rs. 90,000, while The Etikoppaka Co-operative Industrial and Credit Society, Ltd., Vizagapatam and The Vuyyuru Co-operative Industrial and Credit Society Ltd. (1934), were owned and managed by 'bona fide farmers.'

[124] Calculated from written evidence of sugar mills to the Tariff Board in 1937. Indian Tariff Board, *Sugar Industry*, Vol. II, Delhi, 1937. The Government of Madras pointed out in 1934 that 'In view of the importance of substituting other crops for paddy at the present moment, it is encouraging to note that the cultivation of sugarcane . . . has made much headway. The increase in area under sugarcane is marked in Ganjam, Vizagapatam and Kistna districts, especially in localities where sugar factories have sprung up or where there is a prospect for their establishment. G.O. Development No. 1265, 5 September 1934. Also see, *Report on the Administration of the Department of Agriculture in Madras Presidency*, for 1935–36, p. 29.

established tradition of jaggery manufacture, peasants in the context of favourable market price used to produce their own jaggery. But during the 1930s there was a gradual decline in its price, so majority of them sold cane straight to factories.[125] The fall in jaggery prices also enabled the mills to depress the buying price of cane over time. In their replies to the Tariff Board questionnaire (1937), the management of mills mentioned the influence of declining jaggery prices in determining the factory price for peasants' cane. The decrease in the price of jaggery, it appears, must have induced the peasantry to cart their cane to the factories, even at the reduced prices.

Nevertheless, with the expansion of sugar mill industry the cane area, specially in the vicinity of factories, increased year by year.[126] As a result, mills were assured of sufficient and constant cane supply. In Andhra, there was no cultivation of sugarcane on factory lands, as was the case, for instance, with some indigo-plantations-turned-sugar-factories in North Bihar.[127] At the

---

[125] *Ibid.*, p. 417: 'In the locality the manufacture of jaggery from cane is a very old and well established industry. In the years 1934 and 1935 the price of jaggery used to be more than Rs 4–8 per maund and the grower of cane was inclined to make jaggery instead of selling cane to the factory at Rs 10 per ton but now on account of the low price of jaggery at Rs 2–8 per maund he is more anxious to deliver his cane to the factory at Rs 6–8 per ton . . . ' *Answers of the Etikoppaka Co-operative Industrial and Credit Society, Limited, Vizagapatam.* 'In this locality the manufacture of jaggery from cane is a very old and well established industry. In the years 1934 and 1935 the price of jaggery used to be more than Rs 4–8 per maund and the grower of cane was inclined to make jaggery instead of selling cane to the factory at Rs 4–6 per maund but now on account of the low price of jaggery at Rs 2–8 per maund, he is more anxious to deliver his cane to the factory at Rs 4–6 per maund . . . (The Sree Ramakrishna Sugar Mills, Kirlampudi, East Godavari, p. 546). 'The price of one maund of jaggery was Rs 5–8 in the years 1918 to 1924. In the years 1924 to 1929 it was Rs 5; from 1929 to 1933 it was Rs 3; in 1934–35 it was Rs 4–8; in 1936–37 it was Rs 2–8', p. 419.

[126] In their replies to the Tariff Board almost all the mills reported that there was 'a perceptible increase' in the area cultivated and particularly in the immediate neighbourhood of the factories. 'The area under cane has trebled since the advent of the mill' (Bobbili). The management of mills also reported that there were sufficient supplies of cane and there was no competition.

[127] See A.K. Bagchi, *Private Investment in India, 1900–39*, New Delhi, 1980, p. 362ff; Shahid Amin, *Sugarcane and Sugar in Gorakhpur*, New Delhi, 1984.

same time the new mills were not provided with large tracts of land by the government, as happened in the canal area of the Bombay Presidency.[128] The Madras Government's plan to follow the Bombay precedence, however, failed. Since almost all the sugar mills were concentrated in the major cane-growing tracts, a greater proportion of cane came from the surrounding area. Therefore, there was marked preponderance of 'gate cane' over 'rail cane.'[129] Since there was abundant supply of cane, there was little or no competition among the sugar mills. Besides the independent small peasantry, members of the 'Cooperative Industrial and Credit Societies', to whom the factories at Vuyyur, Etikoppaka, etc., belonged, cultivated cane on their own lands. In the Vuyyur Sugar Factory (Kistna district), for example, each shareholder was bound to supply an acre of cane or 20 tons per share.[130] Though none of the factories had enough zerat lands, some of them considered it desirable to produce cane no rented land not only to ensure supply but also, more importantly, to deal with the problems of late and early varieties. However, the mills which had rented land found it uneconomical because the rent was 'exorbitant' Besides, it was also not possible to obtain a contiguous plot (say, of 300 acres) in one block. Moreover, all the sugarcane producing districts did not raise the same variety of cane. The peasants in Ganjam, Vizagapatam and Kistna districts cultivated mostly Co 213 land Co 243 (early variety, average yield 20 tons, sucrose content 12–14 per cent) and J 247 (late variety, yield 30 tones, sucrose content 13–15 per cent). In East Godavari district *ryots* cultivated exclusively J 147. In other words, though the early varieties were

---

Also see Rajat Ray (Bibliography).

[128] A.K. Bagchi, *op. cit.*, p. 385.

[129] 'Gate cane' means the cane which was brought to the factory by peasants at their own cost, whereas 'rail cane' refers to the cane which was bought by factories at distant places and sent in wagons to the factory. While all the mills in Vizagapatam and East Godavari districts were supplied with 'gate cane', the mills in Kistna and Bellary depended to some extent (6–11 per cent of the total requirements) on 'rail cane.'

[130] Indian Tariff Board, *op. cit.*, pp. 450–51.

popular and occupied almost all the area in Kistna and other districts, Godavari districts opted for the late varieties.

Initially, the existing varieties of cane and their cultivation did not satisfy the mills. As there were few varieties of cane in Andhra which ripened at nearly the same time, the mills could not extend the period of crushing. The factories also desired an increase in yield to be able to effect a reduction in the unit price of cane. The factory owners were not satisfied with government research and propaganda to induce the peasantry to grow different kinds of cane the mills needed. More significantly, they wanted an increase in the sucrose content and the dissemination of early and late varieties of cane among the peasantry.

Nonetheless, it can be said that the sugar industry in Andhra evidenced a smooth growth: the sugar factories had an uninterrupted supply of cane; cost of raw material (cane) declined over time with the increased crushing, there was a reduction in overhead and working costs, and the expanding market for sugar assured reasonable profits. In the context of a 'glimmering' prospect of captive market and soaring demand, sugarcane production received attention and aid, though limited, from the government.[131] In fact, three of the important demands of the mill owners were met to a large extent by the government. The latter's 'intensive' and persistent propaganda was largely responsible for inducing the peasantry to cultivate early and late maturing varieties for the benefit of the factories, which meant an extension of the crushing season. Secondly, as a result of the research at Coimbatore, cane varieties possessing high sucrose content and high tonnage were evolved. The new imported 'superior varieties' were high yielding and to a large extent replaced the foreign varieties. Lastly, the available data suggest that with the coming into being of sugar mills, along with sugarcane areas, the area under the improved varieties also increased over time. The area under the improved varieties thus rose from

---

[131] The Sugar mills had no difficulty in finding ready markets for their sugar. Unlike in the United Provinces, in Andhra there was no overproduction. Therefore, sugar was sold in the local market. See G.O. Development No. 2480, 8 November 1937; and also Indian Tariff Board, *op. cit.*, pp. 338–451.

39,597 acres in 1935–36 to 90,573 acres in 1940–41. In other words, 56 per cent of the total cane area was under improved varieties by the early 1940s.[132] Thus the Madras Government remarked: 'the Government is most anxious that the extent under improved varieties of cane of varying maturity should show an appreciable increase... more particularly in the neighbourhood of existing sugar factories.'[133] By and large, the ecological conditions in Andhra were more favourable for sugarcane cultivation. It being a tropical plant, sugarcane could be abundantly grown in this region. Unlike in some parts of the United Provinces, in Andhra cane was cultivated primarily on irrigated and heavy lands (soils) of the delta. Naturally, the irrigated cane contained more sucrose and yielded 'high tonnage.' Therefore, given the suitable ecological conditions and government protection foundations were laid for the sugar industry to grow.

The other agro-industry which began to consume more and more local produce were the jute mills. There were three such mills in Andhra. Until the mid-1930s raw jute was exported to foreign markets, and gunny bags were imported from Calcutta. During the Depression years the trading conditions were unfavourable for raw jute; as in other export-oriented crops. Although the local mills always consumed small quantities of jute, manufacture of gunny bags was unremunerative, particularly in the post-1935 years.[134] Consequently some of them witnessed serious losses. However, during the Second World War large government orders for gunny bags resulted in appreciable price rise. When the mills began war production, local jute found a ready and widening market.

It may also be mentioned that the growing agro-industries, apart from widening local market for agricultural produce, had also an important impact on the process of urbanization and growth of the urban population in Andhra. Along with the expansion of agrarian trade and agro-industries, the urban

---

[132] *Report on the Administration of the Department of Agriculture in Madras Presidency*, for 1940–41, p. 88.
[133] G.O. Development No. 2641, 24 October 1938.
[134] *RADIMP*, for the years 1936 and 1937.

aspect of the region was also changing. Consequently, the post-1920 period witnessed a rapid growth in the factory employment.

Although in some of the agricultural processing industries such as the cotton ginning and pressing factories and groundnut decorticating factories employment was seasonal; rice mills, sugar industries and jute mills employed permanent and semi-permanent labour. In the delta region, where agrarian trade and market were expanding, urban growth was the greatest. Thus the population of such important towns as Guntur, Bezwada ('The nerve centre of deltas trade'), Gudivada, Bhimvaram, Rajahmundry, Ellore, Tenali, etc., had noticed a considerable growth. More significantly, as we shall suggest later (in Chapter V), the growth of ago-industries also meant an expansion of agrarian capital.

	Tons
Total production	9,36,400
Total exports by sea (Cocanada and Masulipatam)	45,211
Total exports by rail to neighbouring provinces	73,369
Total exports by rail to other places within the Madras Presidency but outside the three districts	1,28,860
Urban consumption within the districts	61,210
Total rice marked	3,18,650
Percentage of rice marketed to total production	9,36,400
	3,18,650
	30.20

SOURCE: G.O. Development No. 548, 2 February 1932.

To conclude: in this chapter we have attempted to argue that the rural economy of Andhra was increasingly getting commercialized in terms of a diversified crop structure. We have seen that in the dry districts the area devoted to commercial or industrial crops increased over time. It has been estimated that more than one-third of the total cropped area was occupied by cotton and groundnut. Although the crop statistics served as an useful index to measure the extent of commercialization in the Deccan, where there was a clear distinction between food and cash crops, the same was not true of Coastal Andhra, where a proportion of the staple foodgrain crop (rice) was marketed. Therefore, a distinction between commercial crops and non-commercial crops will not make much sense. In such a context statistics of trade provide an alternative source of evidence on commercialization. An examination of the production and trade figures of rice in the three delta districts of Kistna, East Godavari and West Godavari indicates that nearly 30 per cent of it was marketed in 1930–31.

While the Delta Districts produced and marketed a certain proportion of rice, the other Coastal Districts, viz., Guntur and Vizagapatam produced, in addition to paddy, tobacco and sugarcane respectively for the market. It should be pointed out that the growth of industrial crops in the Ceded Districts, led to exchange not only between agriculture and industry but also between different types of commercial agriculture within Andhra. The crop structure and the pattern of trade demonstrate the division of labour. The growth of commercial crops in the dry region for export resulted in an increasing demand for coastal rice. That is to say, the specialization of industrial crop production gave rise to exchange between the various areas and agricultural products within the region. Therefore, contrary to Hamza Alavi's proposition (of 'external articulation' and 'internal disarticulation'), we argue that the process of commodity expansion led to the widening of the internal sphere of circulation ('internal articulation') along with international trade ('external articulation.') As we have noted earlier, the Coastal Districts exported rice, sugar, provision, etc. primarily to the internal markets. Another indicator of commercialization is the degree of market integration reflected in the movement of prices of agricultural products. We have shown that the convergence of intermarket prices occurred with the construction of railway and roads. Similarly, movement in relative prices of crops induced the peasants to grow 'remunerative' cash crops at the expense of their own food supplies. Cotton and oilseeds cultivation is a case in point. Our study clearly indicates that the peasants' choice of cropping patterns was responsive to shifts in the price movements. The expansion of the transport system reduced the possibility of storing foodgrains for future consumption.

The sharp rise in the relative price of industrial crops led to a spurt in their exports prior to the World Economic Depression in 1928. The Depression led to dislocation in the export market. However, the post-1935 years witnessed a growth in the domestic demand for cotton and groundnut. Certainly the rapid growth in the number of agro-industries (rice, cotton, oil, jute mills and sugar factories) since the second decade of the 20th century would provide evidence for this. Although domestic consumption did not completely replace external demand, to a large extent, it provided an opportunity and impetus for the growth of the internal market.

Chapter Six

Dimensions of Dependence*

SHAHID AMIN

Two conclusions emerge from the discussion so far: first, that the mills created little new marketing, credit or transport networks, but relied on landlords, moneylenders and rich peasants as agencies for the procurement of their supplies; second, that these intermediaries derived their strength from their traditional position of political and economic domination of rural society rather than from connection with the factories or entrepreneurial skill in exploiting new market opportunities. The domination of the cane-growing peasantry of Gorakhpur by the mills in the 1930s was contingent thus on the former's prior subordination to the landlords, moneylenders and the richer peasants.

'MILLERS' AND 'GROWERS'

The question arises as to whether the alliance between the mills and the dominant sections of the agrarian society hampered the capitalists in their attempts to build up direct and more beneficial dealings with the cane-growing peasantry. Representatives of the industry in their official pronouncements certainly gave this impression from time to time. Throughout the 1930s the sugar capitalists argued that there was a basic 'community of interest' between them and the cane grower; that landlords and moneylenders (acting as contractors) were 'parasitic evils' who

* Shahid Amin, *Sugar and Sugarcane in Gorakhpur: An Inquiry into Peasant Production for Capitalist Enterprise in Colonial India* (Delhi: OUP, 1983).

pocketed 'no small a proportion of the legitimate profits of both growers and miller',[1] and that a reduction of interference from the government and the co-operative societies to a minimum would bring the greatest benefit both to the 'grower' and the 'miller.'[2] D.P. Khaitan leading the I.S.M.A. delegation to the Tariff Board in 1937 complained, 'We do not want middlemen, but owing to the force of circumstances middlemen must be employed.'[3]

The capitalists even went so far as to criticize the entire agrarian structure itself on some occasions as needing refurbishing to make Indian sugar competitive on the world market. The most forceful advocate of this line of reasoning was B.J. Padshah of Tata Sona and Company who went against the majority opinion of the 1920 Sugar Committee in advocating active government intervention for the purpose of acquiring land to be cultivated by the factories and a much bigger outlay for irrigation and drainage to extend the geographical spread of the sugarcane belt.[4] Nearer home, some of the bigger sugar magnates of Gorakhpur were calling for state aid to help modify the existing property structure and relations of production in the villages and integrate cane production with the manufacture of sugar. The management of the Ramkola Sugar Mills, Gorakhpur, belonging to the powerful Khaitan business house, demanded that the government should compulsorily acquire about a thousand acres of cane-growing land in the vicinity of the mills and hand it over to the factories concerned. This would have effectively replaced the landlords by the capitalists:

The cultivating tenants should not be dispossessed on their undertaking to cultivate cane of suitable varieties at suitable times. Under this scheme the tenants would remain in possession as heretofore, it would only be the landlords whose proprietary right will be acquired by the mills. The tenants would, we feel sure, receive a kinder and fairer

[1] Noel Deerr, 'Indian Sugar and Protection: An Economic Aspect of the Question', *Capital*, 22 Sept. 1932.
[2] For a general statement of this position see *Khaitan Report* (1939), esp. paras. 22–35.
[3] *I.T.B.*, 1958, iv, pp. 16–17.
[4] *Report of the Indian Sugar Committee*, 1920, pp. 408–62.

treatment from the millowning companies than they have ever received from the landlords.⁵

Here was an attempt by one of the more important business houses to convince the government that the sugar capitalists had a greater complementarity of interests with the cane-growing peasantry than had the traditional landlords, and hence would treat them far better. If the Ramkola capitalists were trying to replace landlordism and integrate the small-peasant cultivation of cane with the specific needs of the mills, Sardar Vivek Singh, big landlord and partner in the Saraya sugar mill advocated a system of capitalist farming which would have eliminated the peasantry altogether.

The main difficulty [Vivek Singh told the Tariff Board in 1937] of cane growers are lack of capital, lack of irrigation facilities, lack of manure and fragmentation of holdings. The only solution is Collective Farming. Each collective farm to be of a thousand acres or more. Cultivators to be given Bonds. Each Bond being of a different value according to the extent of the holding in proportion to the Farm. A committee of cultivators to run the farm under the direction of the zamindar or an official of the Agricultural Department. Preferably under the direction of both. The cultivators to be encouraged to work on the farms but under a system of wages. Farm profits to go to the cultivators in the form of dividends to be declared at the end of each year.⁶

Both these efforts by the two important sugar capitalists of Gorakhpur—replacing the landlords by the factories and transforming the peasantry into share-wage labour—relied on state aid. The sugar policy formulating bodies and the U.P. Government however were not inclined to interfere with the institutional framework of cane cultivation, and therefore none of these schemes could be put into operation. Government policy in the second and third decades of the twentieth century was basically against making huge blocks of land available to the sugar capitalists. The only exception, as we have noted earlier, was in

---

⁵ Memorandum dated 21 July 1937 from Ramkola Sugar Mills Company Ltd., Gorakhpur, *I.T.B.*, 1938, pp. 258–9.
⁶ Evidence of Sardar Vivek Singh, Sardarnagar, *I.T.B.*, 1938, iiiB, p. 509.

western India.[7] In 1920 the Sugar Committee had ruled out acquiring land for the capitalists on the following grounds:

> The argument that it is in the real interest of the cultivator that he should be compelled to grow a crop which is so profitable as cane and to cultivate it in a way which secures the highest return and that such compulsion is also desirable in the best interest of the community ... is one ... which can be carried to very dangerous lengths ... The logical conclusion ... is that the agricultural production of India would be immeasurably increased to the benefit of the country and the world at large, if all land were handed over to be exploited by capitalist enterprise. The truth of this proposition is as undeniable as the impossibility of putting it into practical application.[8]

The Tariff Board which recommended protection to the industry not only ignored the question of land acquisition by the mills, but made its case specifically with reference to the importance of sugarcane to a peasantry caught up in the downward spiral of the depression.[9] Similarly, the U.P. Government doubtless recognized the advantage to the industry of the 'manufacturer ... cultivat[ing] or indirectly control[ling] the cultivation of a sufficiently large area of land suitable for cane growing in

---

[7] In the post-protection period 7 sugar mills controlling 12,000 acres of land in the Bombay canal zones were set up, partially as a result of government help in the procurement of land. See D.M. Waggle, 'The Impact of Tariff Protection on Indian Industrial Growth—with special reference to the Steel, Cotton mill and Sugar industries' (Cambridge University Ph.D. thesis, 1975), p. 473; A.K. Bagchi, *Private Investment in India* (Cambridge, 1972), p. 385.

In a personal communication Dr Donald Attwood has kindly pointed out that this impression, based on my reading of two secondary texts, is mistaken. Land was procured by the Bombay Government, he writes, 'only once, for the Belapur factory in 1920, and not for the 11 new factories started after 1932. There seems to have been much discussion in the Bombay Government for proposals for granting land to sugar factories ... but no action was taken, finally, because it was not necessary after the tariff was enacted. The factories were able to obtain sufficient land through voluntary leasing.' I should here like to retract the suggestion of state-sponsored procurement of land in Bombay Presidency that I made in earlier papers. However, the fact remains that in eastern U.P. sugar capitalists failed to acquire large blocks of land either through private leasing or through state support.

[8] *Report of the Indian Sugar Committee*, 1920, p. 296.

[9] I.T.B. Report on the Sugar Industry (1931), pp. 39–53.

the proximity of his factory.' But it rejected state help in this direction, as 'it would involve ousting the tenants whose sole means of livelihood is the land'[10]

Since the sugar capitalists could not persuade the government to acquire land for them and then give them direct control over cane cultivation, their next plea was that all would be well for the factories and the peasants if only the former were assigned separate 'spheres of influence', or zones, which they could develop to the benefit of all concerned.[11] Thus Noel Deerr, associated with the Begg Sutherland group, remarked immediately after the grant of protection to the Indian sugar industry:

With the adoption of the zone system, that is to say with an area given over to the miller to develop in sympathy with the small holder, there should follow at once an association of agriculture and manufacture for the common benefit of both the interests. It will be the object of the miller to reduce the price of the raw material and this can best be done by increasing the production per acre, and with an increment in the yield the net income of the small holder will increase even with a decrease in the rate paid per unit of raw material.[12]

This was no empty rhetoric on the part of the sugar capitalists. An official memorandum which began with a criticism of the factories for appropriating all the benefits of protection at the expense of the peasantry,[13] ended by supporting the idea of zones with the capitalists cast in the role of benevolent landlords:

In fact the whole idea of the zone is to remove competition [between rival factories] and ill-feeling [between factories and the peasants], and let the factory regard the zone as its own and develop it in the joint interests of both; if either party does anything unfair there would be trouble for both. The very existence of zones would tend to foster good

---

[10] G.U.P. letter dated 2 Sept. 1930, *I.T.B.*, 1930, i, pp. 319–20.
[11] The management of the Ramkola Company, for instance, remarked in their memo quoted above: 'Failing compulsory acquisition of long lease, all cane producing areas should certainly be divided into zones and assigned to different factories on an equitable basis'. *I.T.B.*, 1938, i, p. 258.
[12] Deerr, 'Indian Sugar and Protection.'
[13] Memo submitted by the U.P. Government, para 1, *Progs. of the Sugar Conference held at Simla on the 10th. 11th and 12th July 1933* (Delhi, 1933), Appx. iii, p. 91.

mutual relations. *The factory with its zones would be like a resident landlord who is the guardian of the villagers;* without a zone it is like an absentee landlord who has no interest in the locality save to get his rent.[14]

In similar vein, Gobind Ballabh Pant, the Congress Chief Minister of U.P., remarked in the aftermath of the sugar crisis of 1937 that 'the interests of grower and manufacturers were identical.'[15] Presumably what Pant had in mind was that in the U.P. of the 1930s and more particularly in Gorakhpur sugarcane cultivation had become dependent on the mills, and any recession in the industry—or a natural calamity for that matter—would be equally detrimental to both parties. But to expand this kind of descriptive truism into a general theory of a perfect reciprocity, and to impute an in-built harmony to the capitalist-peasant relationship was a wild distortion of reality; yet both high-ranking Congressmen and sugar capitalists presiding over government-appointed committees indulged in such rhetoric. In 1939 a high-powered committee of sugar capitalists maintained that with the enforcement of minimum prices of cane there no longer was any conflict of interest between the factories and the peasantry.[16]

Capitalists and Dependent Peasants

Despite these pious pronouncements there was a straightforward clash over the pricing of the raw material. The price of cane was roughly half the total cost of sugar depending on the milling efficiency of individual factories.[17] This meant that the

[14] *Ibid.*, para 13 (emphasis added).
[15] 'Progs. of the joint Sugar Conference held at Patna on the 19th and 20th October 1937', p. 6, Rajendra Prasad Papers, xii, 1937, N.A.I. [Hereafter R.P.P.]
[16] *Khaitan Report* (1939), para 35. It is symptomatic of the conflict of interests between the sugar capitalists and the peasantry that the representatives of the 'cane growers' of U.P. and Bihar who were members of the Khaitan Committee, prepared a 7-page long indictment of the factories, and failed to agree with the representatives of the capitalists in almost every respect. I have therefore chosen to call volume I of this committee's report the *Khaitan Report* and volume II the *Saxena Report*, after Shibban Lal Saxena of Gorakhpur.
[17] In the mid-1930s it accounted for 52.58 per cent of the total cost of sugar

mills always tried to get away with as low a price of sugarcane as possible. In fact the very use of the terms 'miller' and 'grower' by the mill-owners tended to mask this conflict by conjuring a picture of harmonious complementarity. The supplier of cane was not just a 'cane grower' in the sense that he produced only sugarcane; he was a peasant cultivating a variety of crops, and his involvement with the mills was not independent of his relations with landlords, moneylenders, rich peasants and the loci of power and authority in the countryside. Thus even if the mills had wanted to deal directly and openly with the 'grower', the latter could not have reciprocated on an equal footing, if only because the cane brought to the mills bore the stamp of economic domination and political authority which determined the conditions of peasant production.

The supposed reciprocity of interest between the 'grower' and the 'miller' was illusory, because these two categories did not exist in any pure form in Gorakhpur. Even if these two had existed as pure economic categories, their interests could not have been identical. In any bargaining over the price of cane, the growers' loss was the millers' gain. The object of the sugar capitalists, as Noel Deerr himself admitted, was to reduced the cost of the raw material. This could be achieved to the mutual benefit of both parties only if there were a substantial increase in yield per acre, so that the total net receipts of the peasantry even at reduced prices exceeded the returns from lower yields at a higher price. Improved varieties of cane did raise yields somewhat,[18] but the net returns to the peasantry were still heavily determined by high rates of interest on cash and consumption loans taken during the fasli year. Only if the 'growers' were organized into a planters' or cane farmers' lobby, like the canegrowers of Queensland or the beet farmers of England, could they have pitted their might against the I.S.M.A., or its branch

---

produced. See *I.T.B. Report on the Sugar Industry*, 1958, table xxvii, pp. 80–1; cf. S. Sivasubramonian, 'Income From the Secondary Sector in India, 1900–47', *I.E.S.H.R.*, 14:4 (1977), pp. 452–5.

[18] Even here the least advance was made in Gorakhpur. See *Report on Cane Supply Unions* (1947), p. 4.

at Gorakhpur; a mutuality of interest arising out of organized bargaining might then have come into being. Such a hypothetical situation did not prevail in Gorakhpur in the 1930s; it has not come about even today.

The rationale for the peasant cultivation of sugarcane, it could be argued, lay in the agricultural sector proper; the mills had not introduced the crop into the agrarian economy, as they did for instance in the New World, but had in fact captured and enlarged an existing market in sugarcane produce. The supposed mutuality between the 'miller' and the 'grower' did not exist because the two did not face each other outside the context of the agrarian economy; in fact, agrarian relations were a primary factor in structuring the relationship between the peasants and the capitalists as they developed through the 1930s.

In this sense the involvement of the Gorakhpur peasantry in the manufacture of cane sugar distinguished it from the other major areas of cane production in the world. In the West Indies, for instance, the sugar industry was built up by using imported slave labour which lacked peasant traditions. After the abolition of slavery these labourers were prevented by the sugar interests from developing a peasant sector independent of the plantations.[19] In Cuba, the Havana sugar oligarchy after 1762 continued to rely on imported slaves, but also invaded the countryside creating huge *haciendos*, ousting the big tobacco farmers and the multicrop raising peasantry alike. Sugar remained 'the foundation of the semi-plantation economy established in the island for nearly two centuries.'[20] In Puerto Rico, the industry forcibly marshalled an embryonic peasantry onto enclave plantations and degraded it into a 'fully proletarianized essentially landless labor force.'[21] A close parallel is to be found in the southern Cauca valley in Colombia.[22] Similarly, if the story

---

[19] Alan H. Adamson, *Sugar Without Slaves: The Political Economy of British Guiana, 1833–1904* (Yale, 1972).

[20] Manuel Mareno Fraginals, *The Sugar Mill: The Socio-economic Complex of Sugar in Cuba, 1760–1860* (New York, 1975), esp. p. 9.

[21] Clifford Geertz, *Agricultural Involution: The Processes of Ecological Change in Indonesia* (California, 1971), p. 89.

[22] M.T. Taussig, 'Rural Proletarianization: A Social and Historical Enquiry

from a village in the Veraguas province in Panama can be generalized, the mills, within a decade of their arrival, had dealt a death-blow to a subsistence economy, converting the *campesinos* into rentiers and rural proletarians.[23] In Java, the 'cane worker remained a peasant at the same time that he became a coolie, persisted as a community-oriented household farmer at the same time that he became an industrial wage laborer. He had one foot in the rice terrace and the other in the mill.'[24]

None of these variations on the plantation theme, which precluded the development of a cane-growing peasantry independent of direct control by the sugar mills, and which forced a peculiar community of interest between the miller and the grower, operated in Gorakhpur. Here peasant production, though independent of the factories, was rooted within an entrenched structure of dependence. For this reason the sale of sugarcane was not a simple transaction confined to the mills and the peasants. As with the previous sale of gur, it was the occasion for a more general extraction of surplus, viz. the realization of rent and interest by the propertied and the moneyed classes in the countryside.

Given the control of the dominant sections of the agrarian population over the disposal of the peasants' sugarcane, there was no question of a mutuality of interest between the mills and the peasants to start with. The sugar capitalists who were interested in a cheap and regular supply of cane relied on the existing credit and marketing networks to secure a continuous flow of raw material. Further, given the fact that their dealings with the peasantry were mediated through the existing structures of economic control and power in the countryside, the mills also dominated the peasantry in a high-handed fashion, viz. by resorting to underweighment and other non-economic ways of effecting

---

into the Commercialization of the Southern Cauca Valley, Colombia (London, University Ph.D. thesis, 1974).

[23] See the important work by Stephen Gudeman, *The Demise of a Rural Economy: From Subsistence to Capitalism in a Latin American Village* (London, 1978).

[24] Geertz, *op. cit.*, p. 89.

a reduction in the price of cane, rather curious examples of 'mutuality of interest.'

Underlying Contradictions

In addition to such fraudulent practices there were other factors which represented a far deeper contradiction between the needs of the mills and the interests of the peasants. First of all there was the clash between the dictates of the peasantry's harvest cycle and the preference of the factories for an extended cane-crushing season so as to minimize overhead costs. The peasants wanted to harvest their cane as early as possible, get the ready money and carry on with other agricultural operations; the factories wanted some early and some late-ripening canes. As the Director of Agriculture, U.P., noted in 1937:

[The cultivator] desires a cane that is harvested between his rabi sowings and harvests and which can be cut at the time of maximum tonnage. The factories however, desire both early and late strands of cane giving profitable recovery at the two ends of the season, and the highest obtainable sucrose content from mid-season purchases.[25]

An additional problem lay in the fact that in the case of early ripening varieties a high sucrose content (i.e. the percentage of sugar per unit weight of cane) was generally associated with long tonnage, and in the case of late-ripening varieties tonnage declined with the advent of the hot season and the consequent drying of cane.[26] Certain improved strains of cane were very good in view of their high sugar recovery but not from the point of view of tonnage, the chief determinant of the peasants' income from cane. Thus 'Coimbatore 214' (Co. 214), an improved variety developed in the late 1920s, was a very fibrous cane with a high sucrose content. However, it was unpopular with the peasants of north Bihar and Gorakhpur because it was not a very good 'tonnage cane.' The difficulty with Co. 214, Wynne

[25] I.T.B., 1958, iiiA, p. 193.
[26] Loc. cit.

Sayer of the Sugar Bureau pointed out, was that 'the raiyat will not grow it.'[27]

In fact an early-ripening cane variety like Co. 214, with a high sucrose content but low tonnage, symbolized the polarity of interests between the factories and the small peasant economy. In U.P., as a rule, the price paid by the mills was based on the weight of cane without taking into account its sugar value, though a system of deductions and premiums for inferior and superior canes was worked out on paper in the late 1930s. Co. 214 raised on some of the factory estates in north Bihar gave a sugar recovery of 8.5 per cent in the early 1930s, while the ordinary factory dependent on the peasants' cane would have the greatest difficulty in getting a 7 per cent recovery. The mills wanted a greater recovery of sugar, and their farms were subsidiary to their milling plants. The small peasant growing cane was not interested in the percentage of sugar in his cane.[28] For him the incentives for taking to improve varieties were three, namely, greater yield per acre, early ripening which would help him get rid of the cane quickly and concentrate on sowing the next year's cane and gathering the rabi crops, and relative immunity from disease and pests.

The factories also wanted an increase in yields so as to effect a reduction in the unit price of cane. But above all they desired an increase in the sucrose content—and the dissemination of early, middling and late ripening varieties of cane among the peasantry. These two requirements were incidental to the peasants' calculations, for now that their weak bullocks did not have to crush the cane for the manufacture of gur, they wanted a weighty as well as an early ripening cane. Here was a clash of interests based on the contradiction between the role of sugarcane in peasant agriculture and its industrial use by the mills. Given the role of sugarcane in the annual crop cycle of the peasantry, there were limits to the peasants growing cane varieties which interfered with this rhythm.

[27] Evidence of Wynne Sayer, *I.T.B.*, 1950, ii, pp. 331–2.
[28] Loc. cit. Earlier, when gur was the commodity form this was not the case. Hence the adage: *preet jo keeje ikh se jame ras ki khaan*: 'Get involved with sugarcane and you shall have a mind-full of juice.'

The capitalists were not satisfied that government propaganda had fully induced the peasantry to grow the kind of cane they preferred at the time they wanted it. The supply of cane of the right maturity was a very important factor in sugar recovery, but this problem, ingrained as it was in the logic of small peasant production, was not easily overcome by the efforts of official agencies like the Cane Development Department. This was conceded even by those very officers who administered these schemes and tried to bridge the gap between the factories and the peasantry. The Report on the Working of the Cane Development Scheme for the year 1939–40 had this to say in reply to complaints received from the factories in the Gorakhpur region about the failure to get cane of the required maturity:

The eastern districts are notorious for their small holdings and for the poverty of the sugarcane growers. A large number of growers have only half an acre of their own land under sugarcane cultivation, and it is difficult for them to grow three varieties—early, middle and late—on this small piece of land. The only alternative therefore is that some cultivators should grow some varieties and others other ... The cultivator has no economic motive to keep the cane crop standing on his field and he wants to supply it as soon as possible.[29]

While the factories wanted cane supplies to be staggered, the pressures of the small peasant economy demanded a quick and speedy disposal of cane, even at the risk of overcrowding at the mill gates. The reasons for the peasants' anxiety to sell their cane as soon as possible are discussed in a note dated 1958 from the Collector of Basti:

The main reasons for the rush to sell the cane are: (1) a perfectly natural desire to see the cash, (2) standing cane means that it has to be watched and looked after against losses accidental or incidental, (3) there is a substantial lightening in the weight of cane from the end of February onwards, so that the grower gets less money for his cane by leaving it standing till March and April, (4) the desire of the grower to have his fields free for ploughing after the cane is cut, (5) the greater costs, as the season proceeds, of clearing the cane (where normally the

---

[29] *Report on the Working of the Cane Development Scheme in U.P., 1939–40*, p. 59; cf. Chapter 9, section IV(B) below.

cleaners are satisfied with taking the tops of the cane, after mid-March I have found that the cleaner expects the tops of the cane together with a small cash payment), and of course, (6) the pressure of the rent demand.[30]

The problem was an entirely different one for the sugar capitalists. The average number of working days in India was around 100 in 1930 and as low as 60 in some cases; the crushing season in Java spanned on an average 126 days.[31] This was because the desi canes of north India—paunda, sarauti, mango etc.—would not mature before the middle of December, and even at this time they were not fully ripe. Consequently, 'the factories had to crush unripe cane in the beginning of the season, over-ripe canes at the end of the season and normal canes during the middle of the season.'[32] The introduction of early-and late-ripening Coimbatore varieties resulted in an economic staggering of the crushing season.

This very fact brought the conflicting interests of the mills and the peasants to the forefront. The capitalists contended that the recovery of sugar from the early crop at the beginning of the season was very low, making crushing operations uneconomic as recovery reached its peak from mid-March to mid-April.[33] However there was a traditional objection in eastern U.P. and north Bihar to cane standing in the fields beyond the end of March.[34] This seems to have been related to the fact that with the advent of hot winds cane lost quickly in weight, and that the crop left uncut in the fields interfered with the other agricultural operations scheduled for the *hast nakshatra* of the month of Chait (roughly the second fortnight of March). Thus, according to a register of agricultural operations in Gorakhpur district, mid-March was the time for sowing the cane and harvesting

[30] Collr. Basti to Commr. Gkp. 1 June 1959, xxx-58–1957–8, Commr. R.R. Gkp.
[31] Evidence of T.S. Venkatraman, Incharge Sugarcane Breeding Station, Coimbatore, *I.T.B.*, 1930, ii, p. 537.
[32] *Loc. cit.*
[33] *Report on Cane Supply Unions* (1947), p. 22; cf. D.P. Khaitan to Rejendra Prasad, 20 Aug. 1937, R.P.P., xiii/1937, N.A.I.
[34] *I.T.B.*, 1930, ii, p. 332.

rabi grains. The two agricultural sayings regarding *Chait Badi* 1st in Gorakhpur district were:

> (a) *Chait kahe kisan chana he mujkmen doona;*
> *Sar uska mati tooteh diyah, rakhiyah mati soona.*
>
> (b) *Ikh paundra bo le tu jo chahe hua nihaal.*[35]

Both these proverbs indicate that the peasants would much rather have sown sugarcane or at least prepared the field for it and gathered rabi crops like gram, than have worried about harvesting the previous year's cane and carting it to the factories. In fact this same argument was put forward by the representatives of U.P. peasants later on in the 1940s. In mid-March, it was pointed out to a committee, the peasants 'find themselves amidst three activities, namely, cane sowing, harvesting the standing cane and carrying it to the factory and gathering the ripened rabi crops. They are so intensively occupied in these activities that they are unable to do justice to any one of these operations.'[36] However, as the capitalists stressed before the same committee, the recovery of sugar reached its peak in U.P. only from mid-March to mid-April, the period when the peasants would rather not supply cane to the mills. The mill owners for their part considered it their prerogative to determine the starting and closing dates of the crushing season, and were unwilling to transfer this privilege even to regional or all-India bodies of the industry itself.[37]

Thus there was a major question at issue between the mills and the peasants regarding the timing of cultivation and the type of cane. The mills wanted a high sucrose cane scheduled

---

[35] (a) 'The month of Chait says: Oh kisan! the yield of your gram would increase two-fold in me provided you do not let the top portions get broken and leave the crop unirrigated.'

(b) 'If you want to be happy, verging on the ecstatic, sow your sugarcane fallow (now).' Gkp. Harvest Calender Register, 1888 Settlement Shelf, G.C.R.R.

[36] *Report on Cane Supply Unions* (1947), p. 22.

[37] Loc. cit.; see *inter alia.* 'A Review of the Activities of the Gorakhpur Sub-office during the Year 1935–36', *I.S.M.A. Report of the Committee*, 1935–6, p. 311.

to mature over a number of months staggered conveniently through the crushing season; the peasants desired an early ripening and high tonnage cane. This contradiction was very much in evidence in the programmes for the dissemination of improved varieties. Swami Sahajanand, an important peasant leader of Bihar, highlighted this conflict in terms that would apply to Gorakhpur as well. Cane varieties with a high sucrose content were distributed even when these were not suitable to peasant needs. For example, cane type no. 453 propagated in south Bihar by the Cane Department was a relatively hard cane with a rough foliage and unsuitable for use as fodder. The mill owners, to quote Sahajanand, 'are interested in propagating this variety, because it has a high sucrose content, and because of its hardness the peasant connot crush it for gur manufacture.'[38] That the improved cane-breeding programme of the government was biased against the small peasant was admitted by Sir R.S. Venkatraman himself, formerly in charge of the Coimbatore breeding centre.[39] Thus in practice even the government sponsored 'cane development' programmes catered to the needs of the sugar capitalists and the bigger landlords rather than to those of the dependent peasantry.

SHORT-CHANGING THE KISANS

But this was merely one aspect of a complex reality where the dice was loaded in all possible ways against the smaller peasants and in favour of the capitalists and the dominant agrarian classes. The sugar capitalists, as we have already noted, did not help the peasantry to function as free agents in the production and marketing of sugarcane. In fact they actually succeeded in defrauding it in the marketing process. The raw material was very important in the total cost of sugar production, and the mills were naturally interested in reducing it. The cost of sugar could

---

[38] *Cane Growers' Co-operative Conference, Patna 7–8 April: Swami Sahajanand Saraswati ka Bhashan* (Patna, n.d., c. 1947), p. 7.
[39] Cited in *ibid.*, p. 8.

be reduced either by increasing milling and boiling efficiency or boiling efficiency or by reducing the price of the raw material. There was some improvement in the efficiency of sugar production,[40] but the main concern of the mills was still to try and pay as little as possible for the cane. A reduction in the unit cost of sugarcane might have come about with the development of cane cultivation along capitalist lines under the aegis of either the rural capitalists or the factories themselves. But no such development took place. Unable to change the agrarian structure and agricultural production to their advantage, the mills attempted to procure a somewhat expensively produced raw material at the cheapest rates possible, even if this involved a certain amount of fraud. This tendency was accentuated in 1934 when the government fixed the minimum price of cane that the mills had to pay to the peasants.

However, the above argument, which seeks to offer an economic rationale for the fraudulent practices of the factories, does not really hold water. The sugar mills of Gorakhpur and Bihar were cheating the peasants with impunity at a time when their profits were as high as 25 to 40 per cent on capital invested.[41] That they persisted in their machinations despite preventive legislation by the provincial government is an indication of the tendency of the capitalists to dominate the peasantry in extra-economic ways as well.

Two further questions need to be answered before the actual methods employed by the factories to cheat the peasants of their legitimate dues are discussed. First, were the mills resorting to anything other than established practice in their dealings with the peasantry, as alleged by some representatives of the industry? Was the peasant not cheated in all marketing networks anyway? Entry into a marketing network for a commercial crop was always difficult and expensive for the smaller peasants. The

---

[40] 'Overall extraction' (i.e. the percentage of commercial sugar produced on the quantity of sugar present in cane) in eastern U.P. mills increased from 77.1 to 80.9 per cent between 1934 and 1937. See *I.T.B. Report on the Sugar Industry* (1938), pp. 72–3.

[41] 'Note on Middlemen in Gorakhpur Sugar Mills', xxx-3–1935–6, G.C.R.R; *Indian Sugar Manual, 1941*, Appx. D, pp. 310–11

gur market in eastern U.P. was in fact characterized by what one source described as 'systematic loot.'[42] It is also well known that even in the grain mandis of U.P. the peasant selling his own produce always had to pay a higher marketing charge than the beopari.[43] Were the mills then doing something exceptional or new in trying to make the extra anna on the side at the expense of the peasants?[44] The only possible answer is negative. In attempting to defraud the peasants in the weighment of cane, etc., the mills set no novel precedents. For the peasantry, the mills represented yet another commodity circuit within the same old marketing network with all its pitfalls and bottlenecks.

Secondly, was the peasant—who spent two to three winter nights in December-January every year at the factory gate waiting for the management to buy his cane and still seriously defrauded at the end of the transaction—not free to sell elsewhere and spare himself all this harassment, if he so wished?[45] The fact is that the peasantry tolerated the harsh treatment in proportion to its dependence on the mills for the disposal of their cane. Sugar capitalists all over U.P. and Bihar tried to defraud the peasants, but their efforts were most successful in Gorakhpur and north Bihar where cane cultivation was more firmly linked to the mills than anywhere else in India. It would seem that there was a direct correlation between the prices that the factories offered for cane and their treatment of the peasantry. It was in areas like Gorakhpur and north Bihar where the minimum price had become indistinguishable from the maximum price (because of the decline in alternative markets for sugarcane) that the peasantry fared the worst at the hands of the mills and their minions. The cane-growing peasants of Meerut or Rohilkhand, who had a strong bargaining counter in

---

[42] Satya Prakash, 'Report of an Economic Enquiry in Village Narainpur Tiwari, Basti District, *U.P.P.B.E.C.* (1931), ii, p. 170.

[43] J.K. Pande, *Price of Cereals in the United Provinces: How Are They Determined* (Bureau of Statistics and Economic Research, U.P., Bulletin no. 4, Allahabad, 1938), table xxxix, para 158.

[44] *I.T.B.*, 1938, iv, p. 16.

[45] For a graphic description of conditions at the mill-gate in the early 1930s, see *Aaj*, 17 Nov. 1933 and 11 Jan. 1934.

the gur and khand sectors respectively, were less likely to fall prey to fraud perpetrated by the factories than the Gorakhpur peasants for whom the sugar mills had virtually come to be the only buyers by the mid-1930s.

In 1940 Shibban Lal Saxena assessed the problem as follows:

> ... A strong gur industry in the neighbourhood of sugar factories is the only safeguard left to the growers. The experience of cane growers in Rohilkhand and Meerut divisions teaches us that they receive far better treatment and much more consideration than the cane growers in Gorakhpur division, where the gur industry is not so well developed and an inferior quality of gur is made.[46]

The maltreatment of the peasantry by the mills in Gorakhpur stemmed from conditions specific to the region—a consequence of peasant dependence which had now acquired a new dimension through the growth of sugar mills.

(A) Malpractices in the Purchase of Cane

We now look more closely at the ways in which the kisans were cheated by the factories. The following is a list of 'malpractices in the purchase of cane' taken from a pamphlet issued by the U.P. Government:

(1) Weighment

    (a) Deliberate recording of lower gross weight.
    (b) Deliberate recording of higher tare weight of carts.
These methods were generally adopted at a large number of purchasing centres and have been generally complained of:
    (c) Deliberate stopping of weighment for some time in the day. The consequent congestion in the afternoon and the anxiety of the growers to sell quickly in order to avoid getting home late were taken advantage of to record false weights. Growers sometimes knew that this was being done but were afraid that their cane would not be taken if they complained.

---

[46] *Saxena Report* (1940), para 209; cf. *I.T.B. Report on the Sugar Industry* (1938), p. 43 for the same point at a more general level.

(d) Deliberate poor lighting arrangements for night weighing so that weighbridge scales were in the dark. This facilitated the recording of false weights.
(e) Taking an assumed weight for a cart known generally to bring cane. This was openly done.
(f) Purchase by appraisement, i.e. without actual weighment, particularly at places where a beam scale is used...
(g) Unloading some of the cane on a cart before weighing.
(h) Incorrect fixing of the beam-scales and manipulation of weights.
(i) Keeping carts waiting for days.

(2) *Unauthorized Deductions and Contributions*

(a) Purchasing agents in some places made growers agree to accept the price of say 70 maunds, for every 100 maunds supplied.
(b) Taking something from each grower, e.g. sticks of cane or a pice as a contribution in charity or for some religious purpose.
(c) Making the growers pay brokerage for changing rupees into small change.
(d) Closing purchasing centres and making growers send cane by rail at their own expense.

(3) *Unauthorized Sale of Requisition Slips for Cane*

There have been numerous instances of this and complaints are frequent. Two kinds of slips are issued, one with the grower's name entered and the second without. In the case of the slips on which the grower's name is entered, growers are left waiting for a long time before being given a slip, particularly in places where the supply of cane exceeds the demand, with the result that they are prepared to pay for slips in order to get their cane taken early. In the case of slips on which the grower's name is not entered they are made to pay for such slips.[47]

---

[47] Cited in *Saxena Report* (1940), para 66.

Some of these fraudulent practices, especially those listed under Section (2) were perpetrated by the purchasing agents or contractors of cane, who were the dominant landlords and prominent moneylenders of the ilaqas from which they arranged for supplies.[48] However, contemporary sources were agreed that it was the short weights which accounted for most of the loss suffered by the peasants; and as weighment was the exclusive responsibility of the factories, which they were unwilling to delegate to any other agency, it should be evident that underweighment of cane was done either at their behest or with the connivance of the managers concerned. The following is an exhaustive list of the ingenious methods employed by the factories in neighbouring Bihar where conditions closely resembled those in Gorakhpur:

(1) Weighbridge is fixed within a room, with small barred windows, and the figures on the arms of the scale are not visible from outside.

(2) No marks of figures are made on the outer side of the arm of the scale for the benefit of the carters.

(3) For cutting weight grease is applied to the arm or a garland is placed (as if casually) on the arm.

(4) In the hollow of the screw at the butt end of the arm a pen knife is inserted to cut the weight. The particular position is not generally visible to the carters and when an inspection officer goes inside the room the knife is immediately removed.

(5) To cut weight, some small bits of paper, lead or some such article is inserted within the cavity of the jockey weight suspended to the arm of the scale.

(6) A few weighment clerks cut the weight of some carts but add the weight thus cut to other carts in collusion with owners of such carts.

(7) When machines cannot be tampered with the clerk usually announces less weight. Those who are clever do not record the weight in writing until a cart is unloaded to avoid being caught in a surprise visit.

[48] See chapter 5, section IV above

(8) There are also favourite carters who earn something by allowing the weight of their employer's cart to be cut.

(9) There is also a system called *chutki* by which an owner, under the stress of deliberate harassment is forced to sell his cane at a nominal price without weighment. For instance, a carter coming early in the morning without wrapper [*kambal*] or money, in the hope of disposal of his cane early in the day would not be attended to till nightfall when cold and hunger would benumb him, and with the prospect of a dreary night he would be compelled to accept a rupee or two for the load in his cart.

(10) *Chukti*—the usual tactic is to threaten that next day the weighbridge would be closed down, and therefore all must come, and when consequently there would be a rush of four or five hundred carts, the contractor would announce that so much was not needed ... He might accept their load at nominal price.

(11) Confidence trick is also played by some clever weighment clerks. They pose as very honest and make the supplier himself weigh his cane and also announce his weight loudly, but actually less weight is noted in the cane receipt which the grower discovers at the time of payment, when he cannot produce any proof.

(12) Some factories have their weighbridges close to the cane carriers, and coolies have been trained to drop some cane into the carrier as soon as an officer comes in sight to avoid the possibility of getting carts re-weighed in a surprise visit by the inspector.

(13) Some factories adopt the most objectionable practice of keeping the main gate closed and guarded by sentries, using whistles to warn all concerned when an officer arrives. By the time the gate is opened, carts are unloaded and every thing is put in order before the inspector covers the distance from the gate to the weighbridge.[49]

The sugar capitalists naturally denied the widespread pre-

---

[49] Pamphlet by Cane Commissioner, Bihar, cited in *Saxena Report* (1940), para 65.

valence of these practices and blamed only the so-called 'black sheep' in the industry for giving them all a bad name. The sugar barons also claimed that the charges against them were politically motivated, and blamed the lower-ranking functionaries on their staff for the alleged malpractices. None of these pleas was entirely valid, however. The source of the complaints mentioned above was not what the capitalists chose to call 'interested parties' but officials of the government who had no reason to run the mills down. Nor were the capitalists entirely innocent of the charge, as we shall presently see. Even the President of the 1938 Tariff Board, who took a reasonable view of the problem, admitted that well-meaning factories were in a minority. 'In the majority of factories', he felt, 'there is either deliberate underweighment of cane and similar malpractices or not sufficient care taken ... to see that the agents ... do not in some way or other cheat the grower.'[50] In Gorakhpur the most damning indictment came from Ajodhya Das, a liberal politician and himself a Director of a mill in Jaunpur district. Das accused the majority of the Gorakhpur factories of underweighing cane and in fact suggested that the government should provide its own weighbridges where the peasants could get the correct weights to check against the factory's vouchers.[51]

Complaints against the mills in Gorakhpur were rife from 1932—the year the capitalists were granted access to a protected home market in the country. That year the Commissioner was bombarded with the 'grievances of the sugarcane growers in regard to low prices, unfair weighment at factories and detention for days involving drying of cane and much damage', as he put it in a confidential letter to the government.[52] During his cold season tour he had indeed found these complaints to be genuine. In 1934 the Collector of Gorakhpur complained at length about the faulty weighment practised by a majority of

[50] *I.T.B.*, 1938, iv, pp. 15–16.
[51] Letter dated 23 June from Mr Ajodhya Das, Gorakhpur, *I.T.B.*, 1938, iiiB, pp. 623–4.
[52] Extract from fortnightly D.O. letter from Commr. Gkp., 25 Oct. 1933, Industries (R.D. Cane) File 563 of 1933, U.P.S.R.R.

the factories in the district. He reported that most weighbridges were so placed that only the clerk could read the scales and that most underweighment was done at night in inadequately lit rooms. The Collector brought the following typical instances to the notice of his superiors in Lucknow:

1. A cartman is frequently paid far less than the full weight of his cane. The mills require canes to be stripped [of leaves etc.]. If a manager considers a cart to be insufficiently stripped he imposes an artificial deduction, e.g. 20 seers or even a maund on a consignment of 17 maunds. Frequently deductions are unfairly high and a natural discontent is caused.
2. Occasionally also deductions are made by way of bargaining. A contractor would induce a cartman, tired of waiting for several days, to take 17 maunds as the weight of his consignment instead of 20 maunds.[53]

The 1934 U.P. Sugarcane Rules, which were modified in 1936, sought to correct some of these abuses. Sugarcane inspectors were appointed and wide powers given to them in matters involving the checking of weighment and weighbridges. Such efforts were not totally ineffectual, but they still fell short of the target. Our data on 'malpractices' are drawn largely from the reports of the very inspectors appointed to curb these underhand dealings, men who claimed no spectacular success for themselves. To quote from the confidential instructions issued to these inspectors:

... A factory is not bound to set up a weighbridge; ... But whether a weighbridge or a balance is used, it must be seen that it weights correctly. Some factories have installed weighbridges which record weights other than that prescribed; in such cases the scale arms should be altered in due course, but as a temporary measure they should be pasted with stout paper showing weights by the maunds of 82 2/7 lbs. *As the inspector cannot carry about weights with him, he will have to use his ingenuity in order to test the weights shown on the weighbridge; knowing one's own weight may be useful in this connection.*[54]

[53] Collr. Gkp. to P.M. Kharegat, 17 Feb. 1934, xxx-159-(xvii)-1933–4, G.C.R.R.
[54] Emphasis added. Industries Dept. File 255 of 1936, U.P.S.R.R.

But the government had underestimated the ingenuity of the factories and their resolve to underweigh the peasants' cane at all cost. An energetic inspector like Rajeshwari Prasad Mathur of Gorakhpur found to his dismay that knowing his own weight or even using the standard two maund weight was no guarantee for the correct weights being used by the factories. In 1936 he was informed that there were certain factories in Gorakhpur district which had so manipulated their weighbridges that the scales read correctly for lower weights up to two or three maunds, but underweighed appreciably for higher weights. Mathur actually detected one such weighbridge which was correct up to two maunds but underweighed by about one maund for weights over twenty maunds. The mechanism of the weighbridge was such that 'by a slight manipulation of . . . the balancing weights on the other side of the scale considerable difference in weights can be obtained.'[55] The inspectors had no effective means of testing the accuracy of the weighbridge other than the use of the standard two maunds weight which was altogether insufficient for the purpose.

There were other problems with the weighment of the cane as well. The cane rules had not outlawed the use of ordinary beam weighing scales which were more open to manipulation. Ram Kar Singh, Sugarcane Inspector of Gorakhpur, called for a ban on these scales in 1935 on two grounds. First, he pointed out that when such a balance-scale was used the cane had to be taken off the cart and then stored on the ground after the weighment. Thus there was no scope for an inspecting officer to reweigh the cane of a particular cart. Secondly, these scales made the task of underweighment easier in many respects. An ordinary cart load of cane weighing about twenty maunds would have to be weighed in at least ten instalments, which meant that even if four seers were cut each time—an easy matter on a

[55] Note by ex-Sugarcane Inspector, Gorakhpur, n.d., c. 1935, xxx-43-(xvii)-1936–7, G.C.R.R.; cf. 'Underweighment has in fact been developed into an art form by using such devices that up to a point the weighment is correct but above that point the machine underweighs.' Explanatory Note by Thakur Phul Singh and Pandit H.B. Tripathi in *Report on Cane Supply Unions* (1947), p. 32.

weight of two maunds—the peasant would have supplied one maund or 10 per cent of his total supply from a cart gratis to the factory.[56] The inaccessibility of the 'weighment room' to the peasants—a general 'malpractice'—was very much in evidence in Gorakhpur. 'In practice the growers either do not dare enter the weighment room or are not allowed inside it', wrote the Sugarcane Inspector in 1937.[57]

The most notorious case of persistent underweighment was that of the Lakshmi Devi Sugar Mills, Chitauni. That mill had been suspected by the Cane Inspector and the Collector since 1934 of 'swindling the growers', but it had been impossible to commit the management for a variety of reasons. The factory was inaccessible by metalled road, and because of the constant and careful vigil kept at the gate the two successive inspectors had been unable to pay a 'really surprise visit' to the factory.[58] The underweighment of cane carried out at Chitauni was revealed in the high recovery figures recorded by the factory, which were out of all proportion to the district average. The 'recovery rate' (that is, the percentage extraction of sugar from a unit weight of cane), in the absence of a superior milling process or extraordinary cane, could be high only if the factory crushed more cane than it recorded as having bought and paid for. And this was precisely the allegation made by the district administration.[59] The Cane Inspector was finally able to catch the factory in the actual act of underweighing cane in late 1937, but he stressed that it was not the lower level staff but the managers and owners who had been responsible for it as a matter of policy. 'I know it for a fact', observed Mathur in early 1938, 'that the man behind all this mischief is Lala Bholi Ram Marwari whose personal gains by swindling must have amounted to several lakhs during the last several years.'[60]

[56] Sugarcane Inspector, Note on the defects found in the practical working of the [U.P. sugarcane] rules, n.d., c. 1936, xxx-43-(xvii)-1936–7, G.C.R.R.
[57] Note in *ibid*.
[58] Sugarcane Inspector to Collr., 1 Jan. 1938, xxx-62-1937–8, Commr. R.R. Gkp.
[59] Commr. Gkp. to Sec. Ind., 13 Jan. 1938, in *ibid*.
[60] Sugarcane Inspector to Collr., 1 Jan. 1938.

The Collector summoned Lala Bholi Ram and Lala Har Narain and made them accept a thorough 'technological' inquiry into the working of the mill at their own expense. However the investigation conducted in April 1938 revealed a percentage recovery in fact higher than what had been alleged by the Cane Inspector to be caused by underweighment. The special report did not fully absolve the factory, and certainly did not attribute the high recovery rate to any technical efficiency. The recovery was high, and was in fact the same as for 1937, but then the factory could very well have arranged for extremely good quality cane to be brought during the course of the investigations.[61] The government did not think it proper to take any action on the report of the technical expert, but the Collector was advised to prosecute the factory on charges of underweighment.[62] The final culmination of this episode is immaterial for our purposes, but the Chitauni case does show the extent to which particular factories would go to cheat the peasants and ensure that the district inspecting staff were kept at bay.

However, the sugar capitalists of Gorakhpur were not alone in their successful attempts to defraud the peasantry. The dealings of American cigarette companies in south China and the jute mills of Bengal, to take two instances, were not above board either.[63] The jute and sugar industries differed significantly enough of course to account for the divergence in their respective pattern of relationships between peasants and capitalists. The jute concerns, unlike the sugar mills, were not 'factories in the field', and the bulking of raw jute meant that the mills did not have to deal with thousands of peasant suppliers at their gates. There was an elaborate hierarchy of middlemen, and a concomitant short-changing at each stage as well. However the machinations of the Jute Mills Association are of interest in themselves. The monopsonist position of the jute mills in the

[61] Report on the working of the Lakshmi Devi Sugar Mill, Chitauni, district Gorakhpur by Abdul Rashid Khan, 26 Apr. 1938.
[62] See. Ind. to Collr. Gkp., n.d., May 1938.
[63] Chem Han Seng, *Industrial Capital and Chinese Peasants: A Study of the Livelihood of Chinese Tobacco Cultivators* (Shanghai, 1939).

home market enabled them to change jute grades with impunity. In theory, the grades were to be fixed in consultation with the Calcutta Jute Dealers Association (a body of non-Indian jute traders) at the beginning of the season and were to be operative at least for the whole season. The I.J.M.A. backed by its stock of raw jute at hand, would however unilaterally downgrade quality jute and, after threatening not to purchase, subsequently buy it up at lower prices as inferior quality jute. To quote an Indian jute dealer:

In the season 1925–26 the grades 4's, R's, and TR's were enforced [R = rejection, TR = terrible rejection]. A lower grade called B.T.R. [bad terrible rejection] was invented later on in the season . . . When I.J.M.A. saw that the grades had reached the lowest point, and they could not possible go beyond B.T.R they invented a quality known as H.J.R (*habijabi* rejection) which is inferior even to B-T-R but is at present the top quality in loose jute trade.[64]

These manipulations no doubt affected the jute dealers more than the peasant producers. The latter sold jute ungraded, 'which meant that they were not directly subject to the manipulation of standards which was common at the higher level of the trade.' However, the jute-growing peasant was affected indirectly, as 'the price he received reflected the merchant's anticipation of further losses.'[65] In Gorakhpur the capitalists short-changed the peasant much more directly, although it seems that the characteristic features of monopsonistic control were evident in both cases.

It has been argued above that there was no necessary complementarity of interest between the mills and the cane-growing peasantry. In Gorakhpur the sugar capitalists and the peasantry faced each other as unequal antagonists; the former had organized themselves into a powerful all-India association and occupied a dominant position in the local economy, while the

---

[64] Extract from the speech of H.P. Bagaria, representative of the East India Jute Association at the Annual General Meeting of F.I.C.C.I, Calcutta, 1928. I am indebted to Omkar Goswami for this and other references to jute in Bengal.

[65] R.G. Heseltine, 'The Development of Jute Cultivation in Bengal, 1860–1914' (University of Sussex, Ph.D. thesis, 1981), pp. 321–2.

latter was a disorganized body of impecunious producers, drawn into a new commodity circuit about which it did not know much and could do little. The interests of the capitalists and the dependent peasantry were mutually opposed. The mills' attempts to cheat the peasants at each step in the marketing process did not improve this relationship, which was one long record of ill-treatment and high-handedness. To be sure, from the mid-1930s certain checks were imposed on the freedom of sugar capitalists. Minimum prices for sugarcane and the growth of a state-sponsored co-operative marketing scheme were the two more important counter-weights. The extent to which these decreased the dependence of the peasantry is the subject of the following chapters.

Chapter Seven

A Typology of Agrarian Social Structure in Early Twentieth-Century Bengal*

SUGATA BOSE

The close intermeshing of revenue-collection structures with land tenures in the pre-colonial agrarian order brought the British face to face with the question of the basic tenure of land during their efforts to set up effective land revenue systems in India. The answer to this question, they reckoned, would help determine the point on the tenurial scale at which a private property right should be lodged. This 'modern' form of proprietary right was to be attached not to the land itself but to its revenues; the existence of a distinct customary right of physical dominion over the soil was never denied. An implicit, though not perfectly clear intention of British revenue law was, if possible, 'to weld the novel property attaching to the revenue-collecting right to this primary right of dominion.'[1] The Permanent Settlement of the land revenue in 1793 with the zamindars of Bengal accorded the property right to a class of people whose role in rural society as territorial magnates and tax-farmers had been to collect revenue and remit it to government, not to hold or exploit land as such. The discrepancy between colonial revenue law infused with its particular concept of property and the manner in which land was actually held and operated in Bengal

---

* Sugata Bose, *Agrarian Bengal: Economy, Social Structure, and Politics, 1919–1947* (Cambridge: CUP, 1987).
[1] Eric Stokes, *The Peasant and the Raj* (Cambridge, 1978), p. 2.

ensured from the very outset that the legal classification of agrarian society was far removed from its working structure.[2]

Yet 'the zamindari system' in its deceptive legal mould has continued to provide a basic framework for the study of Bengal's rural economy and society. Beyond clarifying ideological strands in the formulation of colonial revenue policy and cataloguing the fortunes of the few who occupied the top rungs of the revenue-collecting ladder,[3] such an approach could only mislead. The circulation of revenue rights has often been misinterpreted as a revolution in land. As late as 1980 it has been seriously claimed that as a result of the Permanent Settlement 'the peasant was dispossessed of the land which now became the 'property' of the zamindar . . . The land*lord* became land*owner*. Land was now 'bourgeois landed property' . . . '[4]

Even those who are aware of the problems of making too easy an equation between ownership of the revenue right and actual possession of land have nevertheless seen Bengal's agrarian structure primarily in the context of the colonial land-revenue administration. A recent essay on pre-1947 Bengal opens with the statement: 'We must begin with a consideration of the conditions imposed on Bengal's agrarian economy by the fact of colonialism. The primary and abiding interest of the colonial government in the agriculture of Bengal, or for that matter anywhere else in India, was the extraction of a part of the surplus in the form of land revenue.'[5] The arrangement by which a class of persons designated 'proprietors' were assigned

---

[2] One historian has described the Permanent Settlement with the zamindars as 'a case of mistaken identity.' Ratnalekha Ray, *Change in Bengal Agrarian Society 1760–1850* (Delhi, 1980), p. 73. The compulsory saleability of the new property right meant, of course, that there would have been nothing to prevent the revenue-collecting right and the primary right of dominion, even if initially strung together, from parting ways in course of time.

[3] Excellent monographs on these aspects of Bengal's rural history are Ranajit Guha. *A Rule of Property for Bengal* (Paris, 1963), and Sirajul Islam. *The Permanent Settlement in Bengal: A Study of Its Operation 1790–1819* (Dacca, 1979).

[4] Hamza Alavi, 'India Transition from Feudalism to Colonial Capitalism, *Journal of Contemporary Asia*, 10. 4 (1980), p. 371.

[5] Partha Chatterjee, Agrarian Structure in Pre Partition Bengal in A Sen *et al.*, *Perspectives in Social Sciences* 2 (Delhi, 1982), pp. 113–14.

the property in revenue collection in 1793 together with late nineteenth-century amendments 'to protect as far as possible the predominant organizational form of agricultural production, *viz.* small-peasant farming'[6] is still the main context within which the evolution of Bengal's agrarian structure in the late nineteenth and early twentieth centuries is analysed. The resulting emphasis is on revenue history and rent relations, formal landlordism and tenancy.[7] Yet land revenue and rental demands, important as they were during the nineteenth century, became after 1860 a less significant aspect of the colonial context than the capitalist world economy and the financial policies of the colonial government, and steadily diminished in importance from the late nineteenth century onwards. By the early twentieth century, the agrarian economy of Bengal rested on an extensive subsistence base while at the same time participating in a growing international trade. The history of a subsistence-oriented peasantry engaged in petty commodity production for a well-integrated world market has to be set in the context of the demographic constraint on resources as well as the international and regional dimensions of the colonial political economy.[8] A study of the network of agrarian credit relationships is more relevant in the late colonial period than a focus on ties of revenue and rent. It not only provides the critical linkage along which fluctuations in the world economy were transmitted to the region's agrarian economy, but also forms the most important element in the complex of agrarian relations within the region. This shift of perspective is, therefore, necessary for a better understanding of the connection between changes in the agrarian structure and the nature and course of peasant politics in the final decades of colonial rule.

Trends in the wider economic system were transmitted to the regional economy through the refracting prism of the agrarian class structure and affected various social classes in sig-

[6] *Ibid.*, p. 114.
[7] Asok Sen, 'Agrarian Structure and Tenancy Laws in Bengal 1850–1900' and Partha Chatterjee, 'Agrarian Structure in Pre-Partition Bengal' in A. Sen et al., *Perspectives in Social Sciences 2*, pp. 1–224.
[8] This is attempted in chapter 2 below.

nificantly different ways. The first analytical task is to define the nature of agrarian social structure in early twentieth-century Bengal. Here the dominant 'jotedar' thesis requires serious amendment. The 1970s saw major advances in historical research on agrarian Bengal: the facade of legal tenures was penetrated to offer a glimpse of the economic and social structure. Rajat and Ratna Ray pointed out the important distinction between two structures of land tenure in Bengal—the revenue-collecting structure over the village and the landholding structure within the village. British policy-makers, both proponents and opponents of the Permanent Settlement, often failed to take account of the latter.[9] In consequence of this oversight, the framers of the Permanent Settlement had conferred proprietary rights on a class of zamindars who did not have land in their actual possession. Its critics denied altogether the existence of proprietors of land. But according to the 'jotedar' thesis 'there existed in Bengal, as tenants of the 'revenue-collecting zamindars and talukdars, a class of men known as jotedars who owned sizeable portions of village lands and cultivated their broad acres with the help of sharecroppers, tenants-at-will and hired labourers.'[10] The holdings of these men were believed to run from 50 to 6000 acres. Combining these substantial landholdings with the social authority of village headship and the economic authority of moneylending and grain-dealing, these jotedars exercised supreme control over the labour of the poor peasantry.

The 'jotedar' thesis, which with minor variations became the ruling orthodoxy, has had important implications for the interpretation of social conflict and political upheaval in the Bengal countryside over the last two centuries. The relative quiescence of the first half of the nineteenth century has been attributed to a working alliance between zamindars and jotedars. The agrarian disturbances of the later nineteenth century are explained in terms of a breakdown in this collaborative relationship when the zamindars, attempting to enhance rents in a period of rising

[9] See Rajat and Ratna Ray, Zamindars and Jotedars A Study of Rural Politics in Bengal MAS. 9. 1 (1975), pp. 81–102.

[10] *Ibid.*, p. 82.

prices, could no longer afford special privileges to the favoured jotedars. The colonial government, on this view, now backed tenancy legislation favouring 'the village landlords against the superior landlords.'[11] For the analysis of twentieth-century political developments, the zamindar-jotedar thesis appeared to correspond closely to the revenue-collecting landlord and the village-controlling dominant-peasant pattern of northern India and was quickly assimilated into a broader 'dominant-peasant' thesis. This views the nationalist movement simply as a competition between the British government and the Indian National Congress for the allegiance of these dominant elements in the countryside.[12] It is argued that in Bengal, the nationalist movements were most successful where the jotedars provided the driving force but zamindar-jotedar conflict was patched up by common caste identity, as in the case of the Mahishyas of Midnapur. The origins of the Krishak Praja movement, Muslim 'separatism' and even the partition of Bengal are seen to lie in the conflict between the Hindu zamindar and Muslim jotedar of east Bengal. 'In each area of Bengal', it is asserted, 'the strength or weakness of the nationalist and separatist movements bore a close relationship to the local relationship of zamindar and jotedar.'[13] The tebhaga movement of the sharecroppers on the eve of independence adds another dimension to the 'jotedar' thesis. The 'jotedari system' is once again regarded as central to the agrarian structure but the focus here is on the conflict between

---

[11] Ratnalekha Ray, *Change in Bengal Agrarian Society*, p. 294.

[12] D.A. Low, 'Introduction' in D.A. Low (ed.), *Congress and the Raj* (London, 1977), pp. 1–45.

[13] Rajat and Ratna Ray, 'Zamindars and Jotedats'; p. 101; see also Sumit Sarkar. *The Swadeshi Movement in Bengal, 1903–08* (Calcutta, 1974), pp. 463–4. John Gallagher referred to the 'vitality of agriculture' in east Bengal which spurred the Muslim 'jotedars' to raise their demands against Hindu zamindars; in parts of west Bengal, Gallagher believed the role of Mahishya 'jotedars' to have been crucial: 'it was their domination which gave the resistance the flavour of a mass movement in those areas' See John Gallagher, 'Congress in Decline: Bengal 1930 to 1939' in J. Gallagher *et al* (eds.), *Locality, Province and Nation* (Cambridge, 1973), pp. 269–325, also Hitesranjan Sanyal, 'Congress Movements in the Villages of Eastern Midnapore, 1921–1931', Marc Gaborieau and Alice Thorner (eds.), *Asie du Sud: Traditions et Changements* (Paris, 1979), pp. 169–78.

the jotedars, 'a new category of large land-owners', and the adhiars or sharecroppers employed by them.[14] In short, by virtue of his crucial position in the rent-collecting hierarchy and more importantly, by his place at the head of the village landholding and credit structure, the 'jotedar' is widely seen to have held the key to the social and political destiny of rural Bengal.

The radical exhortation of the Maoist communists of Bengal in the late 1960s and early 1970s to 'bury the jotedar, the reactionary kulak', had scarcely died away when he was resurrected in the historiography of agrarian Bengal. After distinguishing between the revenue and rent structure on the one hand, and the landholding structure on the other, a particular characterisation was offered of the latter in which the ubiquitous 'jotedar' loomed large. Quite apart from the difficulty that 'jotedar' was variously defined as village landlord, rich farmer and dominant peasant, there are two problems with this view. Firstly, a rather rigid and literal distinction between landlords and lords of the land has obscured the role rent-collectors played, at least in some regions, in the village landholding and credit structures. Secondly, and this is more important, the 'jotedar' thesis generalises for Bengal as a whole an atypical form of agrarian organisation that prevailed in the main in north Bengal, but which happened to be graphically portrayed in Buchanan-Hamilton's surveys of Rangpur and Dinajpur and later settlement reports of those districts. Indeed, it may be said that 'jotedars in the sense of *de facto* village landlords or village-controlling dominant peasants had no existence in Bengal except in unusual frontier regions. It is necessary to find a more adequate representation of the agrarian social structure in Bengal than the one turning entirely on the role of the big 'jotedar', thrusting for advantage against the declining zamindar while simultaneously holding down the poor peasantry in debt bondage and by extra-economic coercion in a 'semi-feudal' situation.[15] Using the 'jotedar' thesis as a point

---

[14] Sunil Sen, *Agrarian Struggle in Bengal 1937–47* (Calcutta, 1972), pp. 1–15.

[15] The theoretical underpinning to much of the historical literature on jotedars has been provided by Amit Bhaduri's important article 'A Study in Agricultural Backwardness under Semi-feudalism', *Economic Journal*, 83 (1973), pp. 120–37.

of departure, this chapter seeks to outline a basic typology of the agrarian social structure in early twentieth-century Bengal. The answer to the over-simplifications and distortions of the jotedar thesis is not a retreat into localism which a haphazard wandering through the Bengal districts inevitably entails.[16] The aim here is to extract from the enormous complexity of agrarian Bengal a few broad, general patterns displaying certain enduring characteristics. Without these there can be no meaningful analysis either of the region's agrarian economy and society or its response to trends in the wider economic system. A fuller elucidation of the relations of production and surplus-appropriation and the interlocking conditions of reproduction that held the structure together, as well as the stresses and strains to which it was subjected between the First World War and the end of colonial rule, will form the subject matter of subsequent chapters.

A Typology of Agrarian Social Structure in Early Twentieth-Century Bengal

J.C. Jack and F.O. Bell, two of the more perceptive of settlement officers of Bengal districts, presented totally contrasting representations of socio-economic hierarchy among the cultivating classes of Bengal. Jack wrote of the east Bengal district of Faridpur:

> the cultivators are a homogeneous class . . . It is clear that the agricultural wealth of the district is divided with considerable fairness in such a way that the great majority of the cultivators have a reasonable share. This is no country of capitalist farmers with bloated farms and an army of parasitic and penurious labourers.[17]

Bell, on the other hand, wrote of the north Bengal district of Dinajpur:

---

[16] For the infinite variety of local agrarian structures, see, for instance, Abu Abdullah, 'Landlord and Rich Peasant under the Permanent Settlement', *Calcutta Historical Journal*, 5, 1 (1980), pp. 89–108.

[17] J.C. Jack, *Economic Life of a Bengal District* (Oxford, 1916), pp. 81–2.

The most significant feature of Dinajpur rural life is the inequality in social status and standards of living of different rustic families. Almost every village will reveal some large family of substantial cultivators . . . As elsewhere in North Bengal, this jotedar class is socially supreme in the countryside. The jotedar families may hold several hundreds or even thousands of acres of land in their own possession . . . All these men are of a class which may be described as practising large-scale farming, though it is farming not with any large capital sunk in machinery, but through the traditional methods, employing either labourers or adhiars (sharecroppers).[18]

The contrasting observations of Jack and Bell reveal in a striking manner two different types of agrarian social structure in Bengal. In east Bengal a predominantly peasant smallholding structure was over-laid by various rentier and creditor groups. In north Bengal (and the southern fringe) rich, enterprising farmers who had helped clear the scrub and jungle during the nineteenth century were still the dominant elements in a highly polarised agrarian structure. There were further layers of variation and complexity. The most important of these was the existence of considerable personal demesne of the landlords (revenue-collectors) in west and central Bengal casting a shadow on the lands operated by (tenant) peasant smallholders; the demesne lands were cultivated by attaching a locally available reserve of landless labour. Each of these different types could be tied to the wider economic system in one of two ways: either directly through an agricultural exporting sector which produced commercial crops for the world market or indirectly through the network of credit even where the world market was less significant for the crops produced and rural elites shielded the primary producers from direct exposure to internal market forces. The capitalist plantation would form a distinct type in a broader typology of colonial agrarian societies. It existed only on the northern fringe of early twentieth-century Bengal in the Himalayan foothills where European capitalists had established tea plantations. Its very special features have kept it beyond the scope of this work. In the discussion below, three major types

[18] *Dinajpur SR*, 1934–40, pp. 16–17.

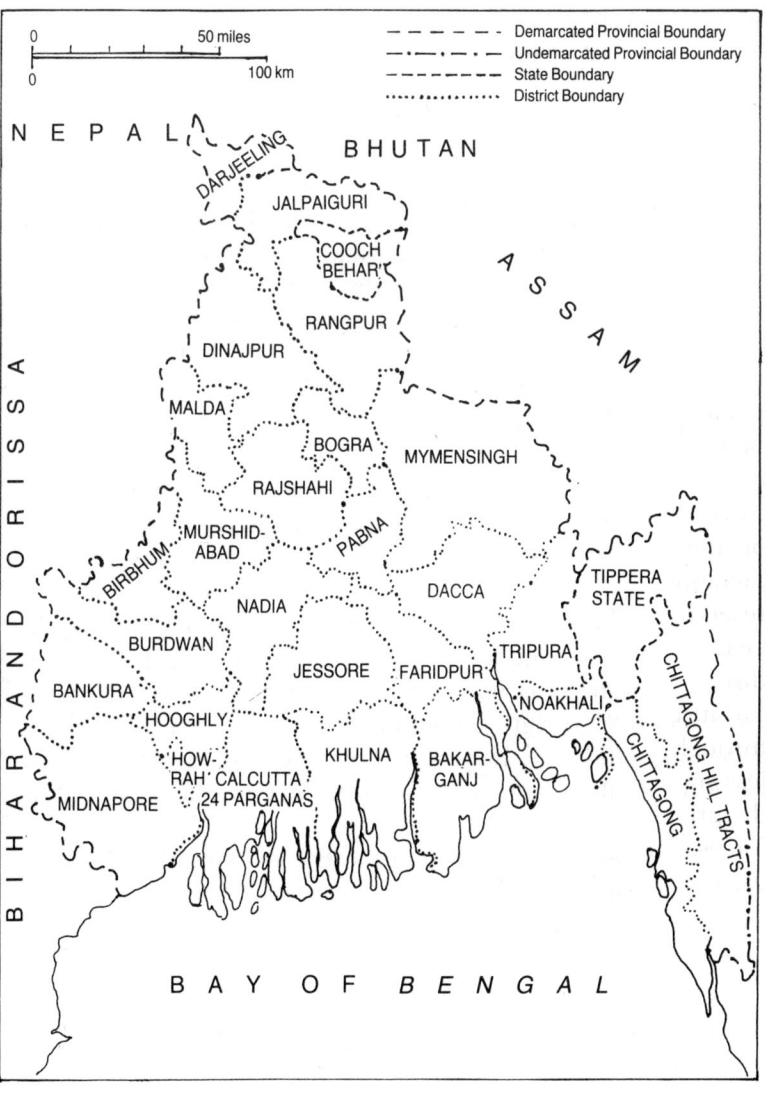

Map 1 Bengal Districts

of agrarian social structure will be analysed: the village landlord/rich farmer-sharecropper system which predominated in north Bengal, the peasant smallholding system which prevailed in east Bengal (and became the principal agricultural exporting sector), and the peasant smallholding-demesne labour complex which was the most common pattern in west and central Bengal.

The Village Landlord/Rich Farmer-sharecropper System in the Frontier Regions

The jotedari-adhiari system of north Bengal grew out of conditions wholly atypical of the older settled regions of west Bengal and east Bengal. Vast areas in this region were uncultivated jungle and settled for reclamation with enterprising tenant-farmers called jotedars. This process was started by the big zamindars prior to the Permanent Settlement and continued in the post-1793 period. Ecological factors had an important bearing on this particular form of agrarian organisation. Large tracts of land were assigned to substantial men of capital at low fixed rents and with permanent and transferable rights to facilitate organisation of large-scale reclamation from jungle. The reserves of labour provided by the semi-tribal Koches and Paliyas—the Rajbansis of later years—and by the immigrant Santal tribes were utilised to conquer the inhospitable wasteland. Once the work of reclamation was completed, they remained as sharecroppers with no right of continued occupancy of the land they tilled and often only in permissive possession of a little homestead on a fragment of the wasteland they themselves had cleared.

During the nineteenth century, the jotedar-adhiar pattern became the dominant feature of the agrarian structure in much of north Bengal. In parts of Ranghpur, Jalpaiguri and the Siliguri subdivision of Darjeeling (which were wrested from Cooch Behar and Bhutan and came under British sway about the middle of the nineteenth century), the practice of leasing out tracts of wasteland to men with capital to organise reclamation had already been resorted to by the Cooch Behar Raj. These leases were recognised by the British and most of the zamindaris included within the permanently settled area. As the District Gazetteer of Jalpaiguri states: 'The tenants in these estates are divided into tenants-in-chief (jotedars), subtenants (chukanidars, darchukanidars and daradarchukanidars) and holders under the metayer system (adhiars). The jotedars are tenants holding immediately under the zamindars; a large number of their rank as tenureholders and others as raiyats under the provisions of the Bengal Tenancy Act.'[19] In Dinajpur where in the aftermath of 1793 the territory of the old Raj was auctioned off in 'lots', the big jotedars held their lands under the new zamindars known as 'lotdars.'[20] The existence of considerable waste in north Bengal usually placed the jotedars in a powerful bargaining position *vis-a-vis* the absentee zamindars and from the later nineteenth century they were also armed with formidable legal rights. Their jotes, either as permanent tenures or as raiyati holdings, became protected interests while they remained free to rack-rent their under-tenants and exact a lion's share of the produce from dependent adhiars.

In north Bengal, big jotedars were the dominant elements in the agrarian structure. But there were also a fair number of small jotedars. It would appear that small and scattered parcels of zamindars' lands were let out to peasants, whereas lands concentrated in large units, usually vast tracts of jungle, were leased *en bloc* to substantial tenant-farmers. The Bengali word, jote, from its Sanskrit original yotra, simply means cultivation or cultivable land. A local officer of Jalpaiguri noted in 1909: 'The term *jote* is applied to any holding, large or small, held direct from a proprietor or from a holder of a recognized tenure, such as a *patni*. I find in Mr Glazier's book on Ranghpur mention of a *jote* paying a rent of Rs 50 000, while the same word is used to describe the peasant's holding of a few *bighas* . . .'[21] The Gazetteers of all the north Bengal districts of the early twentieth

---

[19] *Jalpaiguri DG*, 1911, pp. 83–4.
[20] *Dinajpur SR*, p. 69.
[21] F.J. Monahan to P.C. Lyon, Jalpaiguri, 21 June 1909, Note on the Extension of Survey and Settlement Proceedings under the Tenancy Act to the districts of this (Rajshahi) Division, Progs. of the Conference of Cmsners at Shillong, Oct. 1909, Govt. of Eastern Bengal and Assam (WBSA).

century also refer to the enormous variation in the size of jotes. But where the jote right had not passed to non-agriculturalist moneylenders the jotedar was 'generally a substantial farmer representing the original reclaimer of the soil.'[22]

F.O. Bell in his Dinajpur Settlement Report presents the best profile of the various grades of jotedars and their multifarious moneylending and grain-dealing activities. First, there were the giant jotedars who held several hundreds or even thousands of acres of land in jote or raiyati right and may even have bought into superior proprietary rights in some estates. At the same time, they financed the smaller cultivators and controlled the marketing of the exportable surplus. The famous Chaudhuries of the village of Porsha received 60 000 maunds of paddy each year in their 'golas.' Such men let out some land on cash rents but were more keen on the grain they received from adhiars. Then, there were the jotedars who held some 100 acres and worked on similar methods as the giants but on a smaller scale. 'This well-to-do jotedar class', Bell points out, reminding us of Buchanan-Hamilton 30–100 acre farmers, 'is no new feature of the district, and no creation of the Permanent Settlement, or the Bengal Tenancy Act.' Finally, in the gradation of jotedars came 'the men who possess a raiyati holding which can provide subsistence for the family, and these men who have a few acres add to their produce by working as an adhiar.'[23]

The adhiars, it was found in the 1930s, were, as in the early nineteenth century, usually men with ploughs and cattle and a little land in raiyati right. Yet they were regarded 'more as the servant of the jotedar, or landlord, than as independent landholders. They are mostly the poorer villagers, and the jotedars now finance them through the difficult months. *The jotedar also decides what crop is grown*, and it is frequently divided at the khamar, or other place appointed by the jotedar, a sign of the master and servant relationship.'[24] It is important to note that the jotedar of north Bengal exercised actual possessory domin-

[22] *Darjeeling DG*, (1907), pp. 47–8.
[23] *Dinajpur SR*, pp. 16–17.
[24] *Ibid.*, p. 22 (emphasis added).

ion over the soil and took the major production decisions. An enquiry into the position of adhiars in Jalpaiguri in 1909 is also quite categorical on this point. 'The giri (i.e. the adhiar's employer) selects the crop which is to be cultivated each year. The adhiar cultivates the crop in the best manner possible, but is bound to comply with instructions issued by the giri.'[25] The degree of dependence of the adhiar on the jotedar, of course, tended to vary. It was observed at the time of the tebhaga struggle:

As there are Jotedars of various types there are also adhiars of various kinds ... There are adhiars whose Jotedars have to pay even for the midwife in case of birth and coffin in case of death in adhiars' family [sic] ... In such cases Jotedars apparently give half of the produce to the adhiars but indirectly take back almost the entire [sic] on account of advances, interest etc. The process is repeated annually. Let us consider the question of another type of adhiar. They are those who have land of their own and have their own plough and cattle. Their own land being not sufficient for their ploughs they take a few bighas of land from some other neighbouring jotedars. Such adhiars do not much depend upon their Jotedars ...[26]

In Rangpur the adhiars did not generally form a distinct class but were drawn from the poorer under-tenants who 'quit the land after reaping the harvest'; in the north-western part of the district, however, many adhiars were completely landless and even lived in homesteads belonging to the landlords on their adhi lands.[27] It was in the case of the tied adhiar of parts of Rangpur and Jalpaiguri that conditions approached near-serfdom. The custom of 'hauli—adhiars having to put in unpaid labour on the jotedars' home farm—was fairly common in Jalpaiguri.

To summarise then, the bulk of the land in north Bengal was held in large parcels by substantial jotedars, the rest in small

[25] Beatson Bell's Note on the Position of Adhiars in Jalpaiguri, 21 Feb. 1909, Progs. of the Conference of Cmsners at Shillong, Oct. 1909, Govt. of Eastern Bengal and Assam (WBSA).
[26] Dy Cmsner Jalpaiguri to Addl Secy. Bd of Rev., 11 Mar. 1947, Rev. Dept. LR Br. B. Progs. 15–107, Dec. 1948, File 6M-58/47 (WBSA).
[27] *Rangpur SR*, (1951–8), p. 65.

quantities by small jotedars. The big jotedars monopolised the product market and extracted a large proportion of the adhiars' produce in the form of share rent and loan interest; yet within the type of agrarian structure that had evolved in north Bengal, they played a vital role in the process of reproduction by redistributing part of the surplus both in the form of capital for basic inputs and as grain advances to keep the workers on the soil alive. They sometimes sublet their lands to inferior grades of chukanidars and darchukanidars but it was normal for them to retain large areas as khas (personal demesne) which they cultivated through numerous sharecroppers and labourers. Since the jotedars were also the principal source of credit, they effectively controlled the labour of the poorer cultivators. It was this dual role of landholder-cum-creditor which made the big jotedar of north Bengal a power to be reckoned with.

The only other region which closely approximated the agrarian structure of north Bengal was the abadi (reclaimed) area of the 24–Parganas and western Khulna. The men of substance who had leased government 'lots' in the Sunderbans for reclamation were somewhat similar to the big jotedars of the north. The 'lots' were large blocks of land held at easy rents progressively enhanced as reclamation proceeded, and subject to forfeiture if clearance conditions were not fulfilled. Often speculators and land-jobbers obtained these leases and sublet to smaller lessees for cash payments with the result that the work of reclamation was actually carried out by small peasant cultivators paying rack-rents.[28] The process of dispossession of what was initially a relatively free peasantry and their conversion into near-serfs paid with a share of the crop began in these areas much later than in north Bengal and gathered pace during the first half of the twentieth century. Similar to the government-controlled 'lots' were the 'chaks' in the 24-Parganas and the

[28] The 'lots' were held under the Waste Land Rules of 1853 (99-year grants), the Large Capitalist Rules of 1879 (40-year grants), the Saugor Island Rules of 1847 (40-year grants), talukdari and malguzari leases (in Khulna) and a few under lease to 'small capitalists' and haoladars. GB Rev. Dept. LR Br., B. Progs. 39–45, Sept. 1940, File 41–7 (BSRR). Note the difference with the 'lot-dars' of Dinajpur who were zamindars.

south-east of Midnapur leased out to chakdars for reclamation in the zamindari estates in the Sunderbans. A chak, the 24-Parganas Gazetteer reported 'may be of any size and sometimes in the Sunderbans consists of thousands of bighas of land.'[29] In these areas, the chakdarbhagchashi (sharecropper) sector became dominant in the agrarian structure.

Further east along the Sunderbans in eastern Khulna and Bakarganj, the situation was already slightly different. The hawaladars and ganthidars in these parts are usually considered to be similar to the big jotedar of the north. There were, however, significant differences. The haola (literally, a charge) was 'very largely a small lease for the reclamation of forest granted to a man who was prepared to make his home in the grant and personally supervise its reclamation.'[30] The ganthis were similar in origin but probably earlier assignments. The recipients of these grants were not of the scale of the giant jotedars of north Bengal; they were 'men whose capital was ordinarily small' and who lived in the district of Dacca from where the revenue arrangements of Bakarganj were conducted when the British revenue system was first introduced.[31] More importantly, the physical configuration of this densely forested region, which was cut up into numerous petty blocks by streams and river, did not render feasible the supervision of reclamation and farming in large personal demesne. This led to an extraordinary degree of subinfeudation of tenurial rights. In marked contrast to the north Bengal situation, Jack reported in 1908 that there

---

[29] *24-Parganas DG*, (1914), p. 177. In Midnapur, it was reported, that the chaks 'are large, ranging from 800 to 5300 bighas, practically none of which is now cultivated by the Chakdars.' *Midnapur SR*, (1910–18), p. 45. There was considerable controversy as to whether the chakdar or the bhagchashi should be given the raiyati right. 'Status of Chakdars and Bhagchasis in Pargana Majnamutha and other Khas Mahals in Midnapore', GB Rev. Dept. LR Br., Progs. 17–23, Aug. 1915, File 16-S-22 of 1915 (WBSA).

[30] *Bakargang DG*, (1918), p. 87.

[31] *Ibid*. On the complexity of tenurial patterns in Bakarganj, see Tapan Raychaudhuri, 'Permanent Settlement in Operation: Bakarganj District, East Bengal' in R.E. Frykenberg (ed.), *Land Control and Social Structure in Indian History* (Madison, Wisconsin, 1969), pp. 163–74.

was 'no class of landless labourers in the (Bakarganj) district.'[32] Bargadari (sharecropping) and dhankarari (fixed produce rent) were confined to 5 per cent of the total raiyati area and were resorted to mainly in the area where the Hindu bhadralok gentry had tended to congregate. Operational holdings in these regions were fragmented. While the jotedars of north Bengal had physical dominion over large tracts, the lands held in lease by the hawaladars and ganthidars were possessed, as it were, several times over by numerous degrees of subinfeudatory tenants. Though the tenurial scale spanned a wide range of inequality, the difference between each grade was minute and the dichotomy between classes blurred.

Among the frontier regions of Bengal, the Jungle Mahals on the western periphery and the role of the mandals or clan leaders, often equated with 'jotedars', remains to be considered. The social organisation of the reclaimers and cultivators of the land in these areas exhibited tribal forms. Leaders of tribal communities known as mandals had negotiated the conditions of settlement of an uncultivated tract with the zamindars and undertaken to collect and pay the rent on behalf of the community as a whole. By the early twentieth century, however, the 'mandali system' had virtually disintegrated. Jameson observed in the Midnapur Settlement Report of 1919: 'Mandals exist only in those areas where the aboriginal and semi-aboriginal tribes of Santals, Bhumijs, Mahatos, etc., are, or were until recently, the bulk of the population, and among these tribes the patriarchal village community is the regime under which they lived when the mandali system was evolved, though it has now broken down to a large extent.'[33] The decay of the mandali system was attributed by the settlement officer of Bankura to the introduction of the dikku or foreigner into the Jungle Mahals.[34] These dikkus, Bengalis from the east and Utkal Brahmins from Orissa in the west, who arrived as traders, resorted to the profitable business of moneylending, and eventually obtained a hold on the land. The mahajan first dislodged the mandal and usurped

[32] *Bakarganj SR*, (1900–8), p. 73.
[33] *Midnapur SR*, p. 41.
[34] *Bankura SR*, (1917–24), p. 59.

A TYPOLOGY OF AGRARIAN SOCIAL STRUCTURE / 283

the position of tenure-holder of a village and then went on to encroach upon the best lands of the individual tenants. A special enquiry in 1909 into the conditions of the Santals revealed the alarming rate at which they were losing the most valuable rights in the land.[35] The dispossessed tenant was usually resettled at a very high produce rent or a mixed cash and produce rent. Large-scale dispossession began about the time of the great famine of 1865–66 and proceeded at a rapid pace until an amendment to the Bengal Tenancy Act in 1918 prohibited the transfer of land from aboriginals to non-aboriginals without the collector's permission. The legislation, however, came too late as the superior rights in the rice lands in the jungle area had already passed from the hands of the tribal people into those of the mahajans. It would seem appropriate to treat the tribal settlements in the western fringe as an extreme form of a demesne labour-peasant smallholding complex which is analysed below.

The Peasant Smallholding System in East Bengal and the Peasant Smallholding-demesne Labour Complex in West and Central Bengal

In most parts of Bengal, village-controlling rich farmers even remotely approaching the scale of the big jotedars of north Bengal are scarcely to be encountered. 'Jotedars', as such, were to be found both in east and west Bengal but they were very different from their namesakes in the north. The pattern of settlement often left central villages inhabited by the bhadralok, an elite consisting of the upper Hindu castes as well as aristocratic and learned Muslims who shirked manual labour, surrounded by villages held by the chashis or peasants. The upper-caste gentry in east Bengal did not as a rule enter directly into cultivation as did some of their counterparts in northern India. For the most part, they were simply residential, relying on their profits as petty rent-collectors and increasingly also as moneylenders.[36] For east Bengal, the bhadralok-chashi dichotomy is

---

[35] Mr Alpin's 1909 Report quoted in Bengal Board of Economic Enquiry, *Bulletin District Bankura* (Alipur, 1935), pp. 3–4.
[36] On the talukdars' role as moneylenders, see chapter 4 below.

important and as a broad distinction probably sufficient. The bulk of the Muslim and Namasudra cultivators may be seen to have chashi or peasant status. They held jotes—cultivable lands —owned the implements of cultivation and had solid titles to their homesteads, describing themselves as grihasthi. In west Bengal, it was not unusual for some of the landlords to direct farming on land which they held as khas or personal demesne. In addition to the chashis of the agriculturist castes, such as the Mahishyas, Sadgops and Aguris, there was in west Bengal at the very bottom of the agrarian hierarchy, a distinct layer of landless agricultural labourers drawn from among the low-caste Bagdis and Bauris and the aboriginal tribes, such as the Santals. In east Bengal, with the intensification of demographic pressure, the ranks of a land-poor peasantry swelled after 1920;[37] yet the peasantry here may be seen to merge into the landless category. In west Bengal, a certain discontinuity is apparent; the peasant and the rural proletarian there must be regarded as distinct elements in a pre-existing agrarian social structure.

The typical agricultural work-unit in Bengal was the small peasant family farm. East Bengal was not in as advanced a stage of cultivation at the time of the Permanent Settlement as west Bengal. There was an abundance of fallow land but not the same obstacles in the path of reclamation as in the case of north Bengal. During the latter half of the nineteenth century and the early part of the twentieth, a steady expansion of cultivation took place under the stimulus of a secular rise in population, mainly in the form of a proliferation of atomistic small peasant farms. The relative homogeneity of the chashis of east Bengal, and to a lesser extent in west Bengal as well, in marked contrast to the wide economic differentiation in north Bengal, is striking. Land described in northern India as sir or primary zamindari and cultivated—usually through family and tied labour—by a dominant peasant or a cultivating proprietor, who also has the right to collect tribute, sink wells, plant trees on ryoti or subordinate-cultivator-held land,[38] is virtually non-existent in east

[37] See chapter 2 below.
[38] See Stokes, *The Peasant and the Raj*, pp. 46–62, 205–27.

Bengal. West Bengal provides a sort of transition zone between the upper India and the east Bengal situations. Here, landlords with considerable khas khamar and a segment of rich peasantry with surplus lands which had emerged during the nineteenth century possessed some of the economic power and political clout of the sirdars of northern India. In northern India, the lords of the land derived their position not only from rent-collection and residence but also from the active direction of cultivation. Bengal in most parts did not possess the same nucleated village structure and clan cohesion. It was only in exceptional circumstances (such as the colonisation of new char lands or of pockets of jungle) that Muslim and Namasudra cultivators in east Bengal threw up a form of clan organisation with leaders ready and able to assume the role of dominant peasant. The period of the zamindari rent offensive in the first half of the nineteenth century saw the rise of a sort of seigneurial sergeant class, village leaders known variously as dewanias, mathbars, mandals and pramaniks, who colluded with the zamindars to fleece the peasantry and were allowed to hold land at favourable rates. However, their position had more to do with their ability of manipulate the administrative structure of rent-collection than with the production process itself. With the waning of the rent offensive in the later nineteenth century and the virtual end of the era of high landlordism marked by the Tenancy Act of 1885, their role became less and less relevant. In early twentieth-century rural Bengal there was a mass of peasant smallholders living under broadly similar but splintered conditions of economic existence below a superimposed network of rent-receivers who were proprietors of the land in only legal terms.

In east Bengal, the landlords under the Permanent Settlement consisted of a few big zamindars and numerous petty talukdars. Most of them belonged to the upper Hindu castes but there were also a handful of Muslims. The same social groups filled the ranks of tenure-holders known also as talukdars who were intermediaries in the rent-collecting structure between the superior zamindars and the raiyats. There was not much subinfeudation of rent-collecting rights compared with districts such as Bakarganj and Khulna, and tenures rarely went below two or three degrees. In west Bengal, the dominant tenurial pattern below the level of the zamindars was that of the patni and its derivatives. The patnis had originated after 1793 in the estate of the Maharaja of Burdwan. The revenue assessment of the estate was high and in order to ensure easy and punctual realisation of rent, a number of leases to be held in perpetuity at fixed rents were given to a large number of middlemen. After the Patni Regulation of 1819 legalised and systematised the patni tenure, it gained very wide currency in west Bengal. Patni rights were often held over a whole village or small groups of villages. A considerable number of patnidars sublet part of their interests to darpatnidars, and a few repeated the process with sepatnidars. These interests were mostly in the hands of the upper-caste gentry but it was not unusual for a few members of the agricultural and intermediate trading castes to make their way into these ranks.

Apart from the so-called raiyati land, held and operated by the peasantry, it was a long-established practice of the landed gentry to keep a proportion of khamar land in their personal possession and cultivate it through sharecroppers or hired labourers. In east Bengal, landlords' khas khamar was, generally speaking, minimal. In west and central Bengal, it was considerable. In the 1910s, in the east Bengal districts of Dacca, Faridpur and Tippera, 85 per cent, 86 per cent and 87 per cent of the land area respectively was held by raiyats, a great majority of whom were cultivating peasants. At the other end of the scale, in the west Bengal district of Bankura, 23 per cent of the total area was in the direct possession of proprietors and tenure-holders, 46 per cent in the hands of raiyats and under-raiyats, while the rest was waste and jungle over which the landlords held the primary rights.[39] Consequently, the proportion of labourers and sharecroppers without any rights was much higher among the agricultural population in west and central Bengal than in east Bengal.

In early twentieth-century Bengal agricultural production

[39] *Dacca SR*, (1910–17), p. 70; *Faridpur SR*, (1904–14), p. 29; *Bankura SR*, pp. 66–9 and Appendix.

was predominantly in the hands of peasant smallholders who had recorded rights of occupation. The predominance of the peasant smallholding sector over the demesne labour or khamar sector was much more pronounced in east Bengal than in west and central Bengal. In the words of the Dacca settlement officer, the settled raiyat paying his rent in cash still formed 'the backbone of the agricultural population.'[40] In the battle fought in the late nineteenth century over occupancy right and rent enhancement, it was the raiyats who had scored and won. The Bengal Tenancy Act modified the Settlement of 1793 in important ways. It gave a large body of cultivators a measure of tenurial security on moderate rents.[41] Jack, as we have seen, had the impression that the cultivators of the east Bengal district of Faridpur were a 'homogeneous class.' Bengal was not a land of bloated jotedars. But it would be misleading to suggest that the peasantry was a wholly undifferentiated mass. In east Bengal the scale of inequalities did not produce a class dichotomy within the peasantry; differences in wealth and landholding sizes were fluid within what was basically a peasant smallholding structure. In west Bengal high grain prices during the late nineteenth century had initiated a process of differentiation which had led to the emergence of a small segment of rich peasants. These rich peasants operated in the product and credit markets as partners of

---

[40] *Dacca SR*, p. 71.

[41] The Tenancy Act was the government's response to the wave of agrarian agitation that swept east Bengal in the 1870s and the early 1880s. It provided for a right of occupancy to a tenant who had cultivated any plot of land in a village for the previous 12 years and rent could be enhanced only once in 15 years on clearly defined conditions and by not more than 12.5 per cent of the existing rent.

Many of the protected raiyats could, of course, sublet and become landlords themselves, and there was nothing to prevent non-cultivators from buying into raiyati rights. The extent of subletting until the second and third decades of the twentieth century, when some of the major districts were surveyed, was not very great. The proportion of raiyati area sublet to under-raiyats was in Faridpur 9.1 per cent, Tippera 2.8 per cent, Dacca 1.3 per cent, Mymensingh 4.1 per cent, Midnapur 3.4 per cent, Bankura 4 per cent and Burdwan 2.3 per cent. *Faridpur SR*, p. 33: *Dacca SR*, pp. 70–1; *Mymensingh SR* (1908–19), p. 44; *Burdwan SR*, (1927–34), p. 34.

the moneylending and grain-dealing landlords. The question of a socio-economic hierarchy within the peasantry does not lend itself very easily to a quantitative analysis. Legal categories do not fit real social categories. Nonetheless, the available statistics have to be considered before moving on to a discussion of the qualitative evidence.

It was only in 1938–40 that an attempt was made to establish the number of families in different acreage classes for Bengal as a whole. This was done at the instance of the Land Revenue Commission. Table 1 shows that in the districts of east Bengal, roughly 84 per cent of agriculturist families held less than 5 acres, 11 per cent between 5 and 10 acres and a mere 5 per cent over 10 acres. In the districts of west Bengal, some 72 per cent of the families held less than 5 acres, 19 per cent between 5 and 10 acres and 9 per cent over 10 acres. The figures in table 2 are taken from settlement reports of some west Bengal districts. The reports on Birbhum, Murshidabad, Howrah and Hooghly provide the additional and more significant information about the proportion of the total area covered by different acreage classes of holdings. Howrah and Hooghly appear to show a predominance of small tenancies, while in Birbhum and Murshidabad, rather larger holdings of over 15 acres are of some importance.

More revealing than statistics built around imperfect legal categories are the insights into the relations between real social categories. Differentiation within the peasantry of east Bengal around 1919 appears to be one of subtle and delicate gradation, the scale of inequalities being relatively small. The differences in landholding size were not considerable and did not at any rate allow scope for the development of exploitative class relations through the control of large amounts of surplus lands. Village-controlling landholders so common in north Bengal were conspicuously absent. What then was the position of the 'jotedars' of these regions if they were not rich farmers of the north Bengal type? As Jack discovered in the course of settlement operations in Faridpur. 'The name jot exists throughout the district as a generic name for every tenancy.'[42] In Mymensingh,

---

[42] *Faridpur SR.* P. 27

TABLE 1

DISTRIBUTION OF AREAS HELD BY A FAMILY

District	Number of families enquired into	Average area per family in acres	Proportions (%) of land held by families of different acreage categories					
			Less than 2 acres	2–3 acres	3–4 acres	4–5 acres	5–10 acres	Above 10 acres
East Bengal districts								
Bakarganj	804	2.17	61.8	13.1	9.1	3.9	10.9	1.2
Bogra	464	4.28	34.5	14.2	13.6	12.7	17.9	7.1
Chittagong	690	2.45	60.3	10.1	8.8	5.8	10.7	4.3
Dacca	508	2.13	62.4	11.6	6.1	6.1	5.1	3.5
Faridpur	1104	1.63	81.5	7.6	3.4	1.8	2.6	0.6
Khulna	356	4.78	55.6	7.8	9.0	6.1	13.9	7.6
Mymensingh	931	3.86	34.1	13.9	11.9	10.5	16.9	6.5
Noakhali	502	2.41	65.3	12.1	7.8	3.4	4.2	2.8
Pabna	701	2.39	64.1	9.2	5.8	4.1	7.1	2.4
Rajshahi	1018	5.52	31.8	9.3	9.7	9.1	25.5	14.6
Tippera	950	2.22	63.9	13.7	8.6	4.3	6.6	2.9
East Bengal	8028	3.07	55.9	11.1	8.5	6.1	11.0	4.9

District	Number of families enquired into	Average area per family in acres	Proportions (%) of land held by families of different acreage categories					
			Less than 2 acres	2–3 acres	3–4 acres	4–5 acres	5–10 acres	Above 10 acres
West and central Bengal districts								
Bankura	670	8.17	53.7	8.9	7.8	4.5	14.8	10.3
Birbhum	727	4.64	15.1	10.1	7.4	8.5	19.2	8.2
Burdwan	803	5.63	28.6	10.9	8.9	10.8	26.6	12.8
Hooghly	595	3.74	32.4	13.1	13.0	10.9	18.8	10.2
Howrah	336	3.53	53.2	14.3	5.1	4.5	17.5	5.4
Jessore	1073	4.78	28.5	10.3	9.6	9.8	27.1	13.6
Malda	332	3.34	54.2	7.8	8.4	6.9	15.9	6.8
Midnapur	1110	4.23	38.2	16.1	10.9	10.5	17.6	6.7
Murshidabad	1178	4.30	38.3	10.1	9.3	7.5	16.9	7.7
Nadia	830	4.83	16.8	9.6	10.8	10.1	20.3	11.8
24-Parganas	1174	4.33	56.5	10.7	8.6	4.7	10.9	7.2
West and central Bengal	8828	4.7	37.8	11.0	9.0	8.0	18.7	9.2

SOURCE: Report of the Land Revenue Commission, Bengal, Vol. 2 (1940), pp. 114-15.

A TYPOLOGY OF AGRARIAN SOCIAL STRUCTURE / 291

TABLE 2

ACREAGE CLASSES OF HOLDINGS IN SOME WEST AND CENTRAL BENGAL DISTRICTS

District	0–1	1–2	2–3	3–4	4–5	5–15	15–25	over 25	Total average
1. Midnapur (1911–17)									
Percentage of holdings to total holdings	Under 5 acres:			93.2		4.8	0.7	1.3	100
2. Birbhum (1924–32)									
Percentage of holdings to total holdings	66.8	14.9	7.2	3.7	2.4	4.6	0.3	0.1	100
Percentage of area under holdings to total area	15.2	16.6	13.5	10.0	8.1	26.9	4.8	4.9	100
Average area of each holding	0.29	1.44	2.43	3.46	4.48	7.59	18.52	44.98	1.29
3. Murshidabad (1924–32)									
Percentage of holdings to total holdings	68.9	15.7	6.4	3.2	1.8	3.5	0.3	0.2	100
Percentage of area under holdings to total area	18.6	18.0	12.7	9.0	6.6	21.8	4.6	8.7	100
Average area of each holding	0.33	1.42	2.44	3.46	4.44	7.71	18.73	60.38	1.23
4. Malda (1928–35)									
Percentage of holdings to total holdings	44.0	24.0	2–5 acres:		22.0	5 acres and above:		10	100
5. Hooghly (1930–37)									
Percentage of holdings to total holdings	73.2	15.2	2–5 acres:		9.7	5 acres and above:		1.9	100
Percentage of area under holdings to total area	29.0	23.8	2–5 acres:		31.7	5 acres and above:		15.5	100
Average area of each holding	0.35	1.30	2–5 acres:		2.93	5 acres and above:		7.21	0.89

District	0-1	1-2	2-3	3-4	4-5	5-15	15-25	over 25	Total average
6. Howrah (1934-39)									
Percentage of holdings to total holdings	79.1	13.1	4.1	1.7	0.8	1.1	0.1	0.0	100
Percentage of area under holdings to total area	34.8	24.6	13.3	7.9	4.8	11.0	1.4	2.2	100
Average area of each holding	0.33	1.40	2.41	3.42	4.45	7.42	18.56	48.09	0.74

SOURCE: District Settlement Reports.

apart from a few unusual tenancies, all others were 'almost universally described as jotes.'[43] Anyone in east Bengal with a long-term cultivating right in a piece of land subject to payment of rent was a jotedar. He was more often than not a smallholding peasant. In west Bengal, the term jotedar came to mean sharecropper, almost the opposite of what it usually denoted in north Bengal. The holding of a sharecropper was designated a bhag (share) jote and its holder was a bhagjotedar or simply a jotedar. This local usage was found by sociologists to persist as late as the 1960s even though by that time—in the political dictionary of Bengal—jotedar referred exclusively to the poor peasant's and sharecropper's class enemy.[44]

The terminological question is less important than the substantive issue it deals with. It is not just that the meanings of these indigenous terms varied over time and space or that the features of the agrarian structure remained unchanged. Instead of the 'master-servant relationship' between jotedar and adhiar in Dinajpur, in the Bengal heartland the roles of jotedar and bargadar overlapped and were even interchangeable. The existence of the sharecropping arrangement is too readily seen to indicate relations of dependency bordering on semi-feudalism. No doubt sharecropping and sharecroppers existed all over Bengal. But they did so in a variety of contexts. Dependency in the new areas of reclamation has already been mapped. In east Bengal the sharecropping arrangements both on khamar and raiyati land were to a great extent incidental to the predominating peasant smallholding system. It was a time-worn practice for raiyats to cultivate the khamar lands of zamindars and taluk-

[43] *Mymensingh SR*, p. 43.
[44] Andre Beteille, *Studies in Agrarian Social Structure* (Delhi, 1974), pp. 129–30; A.K. and D.G. Danda, *Development and Change in Basudha* (Hyderabad, 1971), p. 44, O'Malley wrote of the bhag jotedar of Bankura in 1908: 'In such a holding (a bhag jote) the tenant has the use of the land for a year or a season, and pays as rent a certain share on the produce of the land. Ordinarily one half of the produce is so paid, the jotedar cultivating the land with his own plough and cattle, and also finding seed and manure. Occasionally the superior tenant who engages the bhagjotedar finds the manure, in return for which he receives the straw in addition to his half share of produce', *Bankura DG*, p. 102.

dars on a share basis, to add to their income from inadequate raiyati land or even to find full employment for their labour resources. This system survived in the Manikganj subdivision of Dacca. In 1911 it was described by the settlement officer in the following terms:

> It cannot be said on the whole that the bargadars constitute a separate class: in some places especially in the more sparsely populated area around Mahadebpur they consist largely of landless labourers, but in the area as a whole it is the ordinary raiyat who adds to his profits by the cultivation of khamar land ... I doubt whether in a single instance any family gains from its barga lands more grain than is sufficient to support itself; only in a few instances where jute is grown on barga lands is there any monetary gain and for that an extra share is given to the bargadar either in money or in kind.[45]

Similarly, the settlement officer of Mymensingh states quite plainly that nearly all bargadars had jote lands of their own. The bargadars, he wrote, 'do not form a class by themselves ... The bargadar is usually a settled ryot of the village, renting his homestead and one or two plots of arable land from the same landlord on a cash rent.'[46] In a peasant society with inequalities but no sharp differentiation or bi-polar class divisions, the sharecropping relation on a small percentage of peasants' land could in the short term adjust disparate land-labour ratios within the peasant smallholding system.[47]

The dominance of the peasant smallholding system in early twentieth-century east Bengal is undeniable. But there were disquieting reports from some areas of the *de facto* increase of khamar at the expense of raiyati land. *De facto*, because the method often employed by non-peasant rentiers as well as moneylenders and traders was to buy into raiyati rights. The Dacca settlement officer who observed this phenomenon with

---

[45] *Dacca SR*, App. XI, xxvii–xxviii.

[46] *Mymensingh SR*, p. 45.

[47] In Noakhali, where the bulk of the land was held and cultivated by small peasants on money rents, holdings on produce rents were 'only created as a temporary convenience, as for instance when the father of a family dies and a neighbour takes over the cultivation of the land while the children are growing up', *Noakhahi SR*, (1914–19), pp. 91–2.

concern, especially in the regions of new development on the fringes of the Madhupur jungle, made clear the distinction between old barga lands as an adjunct to small peasant cultivation and new barga lands under moneylender-landords at the expense of peasant smallholding.[48] The extent to which non-peasant moneylenders were able to displace small cultivating raiyats and how far the processes of internal differentiation allowed a peasant elite to emerge will be themes we shall return to when we shift from the somewhat static analysis adopted in this chapter to a study of the dynamic processes of the peasants' involvement in the product and credit markets and their impact on the structure of peasants' holdings.

Finally, the role of wage labour in the agrarian economy has to be set into context. This will clarify the differences between the agrarian structures of east and west Bengal. 'The landless labourer', Jack wrote in 1916, 'is unknown in Faridpur and very rare anywhere in Eastern Bengal.'[49] A large number of cultivators, of course, worked for hire at harvest time, but the composition of these hired labourers underlined the relative homogeneity of the peasantry. As a detailed socio-economic survey of Faridpur showed:

The proportion of agricultural labourers amongst the poorer families was naturally much greater than amongst the richer, but not by any means to the extent which might have been expected. Of cultivators in comfort 22 per cent, of those below comfort 31 per cent, of those above want 36 per cent and of the indigent 37 per cent were enumerated as engaged in agricultural labour. It is probable that amongst all these there were none who were exclusively agricultural labourers ... All had their land, some perhaps very inadequate in amount, but others only inadequate because the family contained at the time an undue proportion of young children ... The proportion of the indigent supported by agricultural labour is not larger mainly because this class consists of old men who are unfit for the work or of families whose bread-winner had died before his time.[50]

There was some interdistrict movement of agricultural labour

---

[48] *Dacca SR*, pp. 75–7 and App. XI.
[49] Jack, *Economic Life of a Bengal District*, p. 84.
[50] *Ibid.*

in east Bengal as a consequence of the uneven levels of wealth and the slightly uneven crop cycles. Smallholders from Faridpur, Jessore and Tippera would cut paddy in Bakarganj and strip jute in Mymensingh. Dacca and Tippera display a well-established and elaborate system of labour exchange known as gnata. Labour pools of half a dozen or more smallholders were formed at the time of harvest.[51] In Noakhali as well no clear distinction could be made between the peasant and the landless labourer.[52]

The situation in west Bengal, on the other hand, was entirely different. It was only in the Contai and Tamluk subdivisions of Midnapur that there were no large reserves of landless agricultural labour. Here, as the Midnapur Settlement Report records, the work of cultivation was performed for the most part by the peasant himself using family labour. Owing to the small size of an average holding, this was generally sufficient. If more labour was required it was obtained by 'a system of exchange between neighbours.'[53] Elsewhere in west Bengal, a land of old settlement, high rents, uncertain harvests and a demographic arrest (owing to malaria epidemics from the mid-nineteenth century until about 1920), the Bagdis, Bauris, and tribal people supplied much of the labour on the agricultural lands, invisible to settlement statistics, as bhagdars (sharecroppers without occupancy rights), Krishans (tied labourers paid with a third of the produce), munishes (day labourers) and mahindars (farm servants). These men might perhaps own a garden patch or even

---

[51] *Dacca SR*, pp. 20, 52, S.V. Ayyar and A.K.A. Khan, 'The Economics of a Bengal Village', *Indian Journal of Economics*, 6 (1926), pp. 200–15.

[52] As the settlement officer wrote: 'when the economy of the agricultural classes as a body is considered, it is unnecessary to consider the price of labour. The labour is entirely supplied by the same body. No outside labour is employed and to make an entry on the debit side of the balance sheet of the whole body would merely involve the necessity of including the same figure as an asset of the body on the credit side.' *Noakhali SR*, p. 48. The accounting is dubious, but the passage does suggest the absence of a distinct landless labour class in Noakhali.

[53] *Midnapur SR*, p. 30. This was an area of rising population and expanding cultivation since the withdrawal of the salt monopoly in the mid nineteenth century.

a share in rice fields, yet essentially they constituted a distinct landless element. Robertson, settlement officer of Bankura, described how this large pool of landless labour was put to use:

In the poorer districts of Western Bengal there is a tendency for much of the land to fall into the hands of the tenure-holders. In these districts the crop is uncertain and the people and poor and thriftless. The tenure-holders themselves are the principal moneylenders, and when a tenant has once borrowed money it is only a matter of time before his holding comes to sale. It is then bought in by the tenure-holder, who perhaps retains the more valuable lands for himself settling the less valuable lands with the tenant afresh. In a very poor district such as Bankura there is no lack of hired labourers. Indeed the Bauris as a class are usually landless men who work for others. The tenure-holders, therefore, if they retain lands in their own possession, find no difficulty in hiring labourers to cultivate them.[54]

In 1914, Kaibarta or Mahishya families who formed the bulk of the raiyats in Hooghly district employed Bagdi, Bauri and Santal sharecroppers and labourers.[55] In the wake of the malaria epidemics, the number of working members of raiyat families were often few and far between. The two classes, peasants and agricultural labourers, were consequently brought together in a necessary though unequal collaboration in order to sustain agricultural production. It was not unusual for caste peasants to lease land from the gentry on bhag and employ labourers who were supplied with the necessary plough-team and seeds. For west and central Bengal, the vision of self-cultivation by peasant smallholders has to be modified to take account of the fairly widespread use of tied and hired labour not only on the landlords' and rich peasants' considerable khas lands but also on peasant smallholdings. Table 3 is intended to give a general impression of the much higher proportion of labourers (including sharecroppers without tenancy rights) to the total agricultural population in west and central Bengal than in east Bengal where the great majority of cultivators had some sort of recorded right to the soil.

[54] *Bankura SR*, p. 67.
[55] *Hooghly SR*, (1904–13), p. 37.

TABLE 3

Proportions of Agricultural Population: 'Landlord', 'Tenant', 'Labourer' (per cent)

	Landlords	Tenants	Labourers
East Bengal districts			
Bakarganj	4.26	87.75	7.99
Bogra	1.99	89.24	8.77
Chittagong	9.55	72.53	17.92
Dacca	5.08	89.24	5.68
Faridpur	6.61	88.84	4.55
Khulna	5.73	85.68	8.59
Mymensingh	2.49	92.12	5.39
Noakhali	3.36	82.50	14.14
Pabna	4.12	88.28	7.60
Rajshahi	4.60	84.18	11.22
Tippera	1.93	93.87	4.20
East Bengal	4.52	86.75	8.73
West and central Bengal districts			
Bankura	4.18	68.39	27.43
Birbhum	2.00	62.38	35.62
Burdwan	4.19	68.42	27.39
Hooghly	4.87	68.57	26.56
Howrah	8.89	61.06	30.05
Jessore	6.94	85.34	7.72
Malda	2.12	75.99	21.89
Midnapur	2.57	80.45	17.18
Murshidabad	4.15	68.99	26.86
Nadia	7.85	70.05	22.10
24-Parganas	4.39	76.19	19.42
West and central Bengal	4.72	71.44	23.84

SOURCE: Census of India 1921, Vol. 5, Bengal, Pt 2, Tables 2,3,4 and 5.

This chapter has focused on certain basic distinctions in the agrarian social structure in Bengal. A more complete elucidation of the relations of production and surplus appropriation in the Bengal countryside will become clearer when we study the peasantry's involvement in the credit and produce markets.[56] The impact of these markets on the smallholding peasantry may, however, be sketched here in skeletal form and will be elaborated and developed later. From the later nineteenth century onwards, zamindari rent as a mode of surplus appropriation became less significant, even though in west Bengal districts it continued to rule relatively high. In these parts, the zamindars and patnidars had extensive usurious interests. After the removal of the indigo planters from the agrarian scene by the 1860s, they were clearly the principal source of credit for the cultivating classes. These groups controlled the credit and that critical proportion of surplus lands which enabled the smallholding and landless economies to go on reproducing themselves. It was they who were the chief beneficiaries of the expanding market in grain, and while a shrewder and richer minority of the peasants were able to cash in and become patrons themselves, many peasant smallholders fell into debt and could only carry on by borrowing seed and grain from year to year; in course of time, some were reduced to a position close to that of the landless workers.

The attempt at colonial extraction through capitalist transformation of traditional production regimes ended in Bengal (the Darjeeling tea-plantations excepted) with the demise of the indigo plantations by the 1860s. However, from the late nineteenth century and more emphatically in the early twentieth, the small peasant economy of east Bengal was drawn into the web of an export-oriented colonial economy. Small peasant producers raised jute for the world market on their minuscule holdings. The market and the credit system, which kept the peasant family alive and helped to reproduce the small peasant economy, became more important than rent as the channels of the drain on the east Bengal peasant. The control of the zamin-

---

[56] See chapter 4 below.

dars and talukdars over land was distant and weak. Indeed, the rent charge over which their proprietorship really extended was becoming increasingly difficult to collect. Some were eliminated. Others found a firm niche in the surplus-appropriating mechanism by resorting to lagni karbar (the moneylending business). In addition to the trader-moneylenders, there emerged in the early twentieth century and important section of bhuswami mahajans (landlords who were also moneylenders).[57]

In attempting to raise the value of his product on a diminishing holding, the east Bengal peasant made himself vulnerable to violent and often long-term fluctuations in the world market. Jute was an expensive crop to cultivate, especially since its labour costs were much higher. This in turn enlarged the credit needs of the peasantry. The dispersed nature of peasant production meant that the grower had little bargaining power over prices *vis-a-vis* the highly organised trading sector. An elaborate marketing mechanism involving a long chain of middlemen took away a fair amount of the peasants' due. More importantly, the peasant's lack of holding power compelled him to sell immediately after harvest when prices were at their lowest. Not only did he not have facilities of storage and transport, but he required cash at harvest time to make his exorbitant interest payments to the moneylender and deliver rent to the landlord. Often he would have received a dadan (advance) from a trader-moneylender and had little choice except to sell his produce to the dadan karbari at a price much below the prevailing market rate.[58] In the context of increasing population pressure and diminishing holdings, colonial extraction through the market interacted with and reinforced the credit mechanism. This served to hold the peasant in their pincer grip and perpetuated and impoverished the small peasant economy of Bengal.

Clearly then, possession of the chief means of production—land—did not mean that the peasant smallholders were independent agents in the process of production and reproduction

---

[57] For a detailed discussion of moneylenders and moneylending, see chapter 4 below.
[58] See chapter 3 below.

in agrarian Bengal. Yet the lack of freedom of the small peasant weighed down by scarcity of land and the inequities of highly unfavourable market and credit relations bore no resemblance to the sharecroppers' dependence on the village landlord rich farmer-cum-creditor.

In east Bengal there was no strict equality among the peasantry. But it is still possible to speak of a predominating peasant-smallholding structure with imperceptible gradations from a dwarfholder-cum-bargadar to an owner-tenant-cultivator with a small surplus. While there were no serious class divisions within peasant society, the peasantry in east Bengal found themselves involved in similar sets of tenurial, credit and market relations. In west Bengal, the cleavage between the peasant and the rural proletarian was more obvious. The latter might possess a meagre patch, but he earned his livelihood primarily by hiring out his labour. Here the moneylending landlords and tenure-holders exercised a much greater direct control over the land than in east Bengal. In the dominant structural type of the peripheral regions newly reclaimed from wasteland, the refinement of delicate, graded complexities was of little importance: at the peak of the pyramid stood the giant jotedar flanked perhaps by a few chukanidars and down below were the mass of dependent adhiars.

'For historians, a structure certainly means something that holds together or something that is architectural.'[59] This chapter has identified the major architectural styles in rural Bengal and has given an impression of their contours. We must now examine how the structure held together, the pressures to which it was exposed, the faults that developed within it and the lines along which the edifice of Bengali rural society crumbled.

---

[59] F. Braudel, 'History and the Social Sciences' in P. Burke (ed.), *Economy and Society in Early Modern Europe* (London, 1972), pp. 17–18. Beyond that, in Braudel's definition, a structure 'means a reality which can distort the effect of time, changing its scope and speed.'

Chapter Eight

Malguzars and Peasants: The Narmada Valley, 1860–1920

T.C.A. RAGHAVAN

The Central Provinces of British India, a vast conglomerate of agrarian regions and cultural systems, have been rather inexplicably neglected by historians. The focus of this study is the Narmada Valley or, more specifically, that part of the valley included in the districts of Jabalpur, Narsinghpur and Hoshangabad of the erstwhile Central Provinces.[1] This paper, essentially empirical in nature, documents certain economic aspects of the relations between landlords and tenants. The first two sections are intended as introductory and only briefly discuss the Malguzari settlement and give an overview of economic change in this period. A discussion of the nature of the malguzars of the Narmada Valley is then taken up by documenting their income from rents, the expansion of the *sir* or malguzari home farms, and an examination of contemporary debates about the existence of two functionally different kinds of landlords—'agriculturists' and 'non-agriculturists.' The discussion of the

---

* T.C.A. Raghavan, *Studies in History*, 1, 2, n.s. (New Delhi: Sage Publications, 1985)

** I am grateful to Professor Sabyasachi Bhattacharya and Mr Neeladri Bhattacharya for their encouragement and help in extensively revising an earlier draft of this paper.

[1] This paper is based on material collected for a M. Phil dissertation submitted to the Centre for Historical Studies, Jawaharlal Nehru University, entitled 'Agrarian Change in the Narmada Valley: A Study of Jabalpur, Hoshangabad and Narsinghpur Districts—1860–1925', henceforth 'Agrarian Change in the Narmada Valley.'

peasantry concentrates on certain aspects of peasant differentiation: the differential incidence of rent and the role of ecological factors in the differentiation of the peasantry.

The Malguzari Settlement, 1860s

The valley was acquired by the East India Company (EIC) in 1818 from the Bhonsales of Nagpur. This cession was the final act of a half century of somewhat bewildering array of political change as the original, highly sanskritised Gond rulers under the loose suzerainty of the Mughals yielded to the Maratha sardars of Bilaspur who, in turn, were replaced by the Bhonsales and the Narmada Valley was incorporated, albeit peripherally, into the Nagpur Province.[2]

The early policies of the EIC in the valley are marked by a somewhat credulous faith in the benign effects of enlightened government and imperial law and order. Trusting these to be the sole panacea of all the perceived economic ills of the valley —'a depopulated country and an impoverished and dispirited people'[3]—the EIC inaugurated its reign by increasing the revenue demand to hitherto unprecedented levels. The Company at the same time retained the tenurial and administrative arrangements of the Marathas which were in turn often characterised by relics of the administrative system of the former Gond rulers. The levy of a very high revenue for about a decade-and-a-half led to a near breakdown of the revenue administration in this

[2] Analysis of the socio-political systems and state forms of the Gonds in this region remains at an elementary level. See however Surajit Sinha, 'State Formation and Rajput Myth in Tribal Central India', *Man in India*, 42, 1, 1962, pp. 35–80. Similarly even the broad outlines of continuity and change in the agrarian economy following the transition to Maratha and British rule are as yet unclear. A reading of R. Jenkins, *Report on the Territories of the Raja of Nagpur*, Calcutta, 1827, (reprinted 1923) and J. Malcolm, *Memoirs of Malwa* (reprinted, New Delhi, 1970) brings out some contrasts in the pre-British revenue administration of the Narmada Valley with that of Malwa and the Nagpur Province.

[3] *Hoshangabad Settlement Report*, 1867, p. 45; (hereafter Settlement Reports are cited as *S.R.*)

thinly populated region.⁴ The large arrears of uncollected revenue, the possibility of large-scale desertions by peasants, and the effects of the agrarian depression in the NWP were among the factors which led to a twenty-year settlement on a relatively milder demand.⁵

It was in the 1860s that the government, in a major policy decision for the recently formed Central Provinces, announced a new land revenue settlement. The malguzars were declared to possess full proprietary rights over their estates on which the revenue was to remain fixed for the settlement period of thirty years. There was an undercurrent of official opinion, however, which held that there were many peasants with long-standing rights which would be eroded by a zamindari settlement. The grant of proprietary rights to the malguzars was therefore accompanied—no doubt a concession to this view—by varying degrees of legal protection to certain categories of tenants.⁶ The net result as it emerged was the categorisation of the peasants into those having rights of 'malik makbuza', i.e., 'absolute occupancy' and 'occupancy.' Those without any claims to tenancy protection were classified as 'Ordinary' tenants.⁷

---

⁴ Evidence of over-assessment of revenue in the first decades of British rule is found in Sleeman's correspondence and in the Jabalpur Divisional Records, both of which are preserved in the Central Records Room of the Government of Madhya Pradesh, Nagpur. Some details of over-assessment are also to be found in *Hoshangabad S.R.*, 1867, *Jabalpur S.R.*, 1869, and *Narsinghpur S.R.*, 1866.

⁵ For arrears of revenue see *Hoshangabad S.R.*, 1867, p. 46. For desertions by peasants see for instance, Sleeman's correspondence, *op. cit.*, volume for the year 1846, 16 July 1846 (this correspondence is unindexed), and, Jabalpur Divisional Records, *op. cit.*, Vol. 121, No. 60, 6 October 1832. The depression in the NWP is discussed in Asiya Siddiqi, *Agrarian Change in a Northern Indian State*, London, 1973, pp. 169–78.

⁶ For details of the controversy regarding tenurial rights and the position of the peasant via-a-via malguzar see for instance, Foreign Dept. Rev,, Nos. 39, 40, August 1864 and Nos. 41–44, April 1868, NAI. Baden Powell referred to the malguzari settlement as one which 'combines features from Bengal with features from the North West.' *The Land Systems of British India*, London, 1892, Vol. II, p. 387.

⁷ The status of malik makbuza, in effect a plot proprietary right, was generally granted to those members of the proprietary family who preferred

The grant of proprietary rights to the malguzars has been termed by Peter Harnetty as being a 'revolutionary decision.'[8] His argument is that is one stroke the government converted the malguzar from a revenue farmer into a landlord with almost undisputed authority over his tenants whose customary rights were correspondingly reduced. According to Harnetty this policy was to have disastrous consequences and by the end of the nineteenth century large-scale alienation of land to moneylenders, a growing indebtedness and a steady erosion of peasant rights had become matters of almost continuous official concern.[9]

The precise effect of the grant of proprietary rights is a question intimately related to the larger issue of the status of the malguzar and his position in the agrarian structure under the Marathas and in the early British period. This problem being beyond the scope of this paper, a detailed examination of the effects of the grant of proprietary rights cannot be undertaken. For the sake of completeness, and as a prelude to a study of the landlords and peasants of the Narmada Valley, a few brief comments on the issues raised in Harnetty's paper are made. It should be pointed out that these cryptic comments do not constitute a critique of Harnetty's paper which deals with the Central Provinces as a whole. The remarks which follow merely seek

---

taking plots of land rather than a share in the village; former malguzars who had retained some land for their own use, and holders of *muafi* grants. The malik makbuza paid revenue to the government but no rents to the malguzar. The right of absolute occupancy was conferred at the time of the First Settlement (1860s) to selected tenants and could not be acquired later. This right was hereditary, transferable and ensured fixity of rent for the term of a settlement. The occupancy tenant was entitled to cultivate his plot at a rent to be fixed by a revenue officer and not the landlord. This tenure while heritable was not transferable except with the consent of the malguzar. The rents of the ordinary tenants could be increased at the discretion of the malguzars and they had no rights over the lands they cultivated. During most of the period under study they remained thus essentially as 'tenants at will.' C.P. Settlement Code, 1903, p. vii. See also, *Hoshangabad S.R.*, 1919, pp. 23–24.

[8] Peter Harnetty 'A Curious Exercise of Political Economy: Some Implications of British Land Revenue Policy in the Central Provinces', *South Asia*, No. 6, 19 December 1976, p. 16.

[9] *Ibid.*, pp. 22, 25, 32.

to raise alternative hypotheses as suggested by a study of the Narmada Valley.

By the end of the nineteenth century there was a unanimity of official opinion that the grant of proprietary rights had created a class of landlords who had not existed before. However, examination of evidence relating to the Narmada Valley during the 1820s suggests that the malguzar was considered as being more than a mere revenue farmer, whose sole role lay in depositing the revenue he collected from peasants in government treasuries for a commission.[10] During the 1829s the exigencies of collecting a highly pitched revenue often made district officials embark on a course of action aimed at restricting and eroding the influence of the malguzar. Thus, for example, the 'ryot's rent was fixed as completely as the revenue of the village' and the interest rates for seed grain loans (which the malguzars made to the poorer peasants) was fixed by revenue officers.[11] These, and other similar measures, were not infrequent up to the mid-thirties and it is possible that they did weaken the malguzars. Yet it is interesting to note the consistency with which the initiators of these measures were censored. The most striking evidence for this is found in an inspection note recorded by R.M. Bird in 1834.[12] Bird, known in the North Western Provinces as an advocate of restricting the rights of landlords over their tenants, however recommended, after his tour of the Narmada Valley and the district of Saugar, that official actions intervening between malguzars and tenants cease immediately and that

---

[10] For example there is considerable evidence documenting the dependence of the poorer peasants on the malguzars for seed grain. For instance, Jabalpur Divisional Records (Supplementary List), *op. cit.*, Vol. 28, No. 52, 22 September 1830. Similarly while describing conditions in Jabalpur, W.H. Sleeman rejected the view which saw the malguzar as 'merely as agent for the collection of payment of the government demand', *ibid.*, Vol. 28, No. 2, 22 October 1830.

[11] *Hoshangabad S.R..*, 1867, p. 30. See also, Robert Merttins Bird, *Note on the Saugar and Nerbudda Territories, 31 October 1834*, Nagpur, 1917, for a criticism of the revenue and civil administration in early British rule. Further details of attempts to erode the authority of the malguzars are to be found in Jabalpur Divisional Records, *op. cit.*, Vol. 58, No. 354, 1 April 1827, and Vol. 27, No. 96, 13 August 1830.

[12] Bird, *op. cit.*

MALGUZARS AND PEASANTS: THE NARMADA VALLEY / 307

malguzars be left free to make their own terms with the tenants.[13]

That the malguzar had greater pre-eminence in the agrarian structure than would appear if his role is seen as merely one of a revenue contractor prior to the grant of proprietary rights, is suggested by other evidence also. The economic strength and social influence of the malguzar, testified to by the settlement reports of the early sixties, were clearly the result of more than the few years which had passed since the grant of proprietary rights in 1860. Similarly the large home farms of the malguzars, known as *sir*, suggest a greater control over both resources and produce than would be expected of revenue farmers transformed by a legal fiat into landlords.

What is known about the agrarian structure of the Narmada Valley (and of the Central Provinces) is too limited to boast of many categorical statements. However, some evidence from the Narmada Valley does suggest that the grant of proprietary rights in 1860 was not as 'revolutionary'—either as a decision or in its consequences—as Harnetty has assumed it to be. The process by which the malguzars were to emerge as the dominant class of the countryside was underway by at least the thirties and forties. More intensive research into the Maratha and early British periods may in fact show that the malguzars had occupied a position of dominance from an even earlier period. In any case it appears unlikely that a situation in which 'the individuals who composed the village community' were 'too much at the mercy of the new proprietor' was solely the result of a legal fiat or even 'the natural result of the British system that always dealt with individuals.'[14]

---

[13] Features of Bird's career in the North Western Provinces are found in T.R. Metcalf, *Land, Landlords and the British Raj*, Berkeley, 1919, pp. 67–73. For a summary of his main recommendations, see, 'Agrarian Change in the Narmada Valley', *op. cit.*, p. 62.

[14] Harnetty, *op. cit.*, p. 26. My contention obviously is not that the grant of proprietary rights did not imply any redefinition of rights and selections in the agrarian structure. The changes which took place however have to be seen in conjunction with the patterns of economic change in this period rather than as a result of the changing social composition of the landlords or a charter of rights granted to them by the government.

The Wheat Boom

A study of the 'boom' in the cultivation and trade of wheat and the end of this boom is useful in providing the historical conjuncture within which this study of the landlords and peasantry is located. The second half of the nineteenth century saw the establishment of a trading and transport infrastructure in the valley. While this region had a history of wheat production for export stretching back to at least the seventeenth century, difficulties in navigating the Narmada and lack of fair weather roads were among the factors limiting expansion of trade.[15] The advent of railways brought about a number of dramatic changes which were to significantly affect the course of agrarian change. Apart from bringing about greater internal price integration and linking it to international markets, the railways were also instrumental in bringing about certain internal reorientations in the economy. Among these was a change in the direction of the valley's exports as the flow of wheat to Bombay became the main direction of its trade.[16]

Following the railways were large numbers of merchants and traders. Each railway station thus became a centre of trading operations leading to what one officer called a 'decentralization of trade.' In Hoshangabad by the end of the century there was 'hardly a village of any importance situated more than 12 miles from a railway station of which there are no less than 17 in the district.' At each railway station there would be grain dealers who 'establish themselves during the open season and make

[15] For wheat production for export in the seventeenth century see, Irfan Habib, *The Agrarian System of Mughal India*, Bombay, 1962, p. 73. Problems of transport in the early nineteenth century are mentioned in *Narsinghpur S.R.*, 1866, p. 3, and *Hoshangabad S.R.*, 1867, p. 3. For difficulties in navigating the Narmada, see Foreign Dept. (Pol.), No. 6, 15 March 1841, NAI. The lack of adequate means of transport and stable export markets often led to situations in which local markets were glutted by produce in years of a good harvest. Jabalpur Divisional Records, *op. cit.*, Vol. 27, No. 96, 13 August 1830.

[16] Evidence of lack of market integration before the railways is found in *Jabalpur S.R.*, 1886–94, p. 19, and *Narsinghpur S.R.*, 1866, p. 68. Among the reorientations caused by the railways was the decline of old towns and the emergence of new urban centres, *Hoshangabad S.R.*, 1891–98, p. 9.

their purchases.'[17] The railways, rising wheat prices and the fanning out of merchants and traders brought about striking increases in the cultivation and export of wheat.[18] The potential incomes to be derived from participation in the wheat trade led to what was called an increasing 'competition among exporters.'

Briefly, the mechanics of the trade operated through large firms who established agencies in the larger *mandis* or through smaller local firms who dealt directly with cultivators and in turn sold to the larger firms or consigned the wheat directly. Below these were the village money-lenders and malguzars who recovered wheat after harvests in return for seed grain and money loans made earlier.[19] Not all peasants, however, were in so precarious an economic position that they could not reap benefits from the 'wheat boom' in the nineteenth century and from the continued trade in grains in the twentieth century. In Jabalpur, the Settlement Officer in the 1890s found that a number of tenants

> sow their own seed, while a great many substantial cultivators sow their own seed in part, but borrow some from the malguzar merely to keep up this connection, so that in time of need they may have a *mahajan* ready at hand to assist them.[20]

J.B. Fuller, the Settlement Commissioner, described the operation of the wheat trade as follows:[21]

---

[17] *Hoshangabad S.R.*, 1891–98, p. 10.

[18] Between the 1830s and the 1890s cropped areas increased by about 30 per cent in Jabalpur and Hoshangabad. The area under wheat increased by 62 per cent in Jabalpur and 33 per cent in Hoshangabad. However, as Sumit Guha has argued, some caution is necessary while positing a causal relationship between higher prices or improved transport facilities and expansion of cultivation. See, 'Some Aspects of Agricultural Growth in Nineteenth Century India', in *Studies in History*, Vol. 4, No. 1, 1982. My assumption that the bulk of the expansion in cultivation took place after the railways may not stand ground with further research on the first half of the nineteenth century.

[19] *Memorandum on Rural Conditions and Agricultural Development in the C.P. and Berar*, Nagpur, 1926, p. 10.

[20] *Jabalpur S.R.*, 1886–94, p. 23.

[21] Revenue Dept. No. 212, 1887–88, pp. 14–15. Revenue compilations, 1861–1920, Madhya Pradesh Central Records Office (hereafter MPCRO). More details of the wheat trade and the efforts of the government to increase exports

Competition among exporters has driven them to use every means in their power to procure a rein on the produce and their agents pass from village to village offering money down to any tenant who will forestall his crop and sell it even when yet unsown. It is of course an object with these men to get the tenants into their power and the terms of their agreements are so cast as to bring this about. The stipulation is usually one not for delivery of a crop such as it may be, but for the delivery of a certain amount of produce. The failure of a crop to come up to expectation (and expectation at borrowing time is generally sanguine) will in this case almost certainly make the tenant a defaulter and generally involve him in embarrassment from which he can never manage to escape.

The years from 1880 to 1895 were almost invariably referred to as forming a period of a 'wheat boom.' To many this appeared to be a period of increasing prosperity, of a 'boom' in which all participated. In a stratified society the impact of such economic changes was inevitably differential. This was basically because the boom in the wheat trade in particular, and market forces in general, impinged on the peasantry after considerable modification by a range of intermediaries—the merchant, the malguzar and the moneylender, with these roles often combined in a single individual.

The 'boom' was the result of two exceptionally favourable circumstances. Climatically favourable conditions in the form of well-distributed rainfall combined with a *rising* export demand stimulated production of wheat to hitherto unprecedented levels. The peak years of the 'boom' were from 1888 to 1894, when both wheat acreages and exports touched record limits. From 1894 onwards a long cycle of good monsoons came to an end and the valley had to go through the traumas of either too much rainfall or too little.[22] Second, the cultivation of wheat in the virgin plains of Argentina and its availability at lower prices meant the collapse of the European market for Narmada wheat.[23]

---

are found in Selections from the records of the Government of India, No. CLX, *The Wheat Production and Trade of India*, 1879, pp. 3, 8, 82.

[22] *Jabalpur S.R.*, 1886–94, Settlement Commissioners Review, p. 14.

[23] *Report on the Railborne Trade of the Central Provinces*, 1894–95, Calcutta.

This expansion in wheat acreage or the 'wheat boom' had in it, however, inherent weaknesses. While the valley had always been known as a wheat producing centre, this crop was, in the traditional scheme of things, grown on the richest soils—generally in the rich black soil *haveli* tracts. The *haveli* had been, even by the mid-nineteenth century, almost fully cultivated.[24] The expansion in cultivation which took place in the second half of the nineteenth century was in the remoter regions where soils were thinner and less fertile. What constituted the 'boom' was the cultivation of wheat, under the stimulus of a rising demand, in these newly cleared areas or on plots of land of inferior soils which had previously been devoted to hardier though monetarily less lucrative crops. Yet such cultivation could only be sustained under optimum weather conditions and with well distributed rainfall. The eighties and early nineties had been fortunate, yet the bubble was bound to burst and in 1894 a cycle of good monsoons ended.[25] The following years were of famine whose effects were intensified by the recently concluded settlements which had levied a considerable enhancement of revenue.[26]

Professor Eric Stokes has presented the famine years as forming almost a chronological divide in the agrarian history of the valley. Thus the period upto the nineties is depicted as being one of promise with large wheat farms and mounting exports. The twentieth century, he argues, saw the abandoning of wheat monoculture in favour of mixed farming—a retrograde period compared to the earlier one which was a 'pioneer age.'[27]

---

[24] See for instance *Jabalpur S.R.*, 1886–94, p. 11.

[25] For over-cultivation of wheat see *Narsinghpur S.R.*, 1923–26, pp. 26, 38, and *Hoshangabad S.R.*, 1919, p. 19. One of the consequences of over-cultivation of wheat was a shortage of communal grazing and ristar lands. See *Narsinghpur S.R.*, 1923–26, p. 44.

[26] During the famines, a field for an independent study demonstrating the mutations in the economy under the impact of an ecological crisis, a striking change which occurred in the cropping pattern was a shift from *rabi* to *kharif* crops. See for instance Dept. of Rev. and Agri., Land Rev. Pros., No. 41, F. No. 62 of 1899, July 1899, pp. 2191, 2115, NAI.

[27] Eric Stokes, 'Peasants, Moneylenders and Colonial Rule, An Excursion

This thesis may be discarded as being too simplistic. It was argued earlier that wheat was being over-cultivated to the extent that it was often being grown on land not suited for its cultivation. A decline in wheat acreage could therefore in fact be a return to stabler conditions. More importantly Stokes' contention ignores the fact of market forces influencing acreage. Crops other than wheat—gram and til most prominently—were preferred in the twentieth century because it became more profitable to cultivate them. Wheat had enjoyed a price advantage over other crops in the late nineteenth century but this gap narrowed in the early twentieth century. Although wheat prices remained higher than those of gram yet the differential reduced to the extent that over a number of harvest cycles the cash returns from hardier crops like gram became higher. In other words such a reduction of the price differential implied that it was no longer worth the risk to grow wheat on soils better suited to hardier crops.[28]

The shift to other crops was not in any sense then *symptomatic* of a weakening of market links. Agricultural exports did not fall but what changed was their commodity composition. For instance in Narsinghpur the percentage of wheat to total exports fell from 58 per cent to 22 per cent between 1880 and 1924. In the same period the percentage of gram exports increased from 26 per cent to 56 per cent.[29] The end of the wheat boom and the famines were then in a sense no more than an episode in the agrarian history of the valley. The strong continuity between the period of the 'boom' and the twentieth century was a simple one—of market forces influencing cropping.

---

into Central India', pp. 254–64, in Eric Stokes, *The Peasant and the Raj*, Cambridge, 1979.

[28] Statistical calculations of prices of wheat, gram and til in the periods 1884–94 and 1910–19 show clearly that the price advantage of wheat declined considerably in the latter period. For details, see, 'Agrarian Change in the Narmada Valley', *op. cit.*, pp. 218–27.

[29] Calculated from data contained in *Narsinghpur S.R.*, 1923–26, p. 31.

The Malguzars

Inflation of Income

There may be different views on the original character of the malguzars of Narmada Valley and on the magnitude of the change caused by granting them proprietary rights. Yet by the middle of the nineteenth century their dominance and affluence was clearly visible. Established as a powerful rural gentry, their authority and status were not easily questioned. The house of a malguzar was frequently a structure standing 'well above the other buildings . . . often a handsome two storied building of brick and stone. Inside are larger courtyards, well stocked by cattle and surrounded by dwelling houses and granaries.' Almost all the malguzars of Hoshangabad district, observed its first settlement report, had either a horse or a cart and they all considered 'walking very derogatory to their dignity.'[30]

The economic power and pre-eminence of the malguzar was derived from three principal sources of income: (i) cultivating profits from the home farm or the *sir* lands; (ii) rental payments; and (iii) income from cesses levied on peasants for the use of waste lands, pastures and forests. This income was known as *siwai*.

The financial importance of these sources varied. In money terms the proceeds from *siwai* were much lower than incomes from *sir* cultivation or rental collections. The malguzars, however, enjoyed a broadbased dominance only because of their control over all these three factors. Thus although *siwai* income was relatively small, control over the forests and wastes surrounding the village was in the hands of the malguzar—a powerful tool for asserting his economic and social dominance over the peasants.[31]

---

[30] *Hoshangabad S.R.*, 1867, p. 141.

[31] Notwithstanding the fact that money incomes from *siwai* were much lower than that from rents or *sir*, malguzars realised the importance of the wastelands and forests. The use of these by the peasants and the rate of payment for this purpose was recorded for each village in the *Wajib-ul-arz*. Many malguzars' commercially exploited these forests, and by supplying

It is striking to see the increase in malguzari incomes over the course of a settlement, for although the revenue remained near constant, the rents and *siwai* increased. Moreover, and this is not visible in revenue statistics, incomes from *sir* cultivation mounted with increasing prices. The revenue on each malguzar was fixed at 50 per cent of his assets which were defined as being made up of rental collections, *siwai* income and the rental value ascribed to the *sir*. Ignoring for the moment *sir* cultivation, it is still striking to see the increases in the malguzari incomes in the three decades between the first and second settlements of these districts (see Table 1).[32]

TABLE 1

Siwai and Rental Income of Malguzars and the Revenue Demand
(in Rs Lakhs)

District	Year	Siwai	Rent	Revenue	Revenue as per cent of Rent + Siwai
Jabalpur	1869	0.27	8.86	6.10	66
	1894	0.55	13.08	6.08	44
Hoshangabad	1865	0.16	7.40	4.56	60
	1896	0.54	11.39	4.69	39
Narsinghpur	1866	0.15	6.64	4.29	63
	1894	0.21	9.10	4.30	46

Thus while land revenue increased only marginally, on account of new villages founded, the income of the malguzars increased in this period by about 50 per cent. This excluded, as mentioned earlier, the proceeds from *sir* cultivation.

The rental income of the malguzars and the differential impact of rental demands shall be dealt with in a later section. An

---

timber to railway companies accrued large profits. Dept. of Rev. and Agri., Rev. Pros., 15–18, F. No. 154, p. 151, 1890, NAI.

[32] *Hoshangabad S.R.*, 1891–98, statement VI, VII and XII: *Jabalpur S.R.*, 1886–94, statement VII-XI; and *Narsinghpur S.R.*, 1889–94, statement VI-X.

important aspect of the malguzar's assets, that of income from cultivation of his home farm, which does not figure except marginally in the revenue statistics, will now be examined.

Expansion of the Sir

A feature of the agrarian economy of the Narmada Valley which drew much official comment was the large areas of land held by the malguzars cultivated by hired labour. The history of these farms in the earlier decades of the nineteenth century is obscure but by the 1860s home farms occupied between 15–20 per cent of the total cultivated area.[33] These farms were often very large indeed and for example in Hoshangabad district, Tulsiram Shukal of Harda Pergana had 'no less than 150 ploughs and occupies actually 4,500 acres of cultivated land.'[34] This was no doubt an exception but in general *sir* holdings were much larger than peasant farms. In Narsinghpur in the sixties, for instance, the average size of a malguzar's *sir* was approximately 67 acres while the average size of a peasant holding was 17 acres. In Hoshangabad in about the same period the average *sir* holding was as much as 147 acres.[35]

Despite this large amount of land held directly by the malguzars, the *sirs* increased appreciably over the rest of the nineteenth century in the valley. Thus in Hoshangabad while the total occupied area increased by 18 per cent between the settlements (1867 and 1891–98) the *sir* increased by as much as 33 per cent. In Jabalpur, in approximately the same period, the occupied area increased by 35 per cent yet the *sir* almost kept pace with this expansion showing an increase of 29 per cent. In Narsinghpur, while occupied areas increased by only 12.5 per cent between 1866 and 1889, the expansion in *sir* was as much as 23 per cent.[36]

---

[33] *Hoshangabad S.R.*, 1891–98, statement V, *Jabalpur S.R.*, 1886–94, pp. 48–50, *Narsinghpur S.R.*, 1891–94, statement V.
[34] *Hoshangabad S.R.*, 1866, p. 95.
[35] *Narsinghpur N.R.*, 1866, Appendix 15, *Hoshangabad S.R.*, 1866, p. 10.
[36] Abstracted from *Hoshangabad S.R.*, 1891–98, statement V; *Narsinghpur S.R.*, 1891–94, statement V; and *Jabalpur S.R.*, 1886–94, statement V.

Among the reasons for this increase in the *sirs* is that malguzars tended to increase them to prevent growth of occupancy rights. What was equally important perhaps, with cultivation on a large-scale becoming progressively more profitable due to the rising wheat prices, was that malguzars were induced into getting 'as much land as possible, rather than leave it with protected tenants with a light rent. Throughout the settlement operations nothing has been more marked than the desire of the landlords to get as large an area as possible recorded in their names.'[37]

Sir holdings were almost invariably of the best quality soil and situated in the most favourable positions in each village.[38] Yet despite this the rental value assigned to them was much lower than that of peasant holdings. Thus in Jabalpur ' . . . the home farm has invariably been valued at the rate employed for fixing rents. In fact . . . (in many groups) . . . the assessment placed on the home farm is soil class for soil class below the rate of the ordinary rental while in no group has that rate been exceeded'[39] Similarly in Hoshangabad district during the second settlement of the district:

> The valuation of the *sir* was done at village rates—which was selected with special reference to the existing rental of the protected tenants, and was considerably below the true value of the village lands, as shown by the payments of the ordinary tenants. The valuation of *sir* at village rates is therefore a very great concession to the malguzars . . . [40]

The value of this concession was realised and infringements on the principle of a lower assessment on *sir* lands were generally quickly highlighted.[41] It is easy to see that the expansion of

---

[37] *Hoshangabad S.R.*, 1891–98, p. 33.
[38] *Jabalpur S.R.*, 1869, p. 84.
[39] *Jabalpur S.R.*, 1886–94, p. 56.
[40] *Hoshangabad S.R.*, 1891–98, p. 56. The reason given by the government for this concession was that ' . . . cultivation of *sir* is frequently conducted with hired labour and cannot be so economically carried on as that of the ordinary tenants who till with their own hands.' Rev. Pros., 37 and 38, F. No. 209 lf 1895, p. 170, January 1896, NAI.
[41] For the malguzar's concern with the assessment of *sir* see, *Land Revenue*

the *sir*, and the absorption of the best soil lands into the *sir* had a direct bearing on the peasant economy. This process implied a correspondingly greater vulnerability of the poorer peasants to the vagaries of the monsoon since holdings of inferior soil were most susceptible to sharp falls in output in years of inadequate rainfall.[42] For those peasants not economically strong enough to resist a malguzar's encroachments, expansion of the *sir* meant a continuous loss of plots of land which had superior soil and being made to shift to the marginal lands.

Malguzars: Agriculturists and Non-Agriculturists

The rising income from rents, *siwai* and *sir* cultivation, made the malguzari estates keenly sought after. The Settlement Officer of Hoshangabad wrote in 1896: '... judged by their style of living there is probably no class in the provinces which has a higher standard of material prosperity than the malguzars of Hoshangabad district.'[43] Not very surprisingly then the demand for malguzaris increased and men of capital started buying out the malguzars. That became a matter of almost continuous official concern from the eighties onwards. 'The gift of full proprietary rights to the malguzars of these provinces had hardly been completed when the advisability of curtailing it began to be examined', wrote J.B. Fuller, the leading revenue authority and Settlement Commissioner of the Central Provinces.[44] In Hoshangabad district in 1888 it was seen that 'of 1577 mahals, 264 or 17 per cent transfers have taken place since settlement.' In Nar-

---

Administration in the Central Provinces of India, compiled by the Tenants and Landlords Association of Jabalpur, 1905, p 132, and Revenue and Scarcity, B. Pros., F. No. 229, 1907, 1909, NAI (petitions from the Malguzari Sabhas in Jabalpur, etc.). The under-assessment of *sir was an important issue elsewhere in North India also. Asiya Siddiqi points out that assessment of sir* at the same rate as that of other holdings thus placing the cultivating proprietors on an equal footing 'with peasants who, by custom, were their social inferiors', may have been one of the factors responsible for the revolt of 1857 in the North Western Provinces, Siddiqi, *op. cit.*, p. 109.

[42] This is discussed at some length in a later section.
[43] *Hoshangabad S.R.*, 1891–98, p. 34.
[44] Revenue compilations, Revenue Dept., No. 212, 1887–88, p. 115, MPCRO.

singhpur about 20 per cent of the malguzaris had changed hands in the two decades after its settlement in 1866. In the Murwara tahsil of the Jabalpur district 30 per cent of the malguzaris were transferred to money-lenders in the same period.[45]

Alienation of land by the malguzars was viewed by many officers within a paradigm created by the Punjab and Deccan experience.[46] 'Non-agricultural' castes were thus seen to be increasing their influence at the expense of the traditional rural patriarchs. But was this process of land transfers resulting in any change in the character and manner of functioning of the landlords of the Narmada Valley? Cutting through official rhetoric is difficult. Thus the Settlement Officer (1894) of Jabalpur felt that:

There are two sorts of Malguzars, viz., mahajans (commercial men) and men of the ordinary cultivating classes. Experience has shown that the latter who can most appropriately be termed 'land grabbers' are always inclined to expand their home farm by dispossessing tenants ... The mahajans on the other hand are always aiming at making rent enhancements, so much as, that the whole produce of a tenant's field often comes into their hands in lieu of seed grain advanced to him, interest on the advance and rent.[47]

Yet there were others who were not entirely convinced by

---

[45] *Ibid.*, p. 101.

[46] Harnetty, *op. cit.*, p. 24. There is little doubt that, as Harnetty argues, there was some change in the social composition of the malguzars in the last quarter of the nineteenth century. He does not, however, investigate whether this led to different kinds of economic behaviour. Similarly his assertion that under native rule 'the malguzar or patel was the natural protection of the ryots ... ' by dint of distributing seed grain, providing advances and extending cultivation, ignores the fact that most malguzars continued these practices throughout the nineteenth century and even later (see however fn 81 below).

J.B. Fuller, the leading revenue authority of the C.P. on the nineteenth century, was a major protagonist against the 'non-agriculturist' malguzar. Stokes, *op. cit.*, p. 259, writes that, 'just how much weight must be given to the anxiety of British officials like Fuller is problematical. Fuller joined readily in the chorus orchestrated by Denzil Ibbetson and other Punjab officials who formed the ruling official clique at Calcutta against the nefarious activities of the rural usurer and the danger he posed to continued peasant contentment with British rule.'

[47] *Jabalpur S.R.*, 1886–94, p. 23.

these stereotypes. The Second (1891–96) Settlement Officer of Hoshangabad wrote:

The distinction between the agriculturist and non-agriculturist community has largely disappeared in this district. There is hardly a single moneylender in the district who is not a landlord and many of the landlords combine the business of money and grain dealer with that of cultivator.[48]

Moreover it is clear that most of the malguzars used to lend out seed grain, especially to the smaller and poorer peasants. The with-holding, or, what was more frequent, the threat of with-holding, of these seed grain loans was a time-tested method of dealing with recalcitrant peasants apart from being a source of income in themselves. In Hoshangabad it was seen that ' . . . the malguzar often carries on a money-lending business under the name of a different firm and seeing that the rent law provides a special penalty for the non-payment of rent, his object is to credit all monies received to any debt other than rent, and thus keep the threat of eviction hanging over the tenant.'[49] In other words, any malguzar desiring to check the growth of occupancy rights and continually absorb the best available lands into his home farm would necessarily have to combine money-lending and grain-lending activities with 'traditional agricultural pursuits.'[50]

'Non-agriculturist' malguzars, furthermore, were by no means the creation of the rapid commercialisation which followed the advent of the railways. Official attention started focusing on the growth of these 'non-agriculturist' landlords only

[48] *Hoshangabad S.R.*, 1891–98, p. 23.
[49] Dept. of Rev. and Agri., Major Head IX. Revenue, Minor Head 28, A. Pros. for July 1894, Nos. 21–30, p. 23, MPCRO. See also for details of the practice, *ibid.*, A. Pros. for June 1895, Nos. 1–3. A. Pros. for September 1897, Nos. 1–2.
[50] For example in his evidence to the Banking Enquiry Commission Rai Sahib Pandit Laxminarayan, a prominent malguzar of Narsinghpur, stated that: 'only those malguzars and cultivators have accumulated wealth, who have combined money-lending with their agriculture.' His evidence also stated that practically all malguzars of Narsinghpur carried on money-lending. *C.P. Provincial Banking Inquiry Committee 1929–30*, Calcutta, 1930, Vol. 1, p. 45.

after the boom in wheat prices in the last quarter of the nineteenth century. However, malguzars of 'non-agriculturist' castes had existed in substantial numbers since the very beginning of British rule in this region. ' . . . it is a deplorable act', reads the settlement code (1894) of the Central Provinces

> that under the first 30 years of British administration in the Saugar and Nerbudda territories, a very large number of villages came into the hands of money-lenders . . . hence it came about that the grant of proprietary rights which was intended to benefit and agriculturist castes, was in great measure a concession to the money-lending interests.[51]

The official polemics against the 'non-agriculturist' malguzars, the dichotomies posited between them and the 'agriculturist malguzars', and the consequent clamour over alienation of land, overlooked other evidence which suggested that these distinctions were highly exaggerated. For example, of the eighty-one new villages founded in Hoshangabad district between the 1860s and 1890s, forty-one had been established by non-agriculturist castes, and another twenty-three by castes 'partly agriculturist and partly non-agriculturist.'[52] It thus appears that merchants and money-lenders were not simply extending their trading and money-lending businesses. They also provided incentives to cultivators to occupy and cultivate waste lands.[53] No doubt this was also a boost to their money-lending and trading operations, but, as we shall see, it is doubtful that they would have resisted the lure of profits through cultivation.

There is another indirect way of establishing the weakness of the 'agriculturist' vs. 'non-agriculturist' malguzar stereotype. To an officer like J.B. Fuller, who maintained a basic distinction between the two, there should have appeared to be two contrary processes at work. While on the one hand the number of trans-

[51] *C.P. Settlement Code op. cit.*, pp. iv-v.
[52] *Hoshangabad S.R.*, 1891–98, p. 34.
[53] This is obviously a simplification of the complex factors behind expansion of cultivation. See f.n. 18 also in this regard. For some details of the methods employed to break up waste and establish new villages, see *Hoshangabad S.R.*, 1891–98, p. 184. The increase in cultivated ares is discussed in 'Agrarian Change in the Narmada Valley', *op. cit.*, pp. 168–75.

fers to non-agriculturists was increasing, the home farm areas were also simultaneously increasing. Transfers to money-lenders does not appear to have led to an increase in absentee landlordism and it appears that the money-lenders or traders, apart from expanding their commercial operations, were also organising cultivation of large farms with the help of hired labour.

Closer examination of contemporary debates regarding 'agriculturist' and 'non-agriculturist' malguzars reveals that it was impossible to resolve the basic definitional problem. If at times the distinction between the two was made on a functional basis, on other occasions the criterion chosen was caste. In actual fact money-lending and trading operations were undertaken by malguzars of 'agriculturist' castes, and on the other hand, 'non-agriculturists' also maintained larger farms cultivated with hired labour. A dogmatic insistence on a basic difference between two types of malguzars often led to contradictory postures. A deputy commissioner of Narsinghpur, otherwise a lively opponent of the 'non-agriculturist' malguzar, was surprised to find that 'agriculturists' appear as mortgagees to as large an extent as they do.'[54]

Another officer, this time in Jabalpur, showed the inherent fallacies in making a bipolar division of malguzars:

The term 'agriculturist' embraces landlords who are actively engaged in agriculture as their ancestors were before them. Such men have usually a money-lending business of more or less importance and sometimes they trade as well. *But even if money-lending or trade has become the main source of income the malguzar remains an 'agriculturist.'*[55]

The strongest reason for rejecting an agriculturist vs. non-agriculturist stereotype is the magnitude of income possible for *sir* cultivation. Rough calculations show that income from *sir* would be larger than the rental and *siwai* income.[56] This, of

[54] Rev. and Scarcity Dept., Pros., for February 1913, Nos. 1–12, F. No. 22–1, p. 41, MPCRO.
[55] *Jabalpur S.R.*, 1907–12, p. 18.
[56] Estimates of net income per acre of wheat very widely depending on the quality of soil in the holding, whether labour is hired and on other expenses

course, is not to say that rental income was unimportant to malguzars. In absolute terms it was large enough. The point is that given a low density of population and the consequent low rents, malguzars would have to derive a large part of their income from cultivating profits. The exact proportion of *sir* income to that from trading, money-lending and rents would obviously vary from village to village. What needs to be emphasised is that the malguzars had a broadbased dominance. Control over rural credit, access to wastes and forests, and an increasing proportion of cultivated land, made the economic influence of the malguzar all pervasive. Yet, as will be seen in the following sections, this dominance was not uniform over the whole peasantry. There existed a strata of rich peasants who were able to insulate themselves from the malguzar's control and demands, and consequently participate more freely in market operations.

Change in the Form of Surplus Appropriation by Landlords From the 1890s

The inflation of the rental income of malguzars has already been noted. In later sections the differential incidence of rents on peasants and the landlords' ability to increase rents of poorer peasants will be documented. What is discussed now is the response of malguzars to a situation in which their control over the rental market became increasingly restricted.

---

of cultivation and obviously on price. Estimates in Hoshangabad in the 1890s placed net income per acre of wheat before paying rent at about Rs 10. See *Hoshangabad S.R.*, 1891–98, p. 29; *Hoshangabad S.R.*, 1913–18, Appendix, pp. 3–5; and Dept. of Rev. & Agri., Rev. Pros., F. No. 84 of 1913, October 1913, NAI. Assuming this figure for net income/acre to be a fair approximation of the average return per acre in *sir* cultivation, a rough estimate of the potential income from *sir* cultivation can be made. *Sir* area in Hoshangabad during the years (1891–96) when the settlement was being made was approximately 290,000 acres. This gives a potential *sir* income of Rs 2,900,000 (@ Rs 10 per acre of wheat). The total rent and *siwai* income of the malguzar in this period was estimated as being in the region of Rs 1,193,500. These calculations, although very crude, do suggest the relative magnitude of income from *sir* cultivation.

MALGUZARS AND PEASANTS: THE NARMADA VALLEY / 323

The high rental demand on the 'ordinary'—i.e., without occupancy rights—tenants became a matter of official concern at the time of the second settlement of these districts in the 1880s and 1890s. Efforts by Settlement Officers to persuade landlords to reduce these rents often met with opposition.[57] For instance many landlords of Jabalpur ' . . . considered the step an unwarranted interference with the sanctity of contract.'[58] Similarly in Hoshangabad, as the Settlement Officer wrote: ' . . . the malguzars regarded my proposals with considerable hesitation, feeling that they were being called upon to give up a portion of their rightful income in favour of the tenant.'[59] It was common practice for malguzars often to pitch rents to levels which they themselves had no real expectation of collecting. The reason obviously was to ensure that peasants had a perpetual arrears account on which they would have to keep paying interest. Thus in 1894 when the wheat crop partially failed in Hoshangabad, ' . . . the malguzars did not press their tenants but allowed the arrears to be held over for recovery with interest.'[60]

Rental reductions, however, were not necessarily against the interests of malguzars since a large recorded rental income led to a revenue demand 'based on assets which could not be realised in bad years.'[61] Many malguzars were quick to realise the viability of a high, unrealisable rental. The Settlement Officer of Jabalpur confessed that the plan of reducing ordinary rents had been

unwittingly suggested by Raja Seth Gokuldas who submitted a memorial to the local government in which it was stated the rental of his villages had risen to so high a pitch as to have become in a great part

---

[57] See for instance the covering letter of Settlement Commissioner to *Jabalpur S.R.,* 1891–96, p. 7.
[58] *Ibid.,* p. 8.
[59] *Hoshangabad S.R.,* 1891–98, p. 55.
[60] Dept. of Rev. AGri., Major Head IX, Minor Head 8, Pros. of June 1895, Nos. 1–3, p. 7, MPCRO. For a rental demand so highly inflated that it was incapable of realisation see also Dept. of Rev. and Agri., Rev. Pros. Nos. 53–55, F. No. 413 of 1906, October 1906, p. 2477 (Inception of Settlement Operations in Jabalpur District), NAI.
[61] *Hoshangabad S.R.,* 1891–98, p. 55.

unrealizable and should not be taken as the basis of assessment of revenue.'[62]

Similarly Rao Bahadur Ballabhdas stated in his evidence to the Famine Commission that the malguzars' opposition to rental reduction stemmed from 'misconception and foolishness.'[63]

The early twentieth century also witnessed a broadening of the scope of tenancy laws so as to include an element of protection for ordinary tenants. Malguzars could no longer raise rents of the ordinary tenants without permission of a revenue officer or a law suit. The malguzars' ability to control the rental market thus became restricted—though to a considerable extent, flowing from the recognition of many malguzars that high rents were also a liability, this restriction was voluntary. One of the clearest contrasts in the agrarian history of the valley is thus that between the movement of rent in the nineteenth century with that in the first quarter of the twentieth century. Rents remained largely stationary in the latter period and, for example, in three tahsils of Jabalpur district between 1912 and 1928 the rental income of malguzars increased only by a mere 5 per cent.[64]

The question is whether the malguzars were left with no other option but to reconcile themselves to a stationary (hence in real terms a declining) rental income. In other words were the poorer peasants, the most affected by the landlords' ability to increase rents, able to retain a larger portion of their produce? What happened in fact was that facing rigidities in the rental market the malguzars switched to other methods of surplus appropriation. This transformation took two main forms: levy of *nazarana* and sub-letting portions of the *sir*.

The switch to other forms of surplus appropriation finds its most striking example in Hoshangabad and Narsinghpur where the levy of *nazarana* became a substitute for rental enhancement. By 1913 the importance of *nazarana* was apparent to revenue authorities.

Owing partly to the cumbersome procedure involved in rental enhan-

---

[62] *Jabalpur S.R.*, 1891–96, p. 39.
[63] Evidence to the Indian Famine Commission, 1898, Appendix 189.
[64] *Jabalpur S.R.*, 1927, Appendix B.

cement, partly to the lure of cash in hand, but above all the possibilities of concealment offered, the practice of enhancements during the course of a settlement has given way almost entirely to the levy of *nazarana* on transfers.[65]

It is clear as to why the levy of *nazarana* assumed significance only after the settlement of the nineties:

The settlement taught the landlords a lesson. It made them realise that heavy rental enhancements such as they had forced on the ordinary tenants recoiled at their own heads at the next revision of settlement, when their home farms were, or might be, valued at the enhanced tenants' rate, and when government proceeded to absorb its legal share in the enhancements . . . [66]

Yet as economic conditions returned to normalcy after the famines there was bound to be a substantial difference between the rental payments actually received and what was expected in a situation of rising prices. The method adopted to fill this gap was that of *nazarana*. In Narsinghpur for instance: 'As a general rule the malguzars preferred to take their share of the rise in land values in the form of *nazarana* . . . [67]

The magnitude of the sums collected as *nazarana*, although it was agreed to be very large, is difficult to determine, '... cases in which the true consideration paid is concealed are probably more numerous than those in which it is acknowledged.'[68] In Hoshangabad in 1913 one estimate concluded that ' . . . the ordinary state of *nazarana* for wheat land of fair quality is between Rs 20–25 as acres, equivalent on a cautious estimate of 10–12 years rental.'[69]

There is evidence which clearly suggests that *nazarana*, like rent as will be seen in a later section, had a differential impact.[70] In a large number of cases *nazarana* was levied on a peasant

[65] Dept. of Rev. and Agri., Land Rev. Pros., October 1913, F. No. 84 of 1913, p. 1940, NAI.
[66] *Hoshangabad S.R.*, 1919, p. 33.
[67] *Narsinghpur S.R.*, 1923–26, p. 56.
[68] Dept. of Rev. and Agri., Rev. Pros., October 1913, p. 1940, NAI.
[69] *Ibid*.
[70] The differentials in the movement and incidence of rents is dealt with in a later section.

entering into a new lease. It was 'not necessarily the entry into relations for the first time with a new tenant which creates this opportunity. In the majority of cases, new leases are probably executed by *old* tenants in respect of land not comprised in their present tenancy.'[71] While this in cases would obviously be richer peasants leasing in more land, the bulk of these instances were in conditions of the periodical failure of rainfall or famines when 'a large number of cultivators abandon a portion, not the whole, of their tenancies, and take on a new lease when conditions improve.'[72]

Thus there was a close relation between famines or partial failures of rains, the entering into new leases and the levy of *nazarana*. 'The recurrence of periodical distress is a factor which favours the landlord in his levy of *nazarana*, since an incident of bad seasons is the temporary abatement of cultivation . . .'[73] An extension of this argument implies that peasants with the economic resources to sustain themselves through a bad season, without having to abandon part or the whole of their holding, would not have to face the malguzar's demand for *nazarana*. On the other hand the poorer peasant, more vulnerable as is argued later to the uncertainties of rainfall, would find himself in a situation of having to pay a substantial sum to the malguzar in order to recover his holding.[74]

If the levy of *nazarana* became an attractive alternative to rental enhancement, it was not the only option open to the malguzar to make up for an increasing inelasticity of rents in the twentieth century. As tenancy protection was gradually extended to the 'ordinary' tenants also, the practice of subleasing part of the *sir* farm increased. The peasant cultivating *sir* land would have a legal status no different from that of a tenant-at-will. Subletting of *sir* land increased all over the valley, though this is not to suggest, as Eric Stokes did, that self-cultivation by

---

[71] Dept. of Rev. and Agri., Land Rev. Pros., Part B, F. No. 48, No. 12, January 1916, p. 4, NAI. Emphasis in original.

[72] *Ibid.*

[73] *Ibid.*, p. 3.

[74] See the section on Ecological Factors and Peasant Differentiation.

malguzars gave way to subletting.[75] Table 2 summarises the increase in subletting of the *sir* between the second and third settlements.

TABLE 2

TOTAL SIR AREAS AND AREA LEASED ('000 ACRES)

District		Home Farm Area	Area Leased	Per cent of Sublet Area
Jabalpur	1895	216	14.8	6.8
	1928	230	29.0	12.6
Hoshangabad	1891	286	14.4	5.1
	1918	268	26.0	9.7
Narsinghpur	1891	119	19.8	16.6
	1926	135	26.9	19.8

SOURCE: Abstracted from *Hoshangabad Settlement Report*, 1919, Statement VI, *Narsinghpur Settlement Report*, 1923–26, and *Jabalpur Settlement Report*, 1927.

While subletting was not a new phenomenon what needs to be explained is the increase in its incidence. It is clear that this increase had little to do with any supposed decline in the advantages of wheat cultivation or in fact with any change in the economic conditions which made rentier landlordism more profitable than self-cultivation of the *sir*[76] Increase in subletting had the same general cause as the spread of *nazarana*. Restrictions on the malguzars control over the rental market made them sublet part of a huge home farm at high rents to subtenants who had neither the legal status nor the economic strength to resist future increases in rent.

[75] Stokes, *op. cit.*, p. 261.
[76] Stokes has suggested that the shift away from wheat cultivation implied a decrease in self-cultivation by the malguzars and the rise of rentier landlordism. However, it is seen that the increase in subletting is not large enough to justify such an assertion. Moreover, as was discussed earlier, a decline in the wheat acreage was not symptomatic of any major change but was rather only the response to a changed market situation.

The Peasants

Tenancy Categories and Peasant Differentiation

Absence of research precludes the making of any statement at present about the extent and nature of peasant differentiation in the early nineteenth century. The categorisation of the peasantry into tenants of differing legal status, undertaken by the revenue settlements of the 1860s, yields some information on the existence of clear economic divisions among the peasants. There is sufficient evidence to justify an assertion that there was a broad correspondence between superior economic position and the protected status of absolute occupancy or occupancy granted by the settlement authorities. The Settlement Officer of Narsinghpur, during its settlement in the 1860s, noted that,

... there are probably few districts in which hereditary cultivating tenures are now more welded into the agricultural system than in Narsinghpur. The same prosperity and quietude which has consolidated proprietary tenures has raised up a race of wealthy influential cultivators, who by the common feeling have a firm hold upon the land which they occupy.[77]

The figures in Table 3 show the clear distinction between the protected and unprotected tenants in terms of average size of holding.

While speaking of a correspondence between economic position and tenancy status, caution on two counts is necessary. First, ordinary tenancy land was in cases also held by protected tenants, malik makbuzas and even malguzars of other villages.[78] Second, and it is important to emphasise this, tenancy protection was often granted arbitrarily without due consideration to the actual economic status of the peasant. In such cases the loss of

[77] *Narsinghpur S.R.*, 1866, p. 92.
[78] Thus the settlement report of Hoshangabad 1891–98 says that, 'composite holdings of land in different rights are somewhat numerous, absolute occupancy tenants also possessing land in occupancy and ordinary rights' *Hoshangabad S.R.*, 1891–98, p. 39. Exactly how prevalent this phenomenon was is difficult to say at present and would require detailed research into the statistics of tenancy at the village level.

absolute occupancy rights was almost inevitable and took place most frequently in the case of tribal cultivators such as Gonds.[79]

TABLE 3

LANDHOLDING BY DIFFERENT CATEGORIES OF TENANTS

Hoshangabad District, 1865			Narsinghpur District, 1866		
Pargana	Average Area in Acres		Pargana	Average Area in Acres	
	Occupancy Cultivator	Tenant-at-Will		Occupancy Cultivator	Tenant-at-Will
Rajwara	26	14	Sreenagar	22	13
Sohagpur	24	12	Shahpur	20	14
Hoshangabad	27	15	Narsinghpur	21	14
Seoni	35	20	Buchy	17	12
Harda	44	27	Chawarpatha	16	10
Charwa	35	18			

SOURCE: *Hoshangabad Settlement Report*, 1865, p. 232, A; *Narsinghpur Settlement Report*, 1866, Appendix 15.

A fact which attracted considerable official attention was the decline in area held by absolute occupancy tenants in the three decades following the grant of this right in the first settlement.

---

[79] Loss of land and tenancy rights by tribals remained, throughout the nineteenth and twentieth centuries, an issue of official concern. This along with the alienation of proprietary estates by tribal malguzars formed the crux of the 'tribal problem' in the Narmada Valley as elsewhere in Central India. See W.V. Grigson, *The Aboriginal Problem in the C.P. and Berar*, Nagpur, 1944, especially pp. 21–23, 26–29 and 106–23. Alienation of tribal proprietors was one area in which the changing social composition of the malguzars had deeper sociological and economic consequences. The monetisation and commercialisation of the agrarian economy intensified such alienation of land, yet this process was part of an older historical process of marginalization of the tribal by an expanding peasant economy dating back to at least the seventeenth century which saw the coming of Hindu peasant castes into the valley to large numbers. The situation of the tribal cannot be understood in isolation from the peasant matrix they lived and functioned in, yet their moving and ultimately tragic history demands separate treatment.

Absolute occupancy area declined in this period by about 20 per cent in Hoshangabad, 23 per cent in Jabalpur and 16 per cent in Narsinghpur.[80] What explains this decline is the hostility of malguzars to protected status and fact that in some cases absolute occupancy tenants were without the economic strength to retain this right over a period of time. In Narsinghpur, for example, this fall in absolute occupancy area was thus explained: 'In the vast majority of cases the tenants appear to have become involved in debt and to have transferred their rights and surrendered their land to the *malguzar*.'[81] Yet the bulk of absolute occupancy land stayed intact. Clearly then many of those holding the right of absolute occupancy were peasants economically viable enough to resist the malguzars' attempts to eject them.

Occupancy tenant area increased dramatically after the settlement of the 1860s and, towards the turn of the century, formed approximately 25 per cent of the total cultivated area in the valley.[82] While till 1883 occupancy rights over a holding could be acquired by proving twelve years continuous cultivation, acquisition of occupancy status was rarely so simple.[83] Malguzars frequently pre-empted the possibility of a peasant gaining occupancy status by evicting him after ten or eleven years.[84]

There were several other ways by which malguzars could prevent growth of occupancy rights or erode existing rights. Mention was made earlier of one such method employed by those malguzars who also carried out money-lending operations. Thus all rental payments received would in fact not be recorded as rent but rather as interest payments or repayments

---

[80] Abstracted from data found in *Jabalpur S.R.*, 1886–94, statement V; *Hoshangabad S.R.*, 1891–98, statement V; and *Narsinghpur S.R.*, 1889–94, statement V.

[81] Dept. of Rev. and Agri., Rev. Pros., 25 and 26, F. No. 191, July 1893, NAI.

[82] See f.n. 80 above.

[83] For official dissatisfaction with the 1883 Tenancy Act, see Dept. of Rev. and Agri., Rev. Pros., 14 and 15 August 1896, NAI.

[84] There is a great deal of evidence documenting the efforts of malguzars to decrease occupancy area and their general antagonism to the protected tenants. See *Report on the Land Revenue Administration of the Central Provinces* (henceforth *Revenue Administration Report*), 1870—71, p. 72; 1872–73, p. 40; 1891–92, p. 21.

of loans. In this way a 'threat of eviction hanging over the tenant' would effectively rule out any moves towards protected status.[85] Another potent instrument was the seed grain loan on which the poorer peasants depended heavily. One officer frequently found:[86]

... occupancy tenants who had left their village and gone to a village close by. There is no reason for a man behaving thus. On enquiry I have always found that grain has been withheld and the ryot has been subjected to those petty annoyances which a malguzar knows so well how to inflict.

Yet another method to ensure the perpetual dependence of the poorer peasants, which came to the notice of the revenue authorities, was by manipulation of the *kists* in which payments were made. Revenue and rents were generally payable in two *kists* every year—one for the *kharif* and the other for the *rabi* crop. Legally a malguzar could not demand from his tenants a larger proportion of the rent in a particular *kist* than the proportion of revenue paid in that *kist*. In Narsinghpur district an officer found that it was:

... common practice for the malguzars to realise rents in two equal instalments quite irrespective of the proportions of the revenue *kists*. I have come across numerous villages in which there is practically no *kharif* cultivation, and the first revenue *kist* of which is only 2 annas in the rupee, where nevertheless the tenants two months ago had paid half the rent, generally having borrowed for the purpose. I have found moreover the malguzars commonly charge three months interest on a first *kist* left unpaid.[87]

In brief, acquiring or retaining occupancy status was perhaps rarely as easy as a reading of the legal provisions would suggest. Economic strength was almost invariably a necessary condition

---

[85] See f.n. 60 above.
[86] *Revenue Administration Report*, 1873–74, p. 13.
[87] Rev. and Agri. Dept., Major Head IX, Minor Head No. 43, A. Pros. for April 1892, Nos. 1 and 2, pp. 1–5, MPCRO. For an illuminating discussion of the disjunction between the peasants' harvest calendar and the timing of the *qists* in which rents were paid in Gorakhpur district of the United Provinces, see Shahid Amin, *Sugarcane and Sugar in Gorakhpur*, New Delhi, 1984.

for protected status and categories of tenancy were not simple legalistic abstractions.

Viewed dynamically, the attempts to acquire tenancy protection and the fluctuations in area held by different categories of tenants represents a continuous struggle over rent. To increase rent, the malguzars would act in such a way so as to erode the legal protection of tenants. On the other hand the richer and more substantial peasants would seek to defend or acquire protected status so as to better resist the demands of the malguzars. The bulk of the peasantry, with small holdings and meagre resources, remained at the mercy of the malguzars. To them legal categorisation meant little and they were possibly the vast numbers which make up the statistics of decline in absolute occupancy area. That such was indeed the case is suggested by the differentials in the rental payments made to the malguzars.

Rents: Their Differential Incidence

The three decades between the first and second settlements of the valley saw a differentiated movement of rents. Rents paid by the protected tenants increased only marginally while rents of the ordinary tenants increased substantially (see Table 4).

TABLE 4

PER CENT INCREASE IN RENTS PER ACRE BETWEEN THE FIRST AND SECOND SETTLEMENTS (APPROXIMATELY 1860s TO 1890s)

District	Absolute Occupancy	Occupancy	Ordinary
Jabalpur	4	5	5
Narsinghpur	3	2	62
Hoshangabad	4	6	98

SOURCE: Calculated from *Hoshangabad Settlement Record*, 1891–98, p. 38; *Jabalpur Settlement Record*, 1886–94, p. 5; *Narsinghpur Settlement Record*, 1891–94, p. 3.

For Jabalpur and to an extent Narsinghpur, district-level

averages obscure the extent of the increase.[88] A more accurate picture is obtained if movement of rents in individual assessment groups of these districts are examined (see Table 5). The corresponding increase in Hoshangabad may be gauged from the fact that even the district average shows a near 100 per cent increase.

TABLE 5

PERCENTAGE INCREASE IN ORDINARY RENTS BETWEEN THE FIRST SETTLEMENT AND THE SECOND SETTLEMENTS

Assessment Group	% Increase	Assessment Group	% Increase
JABALPUR DISTRICT		NARSINGHPUR DISTRICT	
Mangela	54	Mugli	97
Northern Garha	49	Jhansighat	70
Patan	104	Chhindwara	89
Shahpura	85	Shrinagar	57
Usna	79	Kaveri Kandeli	142
Pariyal	64	Supla Themi	154
Singaldip	68	Hiranpur Sankal	70
Rangrah	62	Chawarpatha	64
Katangi	62		
Sadar	70		
Lanckana	54		
Majholi	73		
Bachiya	72		

SOURCE: *Jabalpur Settlement Record*, 1886–94, pp. 5, 6; *Narsinghpur Settlement Record*, 1889–94, Statement VI.

[88] It may legitimately be queried as to why district-level averages in Jabalpur show *no* appreciable rise in the ordinary rates. The reason appears to be the prevalence of rental payments in kind over a large part of the district till the end of the nineteenth century when the operations of the second settlement included the conversion of grain rents into cash rents. Payments in kind obscured the real incidence of rent because 'figures are obscured by the extremely perfunctory manner in which patwaris record grain payments and habitually underestimate them.' Dept. of Rev. and Agri., Rev. Pros., Nos. 53–55, F. No. 413 of 1906, October 1906, p. 2474 (Inception of Settlement Operations in Jabalpur), NAI.

Merely considering the incidence of rent in money terms is, however, as one officer wrote, 'entirely misleading in regard to ordinary ryots.'[89] This was essentially due to the fact that lands cultivated by these peasants, a point to be discussed in greater detail later, was generally of inferior soil. In Jabalpur for instance, '. . . of the area held by ordinary ryots at the end of the First Settlement only about a third was identical with the areas held in this tenure at the beginning of the First Settlement, the rest consisting of inferior land broken from waste during its currency.'[90] In other words comparison of ordinary rents at the close of the settlement with those paid at the beginning was misleading because the quality of land held in this right had changed dramatically in this thirty-year period. The real increase in the incidence of rents on ordinary tenants was thus much higher.

This ability to increase rents of legally unprotected peasants explains the general hostility of the malguzars towards any attempt to acquire protected status. 'Except in the rare cases where peculiar ties unite the landowner with his tenants, occupancy tenants are looked on as fair game, or rather as vermin, which it is a duty and privilege to exterminate.'[91] The differential incidence and movement of rent shows peasants with tenancy protection retaining a larger proportion of their surplus than the tenants-at-will. The fact that substantial numbers of peasants were able to retain and take advantage of their protected status in the face of the malguzars' hostility suggests that their superior economic position of original tenancy categorisation was only a recognition of economic differentials among the peasantry. Over time it became a factor reinforcing and intensifying such differentiation.

The dominance of the malguzar was therefore not uniform over all the peasantry. A strata of peasants were able to preserve a greater degree of autonomy from the malguzars. This fact is also suggested by evidence from the second decade of the twen-

---

[89] *Jabalpur S.R.*, 1886–94, Settlement Commissioner's review, p. 5.
[90] Ibid.
[91] *Revenue Administration Report*, 1872–73, p. 18.

tieth century pertaining to the structure of landholding and the differing economic status of peasants. For example in the 1920s it was found that in Narsinghpur 9 per cent of the tenants were 'malguzars who also hold tenancy land; or very big tenants of a financial status equal to that of the malguzars.' Another 20 per cent were 'big tenants considerably above the average with land in more than one village.'[92] Similarly in Hoshangabad almost 5 per cent of the tenants were classified as being 'roughly capitalists' with another 34 per cent 'prosperous tenants with substantial holdings.'[93]

Village studies undertaken in Jabalpur district by the Provincial Banking Enquiry Committee (1927–30) showed a similar skewed land distribution. In Karmetha Village of Jabalpur Tahsil, more than 50 per cent of the holdings were less than five acres and about 20 per cent more than twenty-five acres.[94] In Village Lakhakhera of Katni-Murwara Tahsil, 33 per cent of the tenants held less than five acres but on the other hand about 15 per cent of the tenants held between 25–50 acres of land and about 7 per cent more than fifty acres.[95]

Ecological Factors and Peasant Differentiation

Rental payments and the extent to which a peasant could transform his gross surplus into net income was not the only factor reinforcing peasant differentiations. The growth of a commodity market, following the advent of the railways, became a major factor intensifying rural inequalities to extents not previously reached. As long as produce had glutted local markets in years

---

[92] *Narsinghpur S.R.*, 1923–26, p. 61.
[93] *Hoshangabad S.R.*, 1913–18, p. 29.
[94] *The Central Provinces Provincial Banking Enquiry Committee*, 1929–30, Vol. II, Calcutta, 1930, Appendix E, p. 217.
[95] *Ibid.*, p. 227. This picture of a highly skewed distribution is confirmed by a number of other sources most notably, the annexures to the settlement reports which contain detailed statistics for each assessment group. See also *Memorandum on Rural Conditions and Agricultural Development in the C.P. and Berar*, Nagpur, 1926, p. 3, for statistics relating to the structure of landholding in parts of the valley.

of good harvests in the absence of a steady external demand, there were well defined limits to the profits the richer elements of the agrarian society could accumulate. This changed once the valley established a link with external markets through the export of its principal commodity—wheat. The export demand for wheat led large numbers of merchants and agents of mercantile firms to establish themselves in the grain dealing centres and even go from village to village to hypothecate the crop of any peasant prepared to accept the cash advance offered. This growing commercialisation of agriculture intensified rural inequalities to extents much greater than had been reached in the period before the railways or the boom in production of wheat. What demands closer attention, however, is how these institutional and economic changes in turn reinforced 'traditional' causes of peasant differentiation—rainfall, local price fluctuations and the structure of land-holding in terms of the quality of soil held.

A near constant refrain, of even the most casual traveller, had been comments on the rich black soil of the valley. Charles Elliot describing the black soil of the Hoshangabad district said that 'this is the only soil in the world which will bear wheat crops for forty years in succession without a fallow and without manure.'[96] Correspondingly the view that the Narmada Valley was secure from famine was propagated with near equal vigour.[97] The fallacies in both these views became apparent by the beginning of the twentieth century after the disastrous famines of the late nineties.

There was, for one, considerable heterogeneity in the soils of the valley. In Jabalpur, for instance, in the first decade of the twentieth century it was found that the richest black soils—*Kaber-II, Mund-I and Mund-II*—made up only 31 per cent, and the medium soils—*Demotha and Sehra*—another 36 per cent of the total cultivated area. The poorer soils *Paturua, Bhatua* and *Barra* comprised 32 per cent of the cultivated area.[98] The richest

---

[96] *Hoshangabad S.R.*, 1865, p. 6.

[97] See for instance ibid., p. 5, and *Narsinghpur S.R.*, 1866, pp. 21, 48.

[98] *Jabalpur S.R.*, 1907–12, p. 5. The names of soils mentioned were used by settlement authorities in a graded scale.

soils, as will be discussed later, were able to yield fairly good outputs except in years when the rains failed completely. The continued high yields from these rich soils even in years of a bad (as opposed to a complete failure of rains) monsoon inflated district averages of crop yield and thus was the main factor building up the myth of the 'security' of the valley.

Revenue officials, in their estimates of output per acre of an 'average' peasant farm, used figures of yield valid only for years with exceptionally favourable monsoons. For instance, the estimate of an average peasant budget of Hoshangabad district in the last decade of the nineteenth century assumed output of wheat to be approximately 620 lbs. to an acre for a farm of thirty acres. However in years when the rains failed or were inadequate this figure would have to be scaled down differentially for different types of soils.[99]

What was crucial in determining the outturn of a crop was the distribution of rainfall over the season rather than its total volume. For a crop like wheat, rains in September made all the difference:[100]

The distribution of the rainfall is in fact of infinitely greater importance than its volume in inches; in June, July and August nothing matters so long as the rainfall is enough to soften the ground thoroughly and is not so heavy nor so continuous that ploughing is prevented. In September and October a heavy fall or two is required, followed by light sowing showers—the *Mahavat* on which the cultivator rests all his hopes.

The importance of the September showers however varied from soil to soil. Crops on the richer soils with high capillarity and greater retaining power of moisture would not suffer unduly in the event of a failure of the September showers. Inferior soils, however, in this situation would suffer heavy losses. Evidence from Narsinghpur shows that the outturn of superior soils would fall by only about 100 lbs. per acre when the September rains were inadequate or failed partially. For the inferior soils, however, the difference would be as high as 300 lbs. per acre.

[99] *Hoshangabad S.R.*, 1891–98, p. 29.
[100] *Hoshangabad S.R.*, 1919, p. 14.

The late monsoon was notoriously erratic. Figures collected in the Hoshangabad district showed that the late monsoon was scarce enough to effect the germination of seed every fourth year on an average and there was a near failure of the cold weather rains every second year.[101] Similarly in Narsinghpur ' ... from 1894–95 to 1920–21, the late monsoon has been defective in nearly half the years, to a greater or lesser degree.'[102]

The continued high outturn from the superior soils often obscured the actual impact of a failure of the September rains. Except in those years in which there was a complete failure of the monsoon, the outturn from the rich soils inflated district level indicators of output. Thus even at the height of the famine of the 1890s, district outturns in Narsinghpur did not fall below an 8 anna crop (when an average crop was treated as equal to 13 annas). 'This remarkable result, is due to the security from crop failure of the best soils ... ' wrote the Settlement Officer.[103]

This impact of the monsoon did not operate in an economic vacuum. Inadequate rains and consequently reduced outputs would lead to higher prices. But an inadequate monsoon had an uneven impact. The superior soils suffered from a relatively marginal decline in output while the effect on the poorer soils was more substantial. An increase in prices at this time however meant that the cultivator of superior soils could have an even higher income than in years when rains were adequate, his output higher but prices lower. For those who cultivated poorer soils, even a partial failure of rains could be disastrous as illustrated in the following example from Narsinghpur.[104]

	Year of Adequate Rains	Year of Inadequate Rains
Outturn from superior soils	750 lbs./acre	650 lbs./acre
Outturn from inferior soils	650 lbs./acre	350 lbs./acre

[101] Ibid.
[102] Narsinghpur S.R., 1923–26, p. 10.
[103] Ibid.
[104] Ibid., p. 69.

	Year of Adequate Rains	Year of Inadequate Rains
Price of wheat	16 lbs./Re	12 lbs./Re
Monetary value per acre from superior soil	$\dfrac{490^*}{16} = 31\text{--}10\text{--}0$	$\dfrac{390^*}{12} = 32\text{--}8\text{--}0$
Monetary value per acre from inferior soil	$\dfrac{390^*}{16} = 24\text{--}6\text{--}0$	$\dfrac{90^*}{12} = 7\text{--}8\text{--}0$

* From all the outturns a figure of 260 lbs. was deducted to take care of (a) 130 lbs./acre for expenses of production and (b) 130 lbs./acre for rent and seed for the next season.

There was in effect a tendency inbuilt into the agrarian economy which operated to increase rural inequalities. The erratic nature of the monsoon cycle implied that peasants cultivating inferior soils would almost inevitably be forced into the clutches of a creditor to continue production. Obviously such a process would have operated in earlier periods also, yet it is easy to see that the rapid commercialisation of agriculture in the second half of the nineteenth century would impart to this process a sharper intensity. The 'wheat boom', as seen earlier, had two aspects to it. The first of these was an increase in wheat production and acreage with the spread of cultivation to the thinner, more inferior soils with greater vulnerability to the notoriously unreliable late monsoon. Second, there was a strong tendency—represented by the fanning out of merchants and the growing interest of many malguzars in the grain trade—to get control over the produce of the peasant by a system of hypothecation and cash advances. In such a situation failure of rains and a bad harvest would therefore push the peasant into a nexus of indebtedness from which it would be difficult to escape.

Both these factors impinged more on the poorer peasants cultivating smaller holdings made up of inferior soils. The Settlement Officer of Narsinghpur (1923–26) summed up the position for his district as follows: 'The ordinary tenancy land at the time of the First Settlement was, on the whole, considerably inferior to the land held by the protected tenants, and during the next thirty years its average quality still further deteriorated.'[105] Similarly in Hoshangabad in the second decade of the twentieth century the soil factor value of *sir* and 'absolute occupancy' land was 25, of 'occupancy' land 23, while that of 'ordinary tenancy' land only 18.[106]

Thus if the holdings of the poorer peasants were the ones which became progressively the most vulnerable to rainfall, the expanding credit nexus also increasingly involved them in ties of dependency on the malguzar, the money-lender or the trader. It is possible to assert then that factors intensifying rural inequality tended to reinforce one another. Peasants who cultivated the worst soils also paid the highest rents, were the ones most adversely affected by shortfalls in rainfall, lost the greatest due to seasonal price fluctuations, remained without any legal rights or tenancy protection, and were the ones most dependent on the loans and credit of the malguzar or trader-money-lender.

CONCLUSION

Beginning from the 1850s and 1860s a cluster of economic and institutional changes had a comprehensive effect on the agrarian economy of the valley. The inability of the market forces in colonial conditions to change technological patterns of production and hence overcome constraints on production does not fall within the scope of this paper. The focus of this article has been to empirically detain characteristics of the malguzars and peasants and document some changes which occurred in their relations.[107] Intrinsic to this discussion was the thesis that if viewed

[105] *Ibid.*, p. 64.
[106] To determine the rents which different kinds of lands were to pay, the settlement authorities devised an elaborate method whereby a plot land was assigned a number of 'soil units' to show its relative quality. The soil factor value thus was based on a number of criteria such as depth and quality of the soil. The settlement code defined this operation as follows: 'It suffices to assume that an acre of a given class of soil contains a certain arbitrary number of soil units, and then to slate the relative value of other classes in terms of the number of soil units they contain per acre when compared with the class adopted as the standard or starting point.' *The C.P. Settlement Code*, p. 13.
[107] For a discussion on the technological and ecological constraints on

in abstract terms, the question of landlord-peasant relations may be reduced to one of control over different markets and forces of production. The malguzar of the Narmada Valley exercised a broadbased dominance derived from control over the rental, produce, land and credit markets. It has not been possible to pay more than cursory attention to the nexus of credit relations and the nature of the credit market. The commodity markets similarly were almost entirely ignored. Yet the discussion of the rental and lease markets illustrates the position of the Narmada Valley malguzar and the corresponding relations of subordination of much of the peasantry.

This broadbased dominance of the malguzar was not derived solely from a legal fiat granting proprietary right and the right to collect and fix rents. It was the varying degrees of control exercised by the malguzar over different markets and factors of production which subordinated the bulk of the peasantry and implied that the peasantry's links with the market were themselves dominated and mediated by the malguzar. Such control over different markets and factors of production implied that in situations in which it became difficult to exploit one such market the malguzar could still maintain his levels of income by intensifying surplus appropriation from other markets. A conceptual scheme described by an economist analysing conditions of present-day agricultural production illustrates this point.[108]

Markets become interlocked through price and non-price links given that market and social power is vested in the dominant rural classes and that the dominant party often combines multiple functions thus enjoying a superior position simultaneously in a number of markets . . . Such interlocking of markets increases the exploitative power of the stronger because, while there could be limits to exploitation in any one market due to traditions or conventions—or due to economic factors—the interpenetration of markets allows them to disperse exploitation over the different markets and to phase out exploitation over time as well . . . Thus the ability of the combined functionary to

---

production and how they operated throughout the period of study see, 'Agrarian Change in the Narmada Valley', pp. 10–36.

[108] Krishna Bhardwaj, *Production Conditions in Indian Agriculture*, Cambridge, 1974, pp. 4–5.

exploit in interlocked markets is more than what he could achieve operating in any one individual market.

Thus it was demonstrated how the malguzars dextrously switched to collecting *nazarana* or subletting portions of the *sir* to face a situation in which rigidities emerged in the rental market and as the ability to increase rents gradually declined. It is clear that malguzars understood such interlocking of markets and knew how to manipulate them to their best advantage. 'Some malguzars have reduced the rent to get a bigger *nazarana*', declared one officer.[109] Such a conceptualisation of a broadbased dominance is perhaps of more utility than an enquiry into the character of the malguzar in terms of their being 'agricultural' or 'non-agricultural.' Eric Stokes, while not falling into an agricultural/non-agricultural stereotype trap, nevertheless maintained that the malguzari system was 'an artificially created system of village landlords . . . artificial because the economic conditions for a true landlord rent had not emerged. . . '[110] Such a restricted definition rules out further enquiry along the lines of subordination-domination over different markets and factors of production. Similarly the controversy regarding the 'agriculturist' and 'non-agriculturist' landlords was viewed earlier within this perspective rather than simply one of loss of land by the traditional patriarchs to the scheming banias under pressure of a high land revenue demand. 'There is hardly a single moneylender', wrote an officer in Hoshangabad district, 'who is not a landlord and many of the landlords, even of agricultural castes, combine the business of a money and grain dealer with that of cultivator.'[111]

It was argued that the dominance of the malguzar was not uniform over all the peasantry. There existed a strata of rich and substantial peasants whose status and economic position often rivalled that of the malguzars. By virtue of their economic strength these peasants were able to insulate themselves from

[109] *Central Provinces Legislative Council Debates*, 1917, p. 16, Speech of J.F. Dyer.
[110] Stokes, *op. cit.*, p. 255.
[111] *Hoshangabad S.R.*, 1891–98, p. 23.

the malguzars' dominance over rents, credit and marketing operations.[112] The bulk of the peasantry was, however, subordinate to the malguzar and it is striking to see that individual aspects of this subordination reinforced each other leading to a situation in which the domination of the malguzar would be virtually impossible to shake off. In much the same way as the domination of the malguzar rested on and was reinforced by control over a number of interlocked markets, the subordination of the poor peasantry was intensified by their vulnerability in the rental, credit and land markets.

Apart from this web of socio-economic forces, agrarian life was also conditioned by ecological factors. What was striking to observe was that the process of differentiation of the peasantry and the increasing subordination of the poorer peasants was aided by factors like fluctuations in rainfall. Commercialisation of agriculture and the spread of cultivation to thinner soils imparted to agrarian life a greater insecurity and increased vulnerability to fluctuations in rainfall. It was argued that this insecurity was at its greatest in the poorer soils and thus most deeply experienced by the poorer peasants, as a nexus of dependency led to their continuously being pushed to the inferior soils. For such peasants, a local failure of rains was thus much more than a seasonal shortage of food or a small famine. It was in a very real sense a financial calamity further cementing the bonds of indebtedness which, given their subordinate position in other markets, would be impossible to shake off. On the other hand for the richer peasants or the malguzars cultivating rich black soil, inadequate rain was not necessarily an unmitigated calamity. Yet the very bounties of providence in the shape of

---

[112] In this context Charlesworth's description of the rich peasant is worth recalling, as 'the cultivator who marketed his crop most efficiently; who borrowed, bought and sold most efficiently . . . ' Neil Charlesworth, 'Rich Peasants and Poor Peasants in Late Nineteenth Century Maharashtra', in C.J. Dewey, A.J. Dewey and A.J. Hopkins eds., *The Imperial Impact*, London, 1978, pp. 97–113. Obviously a weakness of such a categorisation is that it does not account for the lack of emergence of technological change in agricultural production by the rich peasants. See, 'Agrarian Change in the Narmada Valley', *op. cit.*, pp. 33–36 and 215–17.

good harvests might be a cause of reproach. The year 1881 saw a good harvest but, as one officer commented, ' . . . the very excellence and abundance of harvests constitute their reproach in the eyes of those who exclusively have regard to the profit of agriculturists.'[113] Thus the forces which moulded the agrarian structure were the result of dynamic interactions between market production, control over different factors of production, varying economic strengths emerging from the processes of peasant differentiation and the uneven impact of certain ecological factors.

---

[113] C.P. *Revenue Administration Report*, 1880–81, p. 16. Similarly in 1871 after a good harvest it was said that 'the money profits of land will not be commensurate with the bounty of nature', ibid., 1870–71, p. 48.

Chapter Nine

Regional Dependence and Rural Development in Central India: The Pivotal Role of Migrant Labour*

CRISPIN N. BATES**

The problem of regional underdevelopment, particularly in tribal India, has long been recognized and more than one political party has campaigned on this issue.[1] The Indian constitution and state and central government development plans have included special clauses aimed at assisting those groups, the tribals or *adivasis*, who are most affected by the problem. Reports have been commissioned and investigations conducted, but rarely have these ended in constructive or relevant action.[2] The work of anthropologists over a number of generations since the

---

* Crispin N. Bates, *Modern Asian Studies*, 19 , 3 (Cambridge: Cambridge University Press, 1985), pp. 573–92.
** This paper is derived from the author's Ph.D. thesis 'Regional Dependence and Rural Development in Central India, 1820–1930', (Cambridge, 1984).
[1] A.B. Bardhan, *The Tribal Problem in India* (Communist Party of India, 1976). The 'Jharkhand' movement in Bihar takes up the same issues, though the current Assam agitation seems increasingly to be developing along communal lines.
[2] A critical review of State development planning is undertaken by Steve Jones, 'Tribal Underdevelopment in India', *Development and Change*, (SAGE, London and Beverly Hills), vol. 9 (1978), pp. 41–70. For details of the post-independence situation in central India, see the Government of Madhya Pradesh's *An Appraisal Report of Economic Problems in the Tribal Areas* (Indore, 1961), and the various reports and bulletins of the Institute of Tribal Research, and the Directorate of Tribal Area Development and Planning, Bhopal, and of the Agro-Economic Research Centre for Madhya Pradesh at Jabalpur.

1920s has perhaps done most to tell us of the real depth of the problem as it has affected central India. Foremost amongst them was W.V. Grigson, the aboriginal tribes enquiry officer of the government of the Central Provinces and Berar, whose 1944 report stands as the most comprehensive study available of the condition of the tribal peoples of this region at the end of the colonial period.[3]

There was a total of thirty-eight million scheduled tribals in India according to the 1971 census and another five million people who 'by their social organisation' might also be regarded as *adivasis*.[4] Next to Orissa, the largest concentration of tribals is to be found in Madhya Pradesh (twenty per cent of the national total), and most of these in the districts of southern M.P. which were formerly under the rule of the Central Provinces government (see map). Even today the economy of this area is only partly monetized, settlements are scattered, communications poor, and production mainly for home consumption. Historically, one of the more visible features of underdevelopment has been the progressive expropriation of tribal lands. W.V. Grigson found a fall in the average size of tribal tenant's holdings from 13.54 to 11.60 acres over the C.P. as a whole between the 1890s and 1939–40, while tribal landlords (who enjoyed much greater freedom of transfer) were almost entirely displaced in some tahsils. Thus the number of Gond villages fell from 205 to 141 in Narsinghpur district between 1866 and 1907 and from 294 to 129 in Raipur district between 1869 and 1912, and no less than 270 villages were transferred in the Mandla district between 1868 and 1888, predominantly from tribal Gonds to mahajan-moneylenders such as Raja Gokuldas, or to members of the Kallar caste of distillers. A comparison between tribal population and tribal landholding as a percentage of district totals also

---

[3] W.V. Grigson, *The Aboriginal Problem in the C.P. and Berar* (Nagpur, 1944) Ameliorative legislation of the colonial period is summarized in appendix R). Reference ought also be made to the pioneering work of D. Symington, *Report on the Aboriginal and Hill Tribes of the Partially Excluded Areas in the Bombay Presidency* (Bombay, 1938), and the anthropological studies of V. Elwin, S. Fuchs and C. Von Furer-Haimen-dorf.

[4] Steve Jones, 'Tribal Underdevelopment', pp. 44–5.

Map 1 Tribals as a Percentage of District Populations
According to the 1872 Census

confirms the impression that tribals tended to be concentrated in the structurally weakest positions in the agricultural economy, as landless labourers or as petty tenants, usually cultivating the poorest quality (*barra*) soils, in some cases using the most primitive *swidden* techniques. This economic backwardness has been compounded by social breakdown of which there is abundant evidence.[5]

All too often the anthropological evidence on which we must rely concentrates on the problematics of ethnology and cultural

---

[5] For example, Edgar Hyde, the district commissioner for Mandla in 1940, concluded that 'there seems to have been a more-or-less complete breakdown in tribal authorities over the greater part of this district', and Verrier Elwin commented of the Baiga village of Binjhwar in Balaghat: 'The souls of the people are soiled and grimy with the dust of passing motor-buses. In the village, you are in the midst not of a living community, but of a collection of isolated units.' E.S. Hyde to W.V. Grigson, Mandla, 23 November 1940, Hyde Papers, South Asian Study Centre, Cambridge, Box. II enclosure 18; Verrier Elwin, *The Baiga* (London, 1939), p. 513.

assimilation rather than on looking for the historical roots or the economic forces behind the changes in tribal society that have been observed. A step forward in the study of the dynamics of regional underdevelopment in India was made by Sachchidanand Sinha.[6] However, Sinha has been criticized for coining the emotive term 'internal colonialism' and for conceiving this as purely a matter of inter-areal exploitation, arguing that Bihar would be better served if its ties with the central government were severed. This argument clearly missed the point that within Bihar the tribals are the most exploited group and that successive Bihar state governments have been amongst the most corrupt and inegalitarian of Indian legislatures.[7] An alternative approach might be to invoke the interpretative value of a notion of a distinct form of economy peculiar to the tribal and more backward parts of India. The responsibility of the State in allowing and even encouraging underdevelopment in these backwaters of the Indian economy has been underlined by Jones and Corbridge, as well as in my own research. Besides the issue of the gradual expropriation of ancestral tribal lands, much more could be said about neglect and the abuse of authority by the colonial administration than there is room to describe. Instead, I wish to stress here the idea that as a modified survival of pre-colonial times, there also persisted a tribal form of economy which in itself acted as an obstacle to the penetration of capital and hence, indirectly, to the measures of reform proposed by capitalist-oriented colonial and post-colonial governments. This backwardness was not merely a product of limited resources, the usual neo-classical argument in favour of the theory of comparative advantage. Rather it was a matter of State, economy and society combining to produce a set of social relations of production that locked the tribal and more backward rice-growing areas of Chhattisgarh into a static non- correspondence to capitalist inputs or (possibly worse) into a negative cycle of

[6] Sachchidananad Sinha, *The Internal Colony: A Study in Regional Exploitation* (New Delhi, 1973).
[7] See Steve Jones, 'Tribal Underdevelopment', and S.E. Corbridge, 'The State and the Transformation of the Tribal Economy of Chota Nagpur (Cambridge Ph.D., in preparation), ch. II.

cumulative causation, bordering on involution, as appeared to be the case in Chhattisgarh.

While the development of the Narmada valley wheat zone in the colonial period was largely State sponsored and fundamentally unbalanced, the development of the Nagpur-Berar cotton zone benefited from a more propitious class structure and land policy, and managed to attract the sort of productive investment sorely lacking from the Narmada valley. This launched it onto a virtuous cycle of cumulative causation and economic growth, but even this growth was subsidized by those resident in the tribal and rice-growing tracts who benefited, surely, in that they were able to export surplus grain and surplus labour, but who did not at the same time enjoy the economic clout to demand something nearer the full value of their product in return and to convert this into productive investment on their own land. This phenomenon perhaps goes some way toward explaining the dynamics of the continuing underdevelopment experienced in these parts. The longer these zones remained in second place and the more the developed lowlands advanced ahead of them, the more locked they became into a subordinate position, dependent on the crumbs falling from the tables of their richer neighbours. Migrant labour was a key element in this relationship as it became integral to the continuing reproduction of the economy of the underdeveloped zones as much as it was integral to the competitive growth of both the wheat and cotton zones. In this sense it was qualitatively more significant than the parallel development of a trade in cheap grain, which was important to the cotton-growing area, but represented a small advance as far as the rice zone was concerned, which was only partly monetized and could equally well have continued to export at the low levels of the early nineteenth century when the only necessity was the need to raise cash to pay the government revenue demand. Few Chhattisgarhi farmers were broken by the decline in this outlet for surplus grain during the depression of the 1930s. This was in marked contrast with the nature of migrant labour. As can be demonstrated from survey and cost of production statistics, migrant labour amongst other sources of off-farm income was a necessity as far as the average smallholding Chhattisgarhi and tribal cultivator was concerned, and the loss of this income could have literally fatal consequences. Furthermore, the phenomenon of cheap, migrant labour not only helps explain the articulation of developed and underdeveloped economies (as has been described in a number of studies of the southern African economy),[8] but it also explains the curiosity of the development that did take place in the wheat and cotton zones of central India, namely the speed with which it progressed (and failed), despite the low average density of population for much of our period, the distance of central India from the main export markets, and the absence of the sort of major state investment that was to be seen in the Punjab.

The nature of the problem can best be understood if we look at the modest output of raw cotton in the year 1890–91 in Berar and the Central Provinces. This came to a total of 3,308,000 maunds, or 2264,640,000 lbs. Most of the crop was reaped by women and children. This being a labour-intensive business, an active woman was able to gather only about thirty lbs of cotton in a day, meaning that a minimum of 8,821,333 labour-days were

---

[8] See, for example, Lionel Cliffe, 'Labour Migration and Peasant Differentiation: Zambian Experiences', *J.P.S.*, 5, 3 (April 1978), pp. 326–46 and H. Wolpe, 'Capitalism and Cheap Labour-power in South Africa: From Segregation to Apartheid', in H. Wolpe (ed.) *The Articulation of Modes of Production: Essays from Economy and Society* (London, 1980). Studies of rural-rural migration in India, of any description, are notable for their absence, mainly due to the difficulty in obtaining appropriate research materials. However, reference may be made to Dietmar Rothermund, 'A Survey of Rural Migration and Land Reclamation in India, 1885', *J.P.S.*, 4, 3 (April 1977), pp. 230–42, K. C. Zachariah, *A Historical Study of Internal Migration in the Indian Subcontinent 1901–1931* (London, 1964) (a study based on census data), and Anand Yang, 'Peasants on the Move: A Study of Internal Migration in India', *Journal of Interdisciplinary History*, xi, ' (Summer 1979), pp. 35–58. J. Connell *et al.*, *Migration from Rural Areas: The Evidence from Village Studies* (Delhi, 1976), draws upon village studies of the past thirty years and is largely concerned with rural-urban migration. Gail Omvedt in 'Migration in Colonial India: The Articulation of Feudalism and Capitalism by the Colonial State', *J.P.S.*, 7, 2, pp. 185–210, also discusses rural-urban migration but in addition has important sections on migration to the Assam plantations and the effects on rural wages.

required to collect the entire crop. This compares with a total population in the cotton zone of Nagpur, Wardha, Amraoti, Akola, Yeotmal, Buldana and Nimar districts of 4,249,242. Approximately one-third of this population were agricultural labourers, a total of approximately 1,373,000 individuals, whilst a sizeable proportion of the remainder (approximately 671,000) dwelt in towns and cities and were thus unlikely to assist much in agriculture. The labour requirements of cotton then had to compete with the combined output of 22,523,000 maunds of other produce being cultivated, and this at a time when cotton covered only about thirty per cent of the Gross Cultivated Area in the cotton zone as a whole. The demand for labour is thus likely to have increased in proportion as the cotton area rose to forty-four per cent of G.C.A. by the 1920s.[9]

A similar situation presented itself in the Narmada valley wheat zone, where the population density was lower than in the cotton zone (about 140 per square mile on average in 1921 compared with 170 per square mile in the cotton zone) and less than thirteen per cent of the total population in the centre of the zone were classed as agricultural labourers for most of our period. The consequence was that as late as 1927, long past the hey-day of the wheat trade, the total number of migrants flocking into the valley to assist in the harvest operations was estimated at 120,000.[10] The scale of this migration must have been proportionately much greater in the 1880s, when wheat exports and acreages were 340 and twenty-five per cent higher, respectively. The 1901 census described the wheat migration as follows:

... in the Jubbulpore haveli there is an immigration of Chaitharas, or those who come in Chait (March–April) to cut the wheat crop. Year by year ... the Gond comes down from the Rewa hills to the Lodhi

---

[9] For details of labour hired-in on farms of the cotton zone in the late 1930s see Imperial Council of Agricultural Research, *Report on the Cost of Production of Crops in the Principal Cotton and Sugarcane Tracts of India* (Delhi, 1938–39), vol. VII, appendix V. Typically, 11.24 man days and 17.74 woman days were hired per acre in the Berar plains, compared with a deployment of 3.88 man days and 1.42 woman days of family labour per acre.

[10] *Royal Commission on Agriculture in India* (London, 1927), vol. VI, p. 5.

in the haveli, the same Gond to the same Lodhi and from father to son. Till the crop is ready to be cut, he occupies himself in roofing the house, building up walls, and doing any other odd job that may be required. Then he assists in the reaping of the crop, and when it is threshed and harvested, he returns home, having received his food while he is there, and taking across his shoulders as much grain as he can get into a *kawar* load.[11]

Of the wheat labourers, or Chaitharas, who came to assist in the harvest of 1921, 38,857 were enumerated as having been born outside of the province. The census was taken in mid-March, and caught a number of districts at the beginning of the wheat harvest. In Hoshangabad 18,812 were recorded as having arrived for the wheat harvest, in Jabalpur 14,862. In Hoshangabad this amounted to twenty-two per cent of the labouring population of the district, 12.5 per cent in Jabalpur. Though these figures will have been affected by the had harvests of that year, it was also the case that many more tribals from the Satpuras had yet to arrive.

As early as 1867 the settlement officer of Hoshangabad district had described the migration as rather like that of the Irish crossing 'the channel' for the English harvest, and many migrants from the Central Indian States penetrated as far as eastern Berar where they occupied themselves weeding and picking during the autumn and winter and either stayed for the spring harvest in Berar (where as many as 9,000 were enumerated in 1911), or returned *via* the Narmada valley, where the wheat harvest gave them employment until they returned home to prepare their own fields for the ensuing agricultural season.[12] The Jabalpur settlement officer in the late 1880s wrote that 'The Gonds flock with their families at the spring harvest time to the wheat fields of the 'Haveli' to eke out their subsistence by working as labourers', and claimed that their earnings gave them food sufficient to life from until the commencement of the rainy season.[13] Jabalpur farmers in fact often grew a crop of the inferior

---

[11] *Census of India*, 1901, vol. XIII, p. 215.
[12] *Census of India, 1911*, vol. X, pt I, pp. 51ff. *Census of India, 1921*, vol. XI, pt I, p. 12.
[13] L.R.S. *Jubbulpore, 1886–94*, p. 24.

kodon grain specifically in order to pay the wages of these labourers—the wheat crop itself being far too valuable to give away in grain wages—and here as well as in Narsinghpur the farmers were said to rely entirely on migrant labour for the embanking of their fields, work which was done at rates much below those demanded by local labourers.[14] From Balaghat district there were a number of reports which spoke of labour shortage due to 'the annual exodus in March and April from the north of the district to the wheat tracts of Seoni and Mandla', and to the fact that the tracts of Katangi and Langi sent large numbers of persons to Berar every year for the cotton picking after Dasahra. Wages were far higher than in Balaghat and it was noted that many of the 'jharis (or jungle folk)' stayed up until the rains in pursuit of casual labour.[15] In Saugor, Damoh, Seoni, Mandla and Chhindwara, the annual migration for harvest operations was common enough to be regarded as *rozgar*, or part of the daily avocation, and in Damoh, at the time of the 1891 census, it was reported that 'the forest tribes ... have actually *decreased*, due, it would appear, to the drain made on the southern or hilly part of the district for labourers to reap the harvest in Jubbulpore[16]

Except when a specific question was asked about labour migration in the census, as occasionally it was, it is not always possible to distinguish between permanent and periodic migration, especially since labourers might not stay merely for a season, but for a whole year or even longer, in which case they were more likely to be enumerated among the permanent workforce of a district. The C.P. and Berar, being traditionally an underpopulated area, had a large number of immigrants. Estimates in 1931 based on a plateau district where there were no massive movements of population suggested that four per cent was a

---

[14] *Ibid.*, p. 12. *C.P. Wages Census, 1912*, memo. by G. Evans, deputy director of agriculture, Northern Circle, J.B. Fuller, *Review of Progress in the Central Provinces in the last 30 years, and of the Present and Past Condition of the People* (Nagpur, 1892), p. 31.

[15] *Balaghat D.G.*, p. 73.

[16] *Report on Famine in the Central Provinces in 1896 and 1897* (Nagpur, 1898), p. 5. *Census of India, 1891*, Vol. XI, pt I, p. 215.

TABLE 1
IMMIGRATION IN THE C.P. AND BERAR

District	Those born elsewhere in C.P. as percentage of district population +		Those born beyond C.P.+		Total immigrants	
	1881	1931	1881	1931	1881	1931
Wheat zone:						
Saugor	3	2	13	7	16	9
Damoh	6	5	10	5	16	10
Jabalpur	4	5	13	10	17	15
Hoshangabad	4	5	10	6	14	11
Narsinghpur	5	6	5	2	10	8
Cotton zone:						
Nagpur	7	11	5	4	12	15
Wardha	14	16	9	3	23	19
Amraoti	9*	11	17*	4	26*	15
Akola	7*	9	11*	6	18*	15
Buldana	5*	3	14*	7	19*	10
Yeotmal	7*	11	21*	6	28*	17
Rice zone:						
Raipur	6	3	1	3	7	6
Bilaspur	6	2	3	4	9	6
Bhandara	7	7	1	1	8	8
Tribal zone:						
Seoni	17	6	1	7	18	13
Chhindwara	10	6	1	2	11	8
Drtul	4	4	3	1	7	5
Malaghat	17	9	1	7	18	16
Chanda	6	4	5	2	11	6

+ those born elsewhere in, or beyond, *Berar* not C.P.
* the figures for 1931 are for the amalgamated province of C.P. *and* Berar.

SOURCE: Berar, C.P., and C.P. and Berar decennial *Census Reports*.

fairly normal proportion of casual migrants compared to the total population. Estimates for many Deccan districts gave combined totals of permanent and casual migration at around the same level, while at the all-India level roughly ninety-one per cent on average were shown by censuses between 1881 and 1931 as having been born in the district in which they were enumerated, six per cent being born in adjacent districts and only three per cent having migrated further. The central Indian censuses, however, give totals of casual and permanent migration significantly in excess of this (see Table 1).

Amongst other explanations of short-distance migration across borders, marriage is usually considered of importance, the bride commonly travelling to the bridegroom's home over a distance of several miles. There were, indeed, a greater number of females recorded as having migrated from Hyderabad and Bombay Presidency into the Berars in 1921 and 1931, but something like one in six of married males in Berar must have married a girl from Hyderabad or Bombay presidency if the marriage explanation were to carry any great weight. More likely this imbalance was tied in with the labour requirements of agriculture in each region. Occupational statistics demonstrated a preponderance of females, at a level around seventy-four per cent, being employed in cotton spinning and other aspects of cotton processing. They also usually constituted around sixty-six per cent of the total number of field labourers and were strongly favoured when it came to employment at cotton harvest time. Only in longer-distance migration and migration to the big industrial centres were males substantially in the majority.[17]

The 1931 census described two main streams of migration by that date as follows:

The one comes in a southwesterly direction from the U.P. and the Central India Agency States, which is caused by the general poverty and periodic scarcity in Central India and the lure of good wages and opportunities for obtaining work in the industrial centres of the C.P. The other comes in a westerly direction from the Chhattisgarh Plain

[17] *Census of India, 1931*, vol. XII, pt I, pp. 95–100. *Census of India, 1901*, vol. XIII, p. 240.

division and Bhandara district, which is caused by the poverty of the not very fertile, land-locked plain and periodic failure of crops in the same region, as also the prevalence of a higher standard of wages in the developed portion of the province.[18]

A survey of the cotton industry labour force at Jubbulpore, Nagpur, Hinganghat, Amraoti and Akola revealed that in fact the largest portion of migrants (thirty-three per cent) came from Chhattisgarh. Similarly, the Berar censuses of the nineteenth century revealed that roughly forty-three per cent of immigrants came from the east, thirty per cent from Hyderabad and eighteen per cent from Bombay presidency, while the 1921 census showed the largest imbalance in migration in the Chhattisgarh districts of Raipur and Bilaspur, which had a total emigration of 322,805 and an immigration of 188,073. Among this total were many out of the 5,000 per annum in the 1900s who went to Assam to work on the tea plantations. But most returned after working out their contracts, and even at its peak, 1918–19, migration to the tea plantations took no more than 31,000 from the Central Provinces as a whole. By comparison, the Berar districts, which were less than two hundred miles away, had a total number of immigrants of 473,559 and 181,155 emigrants recorded in 1921.[19]

Between 1911 and 1921 there is evidence of a definite increase in the number of emigrants from Chhattisgarh to the Maratha plain, partly as a result of bad harvests and the influenza epidemic of 1918 but doubtless also affected by the increased demand from the cotton zone for labour in this period; 27,000 residents of Berar were recorded as having been born in Chhattisgarh in 1911, and 37,000 in 1921, this being over half of the total number of immigrants from the east. In the nineteenth century this total had been running at the level of around 200,000 *plus*.[20] In 1931 emigration statistics were not included in the census, but the total number of immigrants in Berar had in-

[18] *Census of India, 1931*, vol. XII, pt I, p. 100.
[19] *Census of India, 1921*, vol. XI, pt I, p. 30. *Annual Reports on Inland Emigration*. Details of Wardha district migration statistics are to be found in *Wardha D.G.*, pp. 31–4.
[20] This was described as a veritable tide in 1870s—see *Berar Gazetteer*, p. 219.

creased to 449,576. If we assume a mortality rate of forty per thousand on the immigrant population of 1921, this means there may have been as many as 215,000 new arrivals over the period 1921–31: possibly an exaggeration, but the figures give an idea of the continuing scale of the migration phenomenon, and this despite improving employment opportunities in tribal areas and in Chhattisgarh with the construction of irrigation works and the Raipur to Vizianagram railway, Chhattisgarhi irrigation works in 1927–28, to take just one example, employing 8,600 labourers, most of them recruited locally.[21]

The migrant was not always landless, but typically a small-holding tribal or Chhattisgarhi rice cultivator holding only an acre or two of land.[22] For the Chhattisgarhi the prospects were admirably summarized in the late 1920s by Chhotelal Verma, the Bilaspur settlement officer:

In a district like Bilaspur, where the opportunities for agricultural operations are limited only to a part of the year, on account of a large area being devoted to only one crop, paddy, and the pressure of population falling heavy on land, small tenants, whose number is fairly large, take to various other occupations in the non-working season; those residing in the proximity of Government and Zamindari forests engage themselves in extracting and carting forest produce; others find employment on the irrigation works and Public Works Department roads; some do carting of grain, while the more adventurous Satnami emigrates temporarily after each harvest to Calcutta, Kalimati and other centres of trade, to obtain the best market for his labour and returns with the beginning of the year to resume agricultural operations. It is not an uncommon sight to see villages denuded of a large part of their population during January, February, March, April and May each year, on account of the temporary exodus.[23]

Unlike in Maharashtra, where the kharif rice harvest tended to be a little later, in Chhattisgarh harvesting began around the 15th of September and was in all cases finished before the 15th of December. There was no other major crop or winter rice to

---

[21] *Census of India, 1931*, vol. XII, pt I, p. 101.
[22] Memo. by C.E. Low, director of agriculture and industries, *C.P. Wage Census*, 1912.
[23] C.P.P.B.E.C.R., vol. 4, pp. 910–11.

be sown, so until the season for sowing came round again, in the second week of June, there was little work to be done. Hence the attractiveness of the so-called 'jhari' migration to the Berars, helping with the cotton picking, which started in November, or helping with the later wheat harvest in the Narmada valley, which was over rather more quickly, between the second week of February and the month of April.[24] The slum suburbs of Hamalpura, Masanganj and Ratanganj around the town of Amraoti were inhabited very largely by such migrants, and big and small settlements such as these were scattered all about the towns and villages of Nagpur and Berar, Nagpur city itself growing at a rate of 1.4 per cent per annum in the late nineteenth century.[25]

Amongst the Gonds and Bhumias of Eastern Mandla, Stephen Fuchs has described three categories of tribal labourers. These began with the *barsi* or *barkhi*, who was a permanent field servant who worked for a whole year, often in order to pay off a debt. Then there was the *harwaha* or ploughman, usually employed for just one season at a time and paid in kind, though not enough to feed his family as well. In a survey of nineteen villages this category outnumbered the former by three to one. Finally there was the *bani (banihar)* or occasional labourer, who was paid in kind on a daily basis. Fuchs commented that their

---

[24] This skewed structure of demand for labour in the rice zone was one of the factors which also made groundnut an unwelcome addition, despite the efforts of the C.P. agriculture department, and linseed so popular, easily fitting, as it did, into the slack season of the year. *Census of India 1921*, vol. XI, pt I, p. 26. Wardha D.G., p. 137.

[25] *Amraoti D.G.*, p. 359. *L.R.S. Nagpur 1890–95*, p. 21. Mr Balkrishna L. Baput, Deputy Educational Inspector, commented: 'As proof of the gradually increasing demand for labour for agricultural purposes, one may easily observe clusters of little houses which during the last fifteen years or so, have grown up on the outskirts of every village in the mofussil. These are the habitations of immigrant labourers who once came seeking labour for a season, but who, finding enough of what they sought for, and probably, on their terms, have permanently settled in the provinces. In some cases a hundred families of such immigrants have thus been added to the labouring population of a village.' *(Dufferin) Enquiry into the Condition of the Agricultural Classes of India* (Calcutta, 1888), pp. 40–1.

numbers were not great, but that yearly epidemics among cattle reduced many to this status, from which it was extremely difficult to regain the position of tenant or landowner.[26] Similar categories amongst tribal labourers were also described for other of the Satpura plateau districts in the late nineteenth and early twentieth centuries. In Balaghat the permanent farm servants were known as *barsias*, and servants hired on a monthly basis were called *mahinhias*, a similar nomenclature also being found in the Betul district where the terms *harwaha* and *barsalia* were used to describe the permanent servant, *mahantia* those employed by the month. *banihar* or *rozina* being the names for daily labourers.[27] In Seoni, farm servants were sometimes engaged on a *balia* or share basis at a rate of one-fifth of the crop, this form of payment being even more prevalent in Chhattisgarh, the large number of smallholdings here making the roles of tenant and labourer several times interchangeable—crop-sharing labourers here going by the name of *saonjia*.[28] Step by step, more and more Gond cultivators in the Satpura tribal belt were being pushed into these categories of temporary labouring as was explained by R.A.B. Chapman, the Seoni district commissioner in 1906:

The Gond appears unable to retain the good land in his possession and as soon as his holding begins to produce anything like valuable crops, he falls back into the position of a farm-labourer and his fields too often pass to others to whom he has become indebted. The bulk of the Gond population are labourers ... Till lately they were always paid in kind, but with the great rise which has lately taken place in the value of wheat, there is a tendency now for payments in kind to be commuted into cash. At the time of cutting of the wheat harvest, there is always a great movement among the labouring classes. Like the hop-pickers at home, whole families will travel long distances to places where plenty of harvesting is going on ... A considerable number ... have emigrated.[29]

[26] Stephen Fuchs, *The Gond and Bhumia of Eastern Mandla* (Bombay, 1960), pp. 99–100, see also *Mandla D.G.*, p. 140.
[27] *Balaghat D.G.*, pp. 192–4; *Betul D.G.*, p. 152; *Chhindwara D.G.*, p. 124
[28] *Seoni D.G.*, p. 190. *C.P.P.B.E.C.R.*, vol. 4, p. 866 (evidence of W.B. Lakhe, Pleader and Hon. Sec., Co-op. Central Bank Ltd, Raipur). *Bilaspur D.G.*, pp. 157–9.
[29] *Seoni D.G.*, pp. 113–14.

Estimating the remuneration received by these labourers is a difficult matter as the sources are numerous, ill-defined and sometimes contradictory, but there is no doubt that higher wages were obtainable in the lowlands, especially in the cotton zone after 1891, than was available either in Chhattisgarh or the tribal belt. The C.P. administration reports, the series of *Prices and Wages in India* (from 1873), volume B of the district gazetteers and the wages census conducted after 1910 suggest that wages in the Chhattisgarh district of Bilaspur in cash or kind were equivalent to about one anna six pice a day, or much less in rural areas where food and clothing only might be provided—this was about half the rate being paid for unskilled labour in the Nagpur—Berar plain or the Narmada valley in the 1870s and 1880s. By 1910 the wages cited in the C.P. wage census for the principal tahsil of each district were in Bilaspur slightly increased to two annas a day, when paid in cash, but were most often paid in kind, the three seers of grain received being worth no more than one anna six pice. Wages in the central Narmada valley, in Hoshangabad, were unchanged at four annas a day, whereas in the Berar district of Akola, wages as high as five annas might be paid, or six seers of grain worth six annas. By 1923 wages everywhere had doubled, but this was a generally lagged response to grain price inflation. Three seers of grain were still the standard wage in Bilaspur, but now worth three to four annas. In the central Narmada valley, cash wages were higher (from four to ten annas), but grain payments were roughly the same, indeed cash wages would have been lower but for the exceptional demand for labourers to assist the neighbouring cotton cultivators in Nimar. In tribal areas only half these rates might be paid. In Berar, though, the standard wage for a number of years ranged from five annas to as high as twelve annas a day.

However, although wages in the lowlands were higher, it is difficult to find evidence that they increased at a rate anything like comparable to the growth of agricultural surpluses and the extension of cultivation, particularly when rises in the cost of living are taken into account. In Nagpur the cash value of the wages of agricultural labour was thought to be Rs 5 a month in

1873—evidently a hangover from the American Civil War Cotton boom. In 1881 this fell to Rs 4, remaining at this level until 1891. Thereafter it rose to Rs 5 again, reaching Rs 5–12 by 1903.[30] At this date the Nagpur commissioner was able to comment on a real improvement in the labourer's position, but the increase was partly due to the fatal impact of famine at the turn of the century and after 1905 the upward trend ceased, the commissioner of Amraoti expressing his belief that wages in Berar by 1910 were little changed by comparison with 1870, though increasingly paid in cash, rather than in kind.[31]

Women were preferred for cotton picking simply because they could be paid less, it usually taking three to five pickings completely to clear a crop.[32] At the first picking the wage was two annas per diem and at the second and third it was the money equivalent of a wage in kind, roughly four to four and one half annas for every maund of fifteen seers (three hundred lbs) picked. After this wages fell and at the fourth and fifth picking only two annas a day were earned. *Juar* and other crops were sown by men at a rate of pay of from four to six annas a day, cutting, stacking and threshing being done on contract by men on a basis of two to two and one half *kuros* (one *kuro*=sixteen seers) for every *tiffan* (four acres) reaped, or sometimes twenty-four seers for every khandi of 320 seers harvested. The ears were separated from the stalks by women being paid a basketful of ears (value about four and a half annas) per day.[33] J.B. Fuller in his review of progress in the Central Provinces in 1892 also noted a rise in wages in the cotton zone, but he was less sanguine about the rates paid in the Narmada valley for the harvesting of grain, which could go as low as twenty lbs per acre harvested, compared with thirty to forty lbs (the minimum contract) in the Nagpur country. The permanent farm servant he then believed to be paid 1,440 lbs per annum compared with 1,200 lbs sixty

[30] See Wardha D.G., p. 193 and K.L. Datta, *Report on the Enquiry into the Rise of Prices in India*, 5 vols. (Calcutta, 1914), pp. 25, 215 and 235.
[31] *Amraoti D.G.*, pp. 234–5.
[32] The preference for female labour was also unchanged since the American Civil War—see *Berar Gazetteer*, p. 68.
[33] *Amraoti D.G.*, pp. 234–5.

years before at the time of Richard Jenkins' report on the Nagpur kingdom. By the 1890s, he estimated, something like fifty-seven per cent of the wage-earning classes were dependent on daily hire. This class had generally lost ground 'in localities where agriculture has become ancillary to trade, and custom and sentiment are sacrificed to profit. Chief among these localities is the Hoshangabad district, where a casual labourer now receives from four and a quarter to five and a quarter lbs. a day, against six and a quarter lbs. at last settlement.'[34] By way of confirmation, the rate for unskilled labour hired on the railway between Burhanpur and Jabalpur was quoted as being typically as low as two to four annas a day in 1888, and this was a peak year for the wheat export trade.[35]

Despite the doubling of nominal wages by 1923, compared with their level in 1910, this does not compare well with a seventy-eight per cent increase in the wholesale price of wheat and other goods in the Nagpur market, prices which peaked at more than two and a half times the 1910 level between 1917 and 1921 and again at twice their 1910 level in 1924–25. And these were the years when cotton cultivators in the Nagpur—Berar plain were on average making 'profits' of about Rs 1000 per annum. It was also the case that whenever the demand for agricultural exports slackened, the agricultural labourers were the first to feel the pinch, wages falling first in areas (such as the cotton zone) where they had formerly risen most.[36] Typically, employers would only take on groups of labourers on contract whenever there was a slump, the point being that these contracts specified a piece rate rather than payment by head or by the number of hours worked, thus ensuring the farmer an increased

---

[34] J.B. Fuller, *Review of Progress in the Central Provinces in the last 30 years, and of the Present and Past Condition of the People* (Nagpur, 1892), p. 51. Details of off-farm income and the costs of Chaitara labour in Jabalpur district are given in *C.P.P.B.E.C.R.*, vol. II, pp. 217–219.

[35] C.P.R.D. Comps, Mr Riddell, Acting Dist. Engineer G.I.P. Co. to J.B. Fuller, Comm. Sett. and Agric., 20 June 1888, and J.B. Barton, Resident Engineer . . . Harda, to J.B. Fuller, 18 June 1888.

[36] *C.P.P.B.E.C.R.*, vol. I, p. 135. *Census of India, 1931*, vol. XII, pt I, pp. 16–17 and pp. 61–2, *Census of India, 1921*, vol. XI, pt I, pp. 7–9.

return for his money. When famine struck the conditions of employment changed more rapidly:

> The ordinary field labourer depends a great deal on harvest earnings, and if one locality does not offer harvest wages he will move to another which does. The two great movements of the labourers of the province ... are those of the people from the south of the province to Berar in November-December to reap the cotton and juar crops, and the descent of the inhabitants of the upland tracts in March-April to reap the wheat crop of the Nerbudda valley. These movements are in ordinary times most salutary; but when famine comes and there are short crops to be reaped, the migration of labourers is a source of much embarrassment. In 1896 this exodus of people from the Wainganga Districts to Berar in search of harvesting employment resulted in the districts of Wardha and Nagpur being overrun with crowds of wanderers, some pressing on to their imaginary land of promise, others struggling back empty handed and starving unable to support themselves or to return to their homes.[37]

While the major landowning caste of the Brahmins experienced a 0.3 per cent increase in their population over this period between 1891 and 1910, the Gonds suffered a 17.3 per cent decline, and the Mahars (or Mehras), a traditional labouring caste in Nagpur and the Narmada valley, suffered a six per cent decline in their numbers. This was partly due to local differences in the severity of famine: Betul was especially badly hit with the Korku tribal population being cut by forty-three per cent (either by death or migration), and the rice-growing Bilaspur district was also badly affected, the almost complete absence of decent government famine relief in some of these backward areas greatly exacerbating the problem.[38] Berar, by contrast, did not experience anything like the extent of crop failure seen elsewhere.

---

[37] *Report on the Famine in the C.P. in 1899–1900* (Nagpur, 1901), p. 24.

[38] British famine reports, being designed to prove that the administration was not responsible for deaths by starvation (indeed, that these deaths never actually occurred), often disguise more than they reveal. For a good first-hand impression of the reality of famine in underdeveloped parts of central India, see the earlier reports of F.H.S. Merewether in *A Tour Through the Famine Districts of India* (London, 198). For a summary of the census statistics on the impact of the famines on the labouring population as a whole (as nearly as this could be estimated) see *L.R.S. Saugar, 1887–97*, p. 38.

However independent of the vagaries of charity and the local harvest, cultivators in the tribal and rice-growing areas suffered in addition from the failure of harvests in *other* tracts since migrant labour was an integral part of their subsistence base.[39] Hence R.H. Craddock reported:

The aborigines of Betul ... lost not only the millet crops, but also their harvest earnings, for there was no cotton or juar worth mentioning in the neighbouring tracts of Berar. The enormous labouring population of Chhattisgarh, most of whom are usually paid for their work in a share of the produce, received nothing; the labourers of the Wainganga rice districts had little or no transplanting or weeding to carry them through the rains and no rice crop to harvest at the end. The cultivators of these same rice tracts in only comparatively few cases had their seed returned to them ... At a recent census taken of two crowded works in Raipur, twenty per cent were cultivators.[40]

Distress amongst the Chaitharas was similarly observed in the Lakhnadon tahsil in Seoni district, and again when the harvest failed in Jabalpur, Saugor and Damoh districts in 1928 and 1929.[41] Although wage rates sometimes rose as a result of famine, the effect was comparatively short-lived and would hardly compensate a labouring family for the loss of one or more of its earning members. Likewise, the onset of depression in the world economy after 1929 hit the demand for cotton, and the migrant labourer in Berar was among the first to be affected. The Buldana district commissioner noted a 'substantial reduction' in the movements of labourers and observed that many families who had migrated from the Bombay Presidency and Hyderabad, and

---

[39] L.R.S. *Narsinghpur, 1885–94*, p. 26.

[40] R.A.C. Procs., 3 February 1900, no. 5, pp. 40–1, R.H. Craddock to Sec. to G. of I. During these years of distress migration to Assam also greatly increased, 28.2 per cent of recruits for Assam coming from the C.P. in 1896, 37 7 per cent in 1900 and 39.2 per cent in 1901—this compares with only 5.6 per cent in 1894, (children are excluded from these totals). See *Annual Reports on Inland Emigration* and *Annual Reports on Labour Immigration into Assam* (Calcutta), and *Report of the Royal Commission on Labour in India* (London, 1931) (P.P. 1930 21. XI. p. 571, Cmnd. 3883).

[41] *Census of India, 1931*, vol. XII, pt I, p. 39. R.A.C. Procs., Jan. April 1900, no. 5, P. 41, R.H. Graddock to Sec. to G. of I., 3 February 1900.

whose arrival was recorded in the 1921 census, were now moving back again, while the Wardha district commissioner noted the return of migrants to the forests of Chanda.[42] By 1939 the Imperial Council of Agricultural Research report on costs of production revealed the wages of hired male and female labour in Berar (excluding children) to be as low as Rs 2.8, rising to a maximum of Rs 7.5 per month—rates not very different from those paid in 1860–70. However, during the years of depression, 1930–41, the cost of living had also tumbled dramatically, one index of the cost of living of industrial workers in Nagpur and Jabalpur indicating that price levels had fallen back to the level at which they had been in 1914, the cost of living staying at this level for most of the decade. To those in employment this was a great boon after the years of astronomical inflation they had experienced in between, but work was short and wages were cut and if one looked at the total income of the smallholding family, used to every member, men, women and children, contributing something, it is likely that we would find an overall deterioration in their condition.[43]

CONCLUSION

Economic growth in the wheat and cotton zones thus clearly extended the employment opportunities of those resident in the more backward tribal and rice-growing areas, but the rates of remuneration do not seem to have been high enough in them-

[42] *Census of India, 1931*, vol. XII, pt I, pp. 43–4.
[43] Imperial Council of Agricultural Research, *Report on the Cost of Production of Crops in the Principal Sugarcane and Cotton Tracts in India* (Delhi, 1938–39), vol. VII. *Index Numbers Showing the Changes in the Cost of Living of Industrial Workers of Nagpur and Jubbulpore*, C.P. Dept. of Industries (Nagpur, 1919 to 1952). Further confirmation is available from the survey conducted by the C.P.P.B.E.C.R. in 1929, which found the *average* wage in cash and grain of family of farm servants and agricultural labourers was only Rs 151. This sum would typically have to feed five members of the family for 365 days of the year—C.P.P.B.E.C.R., vol. I, p. 89. The income and condition of an evidently fairly well-to-do Chaithara of the Seoni district is described in vol. 2., p. 643. Details of the off-farm earnings of a Betul smallholder are given on p 691.

selves to dent in any way the underdevelopment from which they suffered. It is, of course, unproven what the course of development might have been in tribal central India in the complete absence of this relationship; however, there are other aspects beyond straight remuneration, such as the expropriation of tribal lands by non-tribals, the destruction of forests and the disintegration of social cohesion and formerly effective methods of land control, suggesting that the interaction of developed and underdeveloped went beyond passive symbiosis and had a number of positively damaging effects. In explaining these effects the notion of a distinct tribal form of economy, rooted in an ideology of egalitarianism and which was peculiar to the more backward tribal and rice-growing areas, has a useful role to play. The idea of a conflict of rationalities between this tribal economic system and that of the lowlands, with which the capitalist-oriented colonial government was most at ease, helps explain the failure of colonial administration in these areas, a failure which might otherwise be put down to wilful racial chauvinism. This failure manifested itself in the progressive marginalization of cultivators and the creation of a dependence on migrant labour, despite the presence of other positive developments in the tribal and rice zones, such as the growth of population, the improvement of communications and an increase of trade. The procedure of breaking down the whole of central India into discrete agro-economic zones, analysing the relationship between developed and backward zones as well as the different patterns of growth in each, and the construction of a notion of a tribal form of economy, is also useful because it avoids cliched essentialist debate—such as whether Indian agriculture is locked in a state of Semi-Feudalism (*a la* Bhaduri), or languishes as a peripheral social formation within capitalism viewed as a world system (*a la* Frank). Such categories of analysis are implicit in many a political programme or research proposal, but there are clearly other exploitations beyond the inter-class and the international and these must be looked at afresh if history is to throw off the role of 'sweeper' to the social sciences. In the case of central India the scale of migrant labour we have described here is clearly an important phenomenon of inter-regional exploitation which must be taken into account in the historiography of the region.

As far as the more developed wheat and cotton zones of the lowlands were concerned the presence of backward reserves of cheap grain and labour proved an advantage, assisting rapid growth that otherwise might not have taken place. Colonial land policy itself had very different effects here, dependent partly on the pre-existing class structure of each zone. This point illustrates the second main theme in my research which has concerned the relative merits of what may be loosely termed demand-side and supply-side economics in explaining the historical experience of development. It can clearly be shown that the development of the Narmada valley wheat zone was fundamentally stymied by the colonial government's encouragement of high landlordism in the shape of the *malguzar*. The soil, lesser tenantry and migrant labourers were over-exploited and the absence of any reinvestment caused the late nineteenth-century wheat boom to grind to a halt after only a few decades. The lower level of differentiation among the peasantry of the cotton-growing Nagpur—Berar region, however, made it far less amenable to a *malguzari* settlement policy and the results were more propitious. There were also far greater reserves of capital within the credit system due to the presence of princely kingdoms which held together key mercantile groups during the 'time of troubles' in the early nineteenth century. In the area of Berar there was a fairly successful *ryotwari* settlement which put ownership within the reach of most cultivators, thus also assisting the flow of capital, and in Nimar district a passable *ryotwari* system was established out of the remains of the discredited *malguzari* policy. Markets here were not 'free' but regulated by the State, thus encouraging a more competitive credit system and allowing greater opportunity to cultivators, who learnt to profit by leasing-in and by investment in improved cultivating techniques, and had less need to rely on depressed wages and supra-economic coercion—as was the case in the Narmada valley. The result was a steady growth which was brought to an end not by natural calamity or internal contradictions but primarily by the onset of depression in the international economy after 1930. Thus although tribals, migrants and other labourers were generally in a structurally disadvantaged position, they could be more or less efficiently exploited depending on the local character of the economy with which they were involved. Clearly some did have the chance to improve themselves (as well as others) by migrating to the cotton zone, but by contrast the super-exploitation of the wheat zone proved in the long run to be of little advantage to anyone.

In sum, therefore, any combination of evidence regarding solely resource availability and the growth of opportunities for risk-taking via the development of markets and communications remains unsatisfactory if we are to account for the differential nature of economic growth in different agro-economic zones, as well as the relationship between developed and underdeveloped regional economies. Empirical evidence of changing patterns of distribution (described in other discourses as 'the social relations of production' or simply 'class structure'), and particularly evidence regarding the supply and mobilization of labour—itself a consequence of changing patterns of land control and surplus appropriation—must be an essential ingredient in any economic history that wishes to advance beyond the mere rhetoric of received theory. Above all, the regional character of economic development must be taken into account. It is always possible to find local models of economic success under colonial rule, but no theory of class and development can be adequately constructed without due attention to such wider phenomena as the growth of inter-regional migration and trade.

Annotated Bibliography

This bibliography is anything but comprehensive. It supplements references in the Introduction and Chapters, and it is designed for students and scholars who want to follow up on themes presented in this book. It groups entries according to themes and categories that are useful in research.

1. COLLECTIONS AND REFERENCE

BHALLA, Alok and Peter J. BUMKE, *Images of Rural India in the 20th Century* (New Delhi, 1992). An intriguing collection of ideas from literary studies, the arts, politics and social activists, which begins to indicate the complex role that rural India plays in the culture of the urban intelligentsia inside and outside India.

BOSE, Sugata (ed.), *South Asia and World Capitalism* (Delhi, 1990). A volume on the application of world-system theory to South Asia—including its arguments about development and underdevelopment in agriculture. Includes contributions by economists and historians and essays on the immediate present as well as on the distant past.

CHAMBERS, Robert, Richard LONGHURST et al. (eds), *Seasonal Dimensions to Rural Poverty* (Montclair, 1981). A path-breaking collection. Seasonality merits much more attention by historians. These essays represent a range of subjects to be integrated into the historical understanding of agrarian ecology.

DESAI, Meghnad, Susanne Hoeber RUDOLPH and Ashok RUDRA (eds), *Agrarian Power and Agricultural Productivity in South Asia* (Delhi, Berkeley and Los Angeles, 1984). Essays in economics, anthropology, political science and history that revolve around the linkages that are most relevant today for economic policy makers between power relations in rural society and increases in productivity in the process of agricultural development.

CHAUDHURI, K.N. and Clive J. DEWEY (eds), *Economy and Society: Essays in Indian Economic and Social History* (Delhi, 1979). A wide-ranging set of essays highlighting the role of agriculture in Indian economic and social history.

KUMAR, Dharma (ed.), *The Cambridge Economic History of India, Volume 2: c. 1750–c. 1970* (Cambridge, 1983). A reference work on all development issues, which—except for the absence of bibliography or review of literature on the subjects that it covers—represents the state of the art in growth-oriented economic history at the time.

RAJ, K.N., Neeladri BHATTACHARYA, Sumit GUHA and Sakti PADHI (eds), *Essays on the Commercialization of Indian Agriculture* (Delhi, 1985). The most important collection of essays on this theme, including history and economics, theory and empiricism.

ROBB, Peter (ed.), *Rural India: Land, Power and Society Under British Rule* (London, 1983). A volume not specifically concerned with production that presents case studies of the social and political context of production in various regions.

SCHWARTZBERG, Joseph E. (ed.), *Historical Atlas of South Asia* (Chicago, 1978, 2nd edition, Oxford, 1992). The most important reference tool for agrarian history, which allows the overlay of topographical, ecological, political, social and economic data from different maps to construct a sound geographical understanding of agricultural production from ancient times to the present.

SINGH, Jasbir, *An Agricultural Atlas of India: A Geographical Analysis* (Varanasi, 1974). The most useful geographical presentation of recent data on production at the district level. Excellent for comparative studies of agricultural conditions and development.

SINGH, R.L., *India: A Regional Geography* (Varanasi, 1971). This study compensates for the major weakness of the Schwartzberg *Historical Atlas* and Jasbir Singh *Agricultural Atlas* by providing a systematic analysis of regions. It can be used in combination with Thorner, 'Agrarian Regions'.

STOKES, Eric, *The Peasant and the Raj: Studies in Agrarian Society and Peasant Rebellion in Colonial India* (Cambridge, 1978). The collection from which the essay by Stokes in this volume was drawn, is listed here to emphasize its importance for making connections between agricultural production, social organization and politics. (See Washbrook.)

2. STUDIES THAT FOCUS ON CAPITAL

ADAS, Michael, *The Burma Delta: Economic Development and Social Change on an Asian Rice Frontier, 1852–1941* (Madison, 1974). Describes the role of Chettiyars from Madras Presidency in the financing of rice frontier development in British Burma, and some of its social consequences.

AMIN, Shahid, *Sugarcane and Sugar in Gorakhpur: An Inquiry into Peasant Production for Capitalist Enterprise in Colonial India* (Delhi, 1984). The monograph that represents a subaltern approach to commercial production. All new work on sugar production in India, and most new research on commercial cropping, must take this book into account. (See Attwood.)

ATTWOOD, Donald W., 'Capital and the Transformation of Agrarian Class Systems: Sugar Production in India', in Desai, Rudolph and Rudra, *Agrarian Power and Agricultural Productivity*, pp. 20–50; 'Peasants versus Capitalists in the Indian Sugar Industry: The Impact of the Irrigation Frontier', *Journal of Asian Studies*, 45, no. 1 (November 1985), 59–80; *Raising Cane: The Political Economy of Sugar in Western India* (Boulder: Westview, 1992). Three publications, each with their own points of emphasis and distinct empirical contributions, which describe the economically and socially progressive role of private and public capital investment in sugar-producing regions of Maharashtra. Written by an anthropologist with a primary focus on current conditions, this body of work demonstrates the importance of historical research for contemporary debates in development and economic policy.

BAGCHI, Amiya Kumar, *Private Investment in India, 1900–1939* (Oxford, 1972). The classic study that traces the movement of capital into plantations and agricultural investment in India during the process of overall economic underdevelopment.

BAKER, Christopher John, *An Indian Rural Economy, 1880–1955: The Tamilnad Countryside* (Oxford and Delhi, 1984). Still the best historical study of a large regional economy, showing connections between rural and urban sectors and manufacturing and agriculture.

BANERJEE, Himadri, 'Growth of Commercial Agriculture in the Punjab during the Second Half of the Nineteenth Century', *Punjab Past and Present*, 12, no. 1 (April 1978), 221–56. A study that complements essays by B.B. Chaudhuri and by A. Satyanarayana, in this volume, written on very much the same lines.

BHATTACHARYA, Neeladri, 'Lenders and Debtors: Punjab Countryside, 1880–1940', *Studies in History* (1985). In studies of capital in agricultural development during the colonial period, the scale, distribution, and investment of debt-financing is a critical theme, as indicated by Kaiwar, Bose, and others in this volume. This is an important contribution for Punjab. (See Bose, below.)

BOSE, Sugata, *Agrarian Bengal: Economy, Social Structure and Politics, 1919–1947* (Cambridge, 1986); *Peasant Labour and Colonial Capital: Rural Bengal Since 1770* (Cambridge, 1993). Two monographs that complement one another and together present the argument that debt supplanted rent as a means for surplus extraction from the agricultural production process in Bengal during the twentieth century.

3. STUDIES OF CAPITALISM

ALAVI, Hamza, 'India and the Colonial Mode of Production', *Economic and Political Weekly*, 10, no. 33–35 Special Number (August 1975), 1235–62. A critical essay for scholars who are concerned to theorize capitalist development in agriculture as a distinctive characteristic of India's colonial experience. Much has been written in response to this theoretical intervention.

BAJAJ, Jairus, 'Capitalist Domination and the Small Peasantry: Deccan Districts in the Late Nineteenth Century', *Economic and Political Weekly*, 7, no. 33–34 Special Number (August 1977), 1375–1404, reprinted in Ashok Rudra et al. (eds), *Studies on the Development of Capitalism in India*, pp. 351–428. A theoretical account of the central importance of means for the subordination of peasants rather than workers during the capitalist development, based on data from the Deccan. (See Kaiwar in this volume.)

BHARADWAJ, Krishna, 'A View on Commercialisation in Indian Agriculture and the Development of Capitalism', *The Journal of Peasant Studies*, 12, no. 4 (July 1985), 7–25. All economic argument concerning the levels of commercialization and capitalist development in India, addressing the debate as to the scale and implications of commercial farm production for the Indian economy.

LUDDEN, David, 'World Economy and Village India, 1600–1900: Exploring the Agrarian History of Capitalism', in Bose, *South Asia and World Capitalism*, pp. 159–77. Argues that a form of capitalist development different but comparable to that in early-modern Europe can be seen in India before colonialism, and that a continuity in the distinctive forms of development visible before 1800 continues thereafter.

PATNAIK, Utsa, *Peasant Class Differentiation: A Study in Method with Reference to Haryana* (Delhi, 1987). A detailed empirical and analytical study of degrees and modes of differentiation in the control over means of agricultural production, emphasizing labour exploitation and the conditions of landless agricultural labour.

WASHBROOK, David A., 'Caste, Class and Dominance in Modern Tamil Nadu', in Francine R. Frankel and M.S.A. Rao (eds), *Dominance and State Power in Modern India: Decline of a Social Order, Volume 1*, 204–64 (Delhi, 1989). An account of interactions between development, social change and state politics.

4. STUDIES THAT FOCUS ON LABOUR

ATCHI Reddy, M., 'The Commercialization of Agriculture in Nellore District: Effects on Wages, Employment and Tenancy', K.N. Raj (ed.), *Essays on the Commercialization of Indian Agriculture*, pp. 163–83; 'Official Data on Agricultural Wages in the Madras Presidency from 1873', *The Indian Economic and Social History Review*, 15, no. 4 (October 1978), 451–66. Shows the problems and possibilities of official wage data and the detail that is possible to achieve in studies of wage determination using the private records of individual farmers. Research that is a model to be followed for linking real and actual wages with local conditions.

BARDHAN, Pranab and Ashok RUDRA, 'Labour Mobility and the Boundaries of the Village Moral Economy', *The Journal of Peasant Studies*, 13, no. 3 (April 1986), 90–115. An economic argument and local fieldwork study of labour in the context of rural West Bengal today. The argument is posed in relation to debates about the limits of social change that have been generated by labour markets, in the face of linked markets or land, labour and credit. (See also Rudra in Desai, Rudolph and Rudra, *Agrarian Power and Agricultural Productivity*.)

BHATTACHARYA, Neeladri, 'Agricultural Labour and Production: Central and Southeast Punjab, 1870–1940', in K.N. Raj, *Essays on the Commercialization of Indian Agriculture*, pp. 105–62. A detailed account of labour conditions and their variation over time and space during the growth of commercial agriculture in Punjab.

BREMAN, Jan, *Of Peasants, Migrants and Paupers: Rural Labour Circulation and Capitalist Production in West India* (Delhi, 1985). A kind of rural sociology that has yet to be applied widely enough historically. (See Crispin Bates in this volume.)

DAS GUPTA, Ranajit, 'From Peasants and Tribesmen to Plantation Workers: Colonial Capitalism, Reproduction of Labour Power and Proletarianisation in North East India, 1850s to 1947', *Economic and Political Weekly*, 26, no. 4 (25 January 1986), Review of Political Economy, PE 2–10. Plantation labour and its creation during the last two centuries has not received attention that approximates its historical significance. This essay—like Bates', in this volume—moves in that direction.

KOLFF, Dirk H.A., *Naukar, Rajput and Sepoy: The Ethnohistory of the Military Labour Market of Hindustan, 1450–1850* (Cambridge, 1990). An important monograph on circuits of labour circulation and the seasonal movements into and out of the role of 'peasant' that laid the basis for labour circulation and plantation recruitment during the colonial period.

KUMAR, Dharma, *Land and Caste in South India: Agricultural Labour in Madras Presidency in the Nineteenth Century* (Cambridge, 1965; rpt. Delhi, 1992). The classic study that laid to rest the idea that modern India's massive landless labouring class was the product only of capitalist development under British rule.

THORNER, Daniel and Alice, *Land and Labour in India* (Bombay, 1962). Studies in development economics that forges the link between critical studies of current conditions and the history of colonial capitalism in India.

5. STUDIES THAT FOCUS ON LAND

BANERJEE, Himadri, *Agrarian Society of the Punjab, 1849–1901* (New Delhi, 1982). Describes the basis for agricultural investments in forms of land tenure and rural power relations, and changes in rural society during commercial development.

BHATTACHARYA, Neeladri, 'The Logic of Tenancy Cultivation: Central and Southeast Punjab, 1870–1935', *Indian Economic and Social History Review*, 2, no. 1983 (1920), 121–70. Specifically connects forms of production with forms of tenancy, which needs to be done more widely.

CHAMBERS, Robert, N.C. SAXENA and Tushaar SHAH, *To the Hands of the Poor: Water and Trees* (New Delhi, 1989). Landownership can be detached from the control of necessary production and survival resources that encompasses under most private landed property systems. Historians need to focus on landownership in relation to the control of other material resources in the production process.

KUMAR, Ravinder, *Western India in the Nineteenth Century: A Study of the Social History of Maharashtra* (London, 1968). The classic formulation of changing agrarian social relations during the development of a private property system in colonial India.

PERLIN, Frank, 'Of White Whale and Countrymen in the Eighteenth Century Maratha Deccan: Extended Class Relations, Rights, and the Problem of Rural Autonomy Under the Old Regime', *Journal of Peasant Studies*, 5, no. 2 (1977), 172–237. An important revision of ideas about the isolation of village landownership from regional systems of political power in precolonial India, which has influenced the reinterpretation of agricultural development in the colonial period. (See Bose, *South Asia and World Capitalism*.)

RAY, Ratnalekha, *Change in Bengal Agrarian Society, c. 1760–1850* (New Delhi, 1979). A revision of ideas about zamindars in rural society that argues for the importance of jotedars and rich peasants in production and social relations. (See Bose in this volume.)

SHARMA, H.R., 'Agrarian Relations Since Independence', *Journal of Indian School of Political Economy*, 4, no. 2 (1992), 201–62; 'Evolution of Agrarian Relations in India', *Journal of Indian School of Political Economy*, 4, no. 1 (1992), 80–107. Two recent efforts to summarize trends in agrarian land relations and their interaction with agricultural development in the colonial and national periods.

WASHBROOK, David, 'Law, State and Agrarian Society in Colonial India', *Modern Asian Studies*, 15, no. 3 (1981), 649–721; *The Emergence of Provincial Politics: The Madras Presidency, 1870–1920* (Cambridge, 1976). The best application of ideas developed by Eric Stokes to research in south India. Argues for connections between regional systems of law and power relations in production that determine the direction and effects of agricultural development.

YANG, Anand A., *The Limited Raj: Agrarian Relations in Colonial India, Saran District, 1793–1920* (Berkeley, 1989). As full an account as we have to date of the organization of agrarian society in one zamindari estate over a long period. Emphasizes the relative isolation from one another of layers of power over production resources. (Contrast Washbrook.)

6. Studies that Emphasize Ecology

AGARWAL, Bina, 'Social Security and the Family in Rural India: Coping with Seasonality and Calamity', *The Journal of Peasant Studies*, 17, no. 3 (April 1990), 341–412. The relatively new concern with ecology in agrarian studies needs to embrace the implications of seasonality for rural production and living conditions. This is an excellent place to begin. (Also Chambers, above and Chen, below.)

AHMAD, Nafis, *Economic Geography of East Pakistan* (Dacca, 1958). An early and now dated but still superb regional geography of agrarian ecological conditions in the eastern Bengal region, which can serve as a model for others.

AMANI, K.Z., *Agricultural Land Use in Aligarh District* (Aligarh, 1976). Studies of land use interact with those of ecological conditions by showing the way that producers adapt to and transform the landscape. This is a model of research on contemporary conditions to inform agricultural history.

BARLETT, P.F., 'Adaptive Strategies in Peasant Agricultural Production', *Annual Review of Anthropology*, 9 (1980), 543–73. A bibliographic essay on adaptive strategies that provides a number of enticing propositions applicable to research on India's agrarian past.

BARNETT, Richard, 'The Greening of Bahawalpur: Ecological Pragmatism and State Formation in Pre-British Western India, 1730–1870', *Indo-British Review: A Journal of History XV*, no. 2 (December 1988). A short but instructive and provocative study of agricultural investments and technological adaptations in Rajasthan.

BORKAR, V.V. and M.V. NADKARNI, *Impact of Drought on Rural Life* (Popular Prakashan, 1975). A study that surveys contemporary evidence on an important historical theme, which needs more attention as a feature of agricultural production and livelihoods.

CHEN, Martha Alter, *Coping with Seasonality and Drought* (Newbury

Park, 1991). A case study of contemporary conditions that could serve as a model for micro-regional history.

GUHA, Ramachandra, *The Unquiet Woods: Ecological Change and Peasant Resistance in the Himalaya* (Delhi, 1989). The subaltern study of ecological destruction and politics in the Himalaya. Heavy focus on colonial politics and its continuation after 1947.

LUDDEN, David, *Peasant History in South India* (Princeton, 1985; Delhi, 1990). A history of Tinnevelly district in Madras Presidency based on its agro-technological regions and their distinctive trajectories of development since AD 900.

RICHARDS, John F., James R. HAGEN and Edward S. HAYNES, 'Changes in the Land and Human Productivity in Modern India, 1870–1970', *Agricultural History* (January 1986). A report on a project to analyze large-scale ecological change with district land-use data.

VARADY, Robert G., 'Land Use and Environmental Change in the Gangetic Plain: Nineteenth Century Human Activity in the Benares Region', in Sandria Freitag (ed.), *Culture and Power in Banaras: Community, Performance and Environment, 1800–1980*, pp. 229–45 (Berkeley, 1989). A study that indicates the possibilities for historical studies of ecological change in the Ganga plain.

7. STUDIES OF GROWTH

AGARWAL, Bina, 'Women, Poverty and Agricultural Growth in India', *The Journal of Peasant Studies*, 13, no. 4 (July 1986), 165–220. The implications of growth for different sectors of society represent an important theme for research. Work on women in India today should stimulate history.

ALI, Imran, 'Malign Growth? Agricultural Colonization and the Roots of Backwardness in the Punjab', *Past and Present*, 114 (1987), 110–32; *The Punjab Under Imperialism, 1885–1940* (Princeton, 1988). Power that skews the distribution of social dividends from past development skews reinvestment and distorts future development, so that patterns of backwardness and poverty established in the past reproduce themselves long after their originating political conditions have passed away.

ATHREYA V.B., G. DJURFELDT and S. LINDBERG, *Barriers Broken: Production Relations and Agrarian Change in Tamil Nadu* (New Delhi, 1990). Political pressures and movements that arise from past agricultural development can, however, alter the conditions and means of resource distribution in the political system, thereby altering the dynamics of development.

BAGCHI, Amiya Kumar, 'Reflections on Patterns of Regional Growth in India During the Period of British Rule', *Bengal Past and Present*, 95, no. 1 (1976), 247–89. A path-breaking study of regional diversity that links land control, social formations and economic growth.

BAJPAI, Gita, *Agrarian Urban Economy and Social Change: The Socio-Economic Profile of Select Districts of Gujarat, 1850–1900* (Delhi, 1989). A statistical and historical sociology that indicates a direction for future research into the combined growth of urban and rural economies. (See Baker, above.)

BANDOPADHAYAY, Arun, *The Agrarian Economy of Tamilnadu, 1820–1855* (Calcutta, 1992). A study based on district comparisons of Company policy as a feature of growth and development in Tamil districts of Madras Presidency.

BATES, Crispin, 'The Nature of Social Change in Rural Gujarat: The Kheda District, 1818–1918', *Modern Asian Studies*, 15, no. 4 (1981), 415–54. A study that complements the author's essay in this volume, linking changes in the organization of production with social relations.

BHADURI, Amit, 'Agricultural Backwardness Under Semi-Feudalism', *The Economic Journal* (March 1973), 120–37; *The Economic Structure of Backward Agriculture* (London, 1983). Influential works of theory on the connection between inadequacies in the transformation of social relations and developmental backwardness in agricultural production.

BOSERUP, Ester, *The Conditions of Agricultural Growth* (Chicago, 1965). A path-breaking formulation of positive connections between demographic and economic growth, with important implications for long-term agricultural history in India.

BOYCE, James K., *Agrarian Impasse in Bengal: Institutional Constraints to Technological Change* (New York, 1987). The most compelling and detailed account of how inequalities of power and wealth and their institutionalization in society and politics obstruct agricultural development.

BRAY, Francesca, *The Rice Economies: Technology and Development in Asian Societies* (London, 1986). The argument in this book that rice economies in Asia have distinctive trajectories of development has broad implications for agricultural history in India.

CHAUDHURI, Binay Bhushan, 'Rural Power Structure and Agricultural Productivity in Eastern India, 1757–1947', in M. Desai et al., *Agrarian Power and Agricultural Productivity*, pp. 100–71. A sweeping, long-term survey of the theme pursued in Boyce, *Agrarian Impasse*.

GUHA, Sumit, *The Agrarian Economy of the Bombay Deccan, 1818–1941* (Delhi, 1985). A detailed study of economic trends in the Bombay Deccan, describing immiseration and ecological destruction—a necessary complement to Kaiwar, in this volume.

LUDDEN, David, 'Productive Power in Agriculture: A Survey of Work on the Local History of British India', in Desai et al. (eds), *Agrarian Power and Agricultural Productivity*, pp. 51–100. A useful bibliographic essay on locality-focused research concerned with social power and agricultural development.

LUDDEN, David, 'Asiatic States and Agrarian Economies: Agrarian Commercialism in South India, 1700–1850', *Calcutta Historical Journal*, 1–2, 13, 112–37; 'Agrarian Commercialism in Eighteenth Century South India: Evidence from the 1823 Tirunelveli Census', *Indian Economic and Social History Review*, 25, 4 (1988), 493–519. Rpt. *Merchants, Markets and the State in Early Modern India*, Sanjay Subrahmanyam (ed.) (Delhi, 1990), pp. 215–41. Two studies that argue for a reconsideration of the precolonial political economy of south India as being characterized by dynamic commercial production, which laid the basis for early-modern states and established the political economy of early colonialism.

PANDIAN, M.S.S., *The Political Economy of Agrarian Change: Nancilnadu, 1880–1939* (New Delhi, 1990). An excellent study of a micro-region at the tip of the Indian peninsula. A model for future work.

WASHBROOK, David A., 'Progress and Problems: South Asian Economic and Social History, c. 1720–1860', *Modern Asian Studies*, 22, no. 1 (1988), 57–96. A challenging survey of revisions to our understanding of social and economic conditions before British rule and their implications for colonial history.

8. STUDIES AND REPRESENTATIONS OF IDEAS

ADAS, Michael, *Machines as the Measure of Men: Science, Technology and Ideologies of Western Dominance* (Ithaca, 1989). Describes the rise of science as an ideology of growth and development, and its deployment by colonial regimes.

ALTEKAR, A.S., *A History of Village Communities in Western India* (Madras, 1927). A fundamental book in the history of modern ideas about rural India.

BHALLA, Alok, 'Realms of Desire: Some Gandhian Reflections on the Rural India in Modern Times', in Bhalla (ed.), *Images of Rural India in the 20th Century*, pp. 7–16. A representation of political romanticism grounded in a Gandhian imagination of rural India.

BHATTACHARYA, Mirmal Chandra (ed.), *Some Bengal Villages: An Economic Survey* (Calcutta, 1932). A classic early village economic study, along lines pioneered by Gilbert Slater.

BHATTACHARYA, Swapan Kumar, *Farmers, Rituals and Modernization* (Calcutta, 1976). An intriguing account of farmers' ideas about efficacious techniques for increasing productivity.

BHUTANI, V.C., *The Apotheosis of Imperialism: Indian Land Economy Under Curzon* (New Delhi, 1976); 'Lord Curzon's Agricultural Policy in India', *Punjab Past and Present*, 2, no. 2 (October 1968), 366–99. Excellent examples of work on colonial ideas that influenced agrarian India and its political representation in the early nationalist phase.

CHANDRA, Bipan, *The Rise and Growth of Economic Nationalism in India: Economic Policies of the Indian National Leadership, 1880–1905* (New Delhi, 1966). The single most important work for understanding Congress agrarian politics through the 1920s, and for appreciating the connection between early nationalism and scholarship in the field of agrarian history.

CHARLESWORTH, Neil, *Peasants and Imperial Rule: Agriculture and Agrarian Society in the Bombay Presidency, 1850–1935* (Cambridge, 1985); *British Rule and the Indian Economy, 1800–1914* (London, 1982). Two volumes that summarize ideas and debates about colonial rule and its impact on agrarian India.

LUDDEN, David, 'Orientalist Empiricism and Transformations of Colonial Knowledge', in C.A. Breckenridge and Peter Van der Veer (eds), *Orientalism and the Post-colonial Predicament* (Philadelphia, 1993), pp. 250–78. A discussion of ideas about village society integrated into orientalism as a body of scientific and political knowledge about India.

LUDDEN, David, 'Archaic Formations of Agricultural Knowledge in South India', in Peter Robb and Utsa Patnaik (eds), *Meanings of*

Agriculture in South Asia (Cambridge) (forthcoming). An effort to reconstruct ideas about agricultural production and technique, over the long term, at the village level, in south India.

LUDDEN, David, 'India's Development Regime', in Nicholas Dirks (ed.), *Colonialism and Culture* (Ann Arbor, 1992), pp. 247–87. An account of ideas about development, especially in agriculture, as they were institutionalized under British rule and adapted by the state in independent India.

PARASHER, Aloka, ' "The Village Given to Us": Intellectual Constructions of the Village Community in Historical Writing', in Bhalla (ed.), *Images of Rural India*, pp. 17–43. A skimpy but useful summary of ideas about the village as the basis of rural India and Indian society.

SONTHEIMER, Gunther Dietz, *Pastoral Deities in Western India* (Delhi, 1993). The role of pastoral peoples and cultures in the history of agrarian India has been poorly represented in the literature: the ideas in this book should change that.

THORNER, Daniel, 'Agrarian Region', in A.R. Desai (ed.), *Rural Sociology* (Bombay, 1959), pp. 152–60. The original effort to create a regional framework for agrarian history, which still has fresh insights and arguments to be developed by future scholars.

9. STUDIES OF IRRIGATION

ALVARES, Claude and Ramesh BILLOREY, *Damming the Narmada: India's Greatest Planned Environmental Disaster* (Natraj, 1989). A recent attempt to put massive irrigation works into environmental perspective.

ATTWOOD, Donald W., 'Irrigation and Imperialism: The Causes and Consequences of a Shift from Subsistence to Cash Cropping', *Journal of Development Studies*, 23, no. 3 (April 1987), 341–66. A positive view of the creative role that irrigation can play in the advance of commercial production, based on the case of Maharashtra sugar cultivation. (See Attwood, above.)

BHARADWAJ, Krishna, *Irrigation in India: Alternative Perspectives* (Delhi, 1990). A discussion of contemporary debates with important historical implications.

BHARATA, L.P., S.P. MALHOTRA and F.C. PATWA, 'Impact of Irrigation on the Changes in the Population Characteristics of Beneficiary and Non-Beneficiary in a Desert Region', *Indian Journal of Social Research*, 14, no. 2 (1973), 137–45; 'Some Socio-Agricultural Changes as a Result of the Introduction of Irrigation in a Desert Region', *Annals of the Arid Zone*, 13, no. 1 (1974), 1–10. Small studies of small-scale effects that could be replicated and adapted by historians for many localities.

CLAY, Edward J., 'Equity and Productivity Effects of a Package of Technical Innovations and Changes in Social Institutions: Tubewells, Tractors and High-Yielding Varieties', *Indian Journal of Agricultural Economics*, 30, no. 4 (1975), pp. 74–87. A careful study that indicates the kind of social impact that irrigation can have at various levels of scale and also the kind of measures that are available to document such impact historically.

LUDDEN, David, 'Patronage and Irrigation in Tamil Nadu: A Long-Term View', *Indian Economic and Social History Review*, 8, no. 4 (Fall) (1979), 347–65. Argues that irrigation be understood as a social technology organized by state power, whose character and impact on production it significantly represents. Describes the organizational changes that have favoured large over small-scale state irrigation investments in modern times.

STONE, Ian, *Canal Irrigation in British India: Perspectives on Technological Change in a Peasant Society* (Cambridge, 1984). The book that continues the line of argument developed in the essay reprinted in this volume.

WADE, Robert, 'Irrigation Reform in Conditions of Populist Anarchy: A South Indian Case', *Journal of Development Economics*, 14, no. 3 (April 1984), 285–303; 'The System of Administrative and Political Corruption: Canal Irrigation in South India', *Journal of Development Studies*, 18, no. 3 (May 1982), 287–328. These essays represent contemporary history and should be emulated by historians to improve scholarship on the political economy of agriculture in the twentieth century.

WHITCOMBE, Elizabeth, *Agrarian Conditions in Northern India: The United Provinces Under British Rule, 1860–1900, Volume 1* (Berkeley, 1972). The classic that set the tone for later research with its argument against the development ideology of the Raj as articulated in its irrigation policies.

3951-
p
25
060896
tps